Targeting of Drugs 4

Advances in System Constructs

NATO ASI Series

Advanced Science Institutes Series

A series presenting the results of activities sponsored by the NATO Science Committee, which aims at the dissemination of advanced scientific and technological knowledge, with a view to strengthening links between scientific communities.

The series is published by an international board of publishers in conjunction with the NATO Scientific Affairs Division

A	**Life Sciences**	Plenum Publishing Corporation
B	**Physics**	New York and London
C	**Mathematical and Physical Sciences**	Kluwer Academic Publishers
D	**Behavioral and Social Sciences**	Dordrecht, Boston, and London
E	**Applied Sciences**	
F	**Computer and Systems Sciences**	Springer-Verlag
G	**Ecological Sciences**	Berlin, Heidelberg, New York, London,
H	**Cell Biology**	Paris, Tokyo, Hong Kong, and Barcelona
I	**Global Environmental Change**	

Recent Volumes in this Series

Volume 267 — Bioelectrochemistry IV: Nerve Muscle Function—Bioelectrochemistry, Mechanisms, Bioenergetics, and Control
edited by Bruno Andrea Melandri, Giulio Milazzo, and Martin Blank

Volume 268 — Advances in Molecular Plant Nematology
edited by F. Lamberti, C. De Giorgi, and David McK. Bird

Volume 269 — Ascomycete Systematics: Problems and Perspectives in the Nineties
edited by David L. Hawksworth

Volume 270 — Standardization of Epidemiologic Studies of Host Susceptibility
edited by Janice Dorman

Volume 271 — Oscillatory Event-Related Brain Dynamics
edited by Christo Pantev, Thomas Elbert, and Bernd Lütkenhöner

Volume 272 — Photobiology in Medicine
edited by Giulio Jori, Roy H. Pottier, Michael A. J. Rodgers, and T. G. Truscott

Volume 273 — Targeting of Drugs 4: Advances in System Constructs
edited by Gregory Gregoriadis, Brenda McCormack, and George Poste

Series A: Life Sciences

Targeting of Drugs 4

Advances in System Constructs

Edited by

Gregory Gregoriadis and
Brenda McCormack
School of Pharmacy
University of London
London, England

and

George Poste
SmithKline Beecham Pharmaceuticals
King of Prussia, Pennsylvania

Springer Science+Business Media, LLC

Proceedings of a NATO Advanced Study Institute on
Targeting of Drugs: Advances in System Constructs,
held June 24–July 5, 1993, in Cape Sounion, Greece

NATO-PCO-DATA BASE

The electronic index to the NATO ASI Series provides full bibliographical references (with keywords and/or abstracts) to more than 30,000 contributions from international scientists published in all sections of the NATO ASI Series. Access to the NATO-PCO-DATA BASE is possible in two ways:

—via online FILE 128 (NATO-PCO-DATA BASE) hosted by ESRIN, Via Galileo Galilei, I-00044 Frascati, Italy

—via CD-ROM "NATO Science and Technology Disk" with user-friendly retrieval software in English, French, and German (©WTV GmbH and DATAWARE Technologies, Inc. 1989). The CD-ROM also contains the AGARD Aerospace Database.

The CD-ROM can be ordered through any member of the Board of Publishers or through NATO-PCO, Overijse, Belgium.

Library of Congress Cataloging-in-Publication Data

Targeting of drugs 4 : advances in system constructs / edited by
Gregory Gregoriadis and Brenda McCormack and George Poste.
p. cm. -- (NATO ASI series. Series A, Life sciences ; v.
273)
Proceedings [of] a NATO Advanced Study Institute on Targeting of
Drugs: Advances in System Constructs, held June 24-July 5, 1993, in
Cape Sounion, Greece"--T.p. verso.
Published in cooperation with NATO Scientific Affairs Division.
Includes bibliographical references and index.

1. Drug targeting--Congresses. I. Gregoriadis, Gregory.
II. McCormack, Brenda. III. Poste, George. IV. North Atlantic
Treaty Organization. Scientific Affairs Division. V. NATO Advanced
Study Institute on Targeting of Drugs: Advances in System Constructs
(1993 : Ákra Soúnion, Greece) VI. Title. Targeting of drugs four.
VII. Series.
[DNLM: 1. Drug Carriers--congresses. 2. Drug Delivery Systems-
-congresses. 3. Drugs--administration & dosage--congresses. QV
785 T185 1994]
RS200.T37 1994
615'.7--dc20
DNLM/DLC
for Library of Congress 94-41747
CIP

ISBN 978-1-4899-1209-1 ISBN 978-1-4899-1207-7 (eBook)
DOI 10.1007/978-1-4899-1207-7

PREFACE

The NATO Advanced Studies Institute series "Targeting of Drugs" was originated in 1981. It is now a major international forum, held every two years in Cape Sounion, Greece, in which the present and the future of this important area of research in drug delivery is discussed in great depth. Previous ASIs of the series dealt with drug carriers of natural and synthetic origin, their interaction with the biological milieu, ways by which the latter influences such interaction, strategies by which milieu interference curtailing the function of drug carriers is circumvented and, more recently, with the application of drug carriers for the delivery of peptides and proteins. The present book contains the proceedings of the 7th NATO ASI "Targeting of Drugs: Advances in System Constructs", held in Cape Sounion during 24 June - 5 July 1993. As the title implies, the book deals with a variety of approaches to carrier design or modification that contribute to optimal carrier function.

We express our appreciation to Mrs Concha Perring for her assistance with the organization of the ASI. We thank Dr. G. Deliconstantinos who, as chairperson of the Local Committee, contributed to the success of the Institute. The ASI was held under the sponsorship of NATO Scientific Affairs Division and co-sponsored and generously financed by SmithKline Beecham Pharmaceuticals (King of Prussia). Financial assistance was also provided by Liposome Technology Inc. (Menlo Park), Vestar Inc. (San Dimas) and Zeneka (Macclesfield).

Gregory Gregoriadis
Brenda McCormack
George Poste

August 1994

CONTENTS

DRUG TARGETING WITH GLYCOPROTEINS AND OTHER PEPTIDE CARRIERS: AN OVERVIEW

Dirk K.F. Meijer

Dept of Pharmaceutical Pharmacology and Clinical Pharmacy, section Pharmacokinetics and Drug Delivery, Groningen Inst for Drug Studies, University Centre for Pharmacy, Ant. Deusinglaan 2, 9713 AW Groningen, The Netherlands

INTRODUCTION

Drug targeting is aimed at the manipulation of drug distribution in the whole body in order to increase the therapeutic efficiency and decrease side effects and toxicity. In vitro studies with potential target cells can be helpful to study the relative rates of endocytosis, carrier degradation and release of the targeted drug. However, studies in the intact organism, also in the diseased state, should provide more definite clues as to the cell specificity of the chosen carrier system, possible toxicity and immunogenicity of the carrier as well as the potential to pass anatomical barriers en route to the target cells (for some general guidelines in drug targeting research, Fig. 1).

Currently available carrier systems include particle type of carriers such as liposomes, lipid particles (LDL, HDL) and nanoparticles. Soluble carrier systems are prepared from monoclonal antibodies, modified plasmaproteins, polyamino-acids, polysaccharides and other bio-degradable polymers (see Fig. 2). In some cases the peptide carrier can be designed such that it contributes to the therapeutic effects (Meijer et al, 1992)

The mechanisms responsible for the disposition of peptides in the body include carrier-mediated transport (liver, CNS), filtration and reabsorption (kidneys), as well as receptor mediated endocytosis and phagocytosis. Important features determining the fate in the body are amino-acid composition, protein conformation and folding, overall size, net charge, presence of sugars, aggregation state and potential opsonization by blood components.

Small peptides (up to 12 amino-acids) can be efficiently cleared by the liver depending on hydrophobicity and charge through four different carrier-systems, some of which are bile acid transporters (see Fig. 3). Time and dose dependent kinetics, extremely short half-lives and low bioavailability can therefore result (Meijer and Ziegler, 1993).

Larger peptides (up to 60 kD) are easily filtrated by the kidneys but often fully reabsorbed and degraded in renal lysosomes. Such low molecular weight proteins (LMWP's) can be used as drug carriers for renal-specific drug delivery. Targeting to the kidney of anti-immflamatory drugs, antimicrobial agents, ACE inhibitors and dopaminergic compounds have been studied.

Targeting of Drugs 4, Edited by G. Gregoriadis *et al.*
Plenum Press, New York, 1994

a. It is preferable to test drug-delivery preparations as soon as possible in the whole body.

b. It is advisable to test dug-delivery preparations with regard to possible immunogenicity in an early stage of development.

c. Cell-specific distribution of the drug-targeting preparations as well as the rate of drug release from the carrier should be studied in vivo both in the normal and the pathological situation.

d. Drug loading of the carrier should be carefully balanced: sufficient drug molecules should be internalized to obtain therapeutic levels however excessive loading may corrupt cell specificity of the carrier.

e. A type of carrier should be chosen that is relatively non-toxic also with regard to its degradation products.

f. The chosen carrier should be able to pass anatomical barriers in the body en route to the target tissues.

g. Since parental administration is required, drug targeting formulations should have major advantages compared with the parent drug.

Fig. 1 General guidelines in drug delivery research.

Larger (glyco-)proteins are often recognized by sugar specific receptors or scavenger receptors in various cell types in the body. This can occur directly or after opsonization with circulating plasma components. For instance the extent and type of glycosylation can be crucial for the elimination of monoclonal antibodies while complex formation with circulating antigens can largely determine the fate in the body (Fc receptors). Some

• **Viral receptors**	sCD4; CD_4-immunoadhesins; CD_4-toxins (viral inf.)
• **Immunotoxins**	(tumors, rheumatic dis., viral inf.,graft v host reactions)
• **Liposomes**	(solid tumors, lymphomas/leukemias, dermal dis., fungal inf., bact. /viral inf., vaccination)
• **Polymeric matrices**	(lymphomas/leukemias, bact.infect., solid tumors, brain tumors)
• **Monoclonal antibodies**	(tumor imaging, solid tumors, lymphomas/ leukemias, cardiovasc. dis, autoimmune dis.)
• **Transfected cells**	(tumors, HIV inf., genetic deficiences)
• **Other potential carriers**	Glycoproteins, LMWPs, HDL/LDL, bispecifc ABs catalytic ABs, pegylated proteins, poly-amino acids(sugar deriv.), nanoparticles

Fig. 2 Carrier modalities currently employed in drug delivery research for the treatment or prevention of disease.

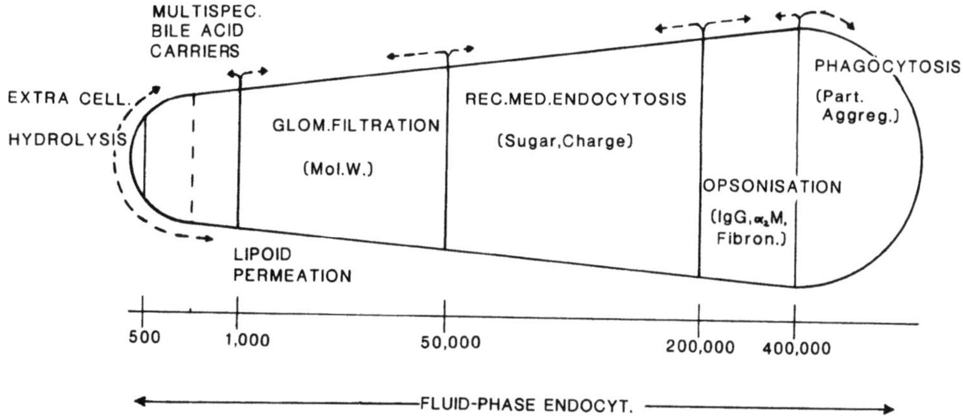

Fig. 3 Clearance mechanisms for oligopeptides and proteins in the intact organism. Dependent on size, charge, sugar recognition, aggregation to particles, and formation of complexes with opsonication factors, peptides can be filtered in the kidneys, or hydrolyzed in the blood or at cell surfaces, or recognized by carrier- or receptor-mediated transport systems. Molecular weight as indicated on the x axis roughly determines the clearance mechanisms. However, the indicated mechanisms largely overlap and are dependent not only on size but also on functional groups (sugars), charge, lipophilicity, and vulnerability for circulating or fixed proteases. Fluidphase endocytosis in principle occurs over the entire molecular weight range

glycoproteins can be used for cell specific targeting of antiviral drugs to the liver (viral hepatitis) (Fiume et al, 1982; Meijer and Ziegler, 1993; Meijer et al, 1990) and blood cells (HIV infections) (Meijer et al, 1992). Such drug targeting devices can also be used for delivery of antisense nucleotides and genes (+ promotors) to reprogram various cell types in the body in vivo to produce therapeutic proteins (Wu et al, 1989; Keegan-Rogers et al, 1991; Meijer et al, 1992; Meijer and Ziegler, 1993; Meijer et al, 1990).

DRUG TARGETING TO THE LIVER WITH GLYCOPROTEINS AND OTHER HEPATOTROPIC CARRIERS

Glycoproteins represent a wide variety of macromolecules which, depending on their charge and carbohydrate structure, can be more or less specifically recognized by certain tissues and cell groups within these tissues. They potentially can be used as drug carriers targeting drugs to their targets in the body (Poznansky and Juliano, 1984; Monsigny et al, 1988; Meijer and Van der Sluijs, 1989; Shen, 1989; Fallon and Schwartz, 1985) through receptor mediated endocytosis and subsequent proteolytic degradation.

Fluid phase endocytosis takes place for any material in the blood stream but is relatively slow. Adsorptive endocytosis is important for ligands that have a multivalent interaction with the plasma membrane due to charge and is quantitatively important after large doses of macromolecules. Such processes may to some extent corrupt specific receptor mediated drug delivery. However, due to the high affinity of the drug carriers for the receptor systems actually aimed at, receptor mediated processes often can easily compete with these more aspecific mechanisms, that is if their maximal uptake capacity is not exceeded.

Drug targeting to the liver is performed for diagnostic and therapeutic purposes. Liver disorders or liver metabolism can be influenced by various agents: agents for antiviral, antiparasitic, antineoplastic or antilipidemic effects, drugs counteracting liver fibrosis, or hepatotoxic reactions, and finally modified enzymes to correct genetic enzyme deficiencies.

POTENTIAL CARRIER SYSTEMS FOR LIVER TARGETING

Both soluble and particle carrier materials have been employed for the purpose of drug targeting to the various cell types in the liver. Relevant in this respect is the endothelial barrier: particles of > 100-150 nm cannot pass the fenestrae (pores) in the endothelial lining (Brouwer et al, 1988). This is true for most of the normal size liposomes, and such carriers can only reach Kupffer-, pit- and endothelial cells. Small sized liposomes and HDL particles can, in principle, be used for delivery of lipophilic (pro-)drugs to hepatocytes and lipocytes (Van Berkel et al, 1986; Van Berkel et al, 1987; Roerdink et al, 1988). Such agents can also be enclosed in the lipid core of LDL particles or in liposomes and be delivered to the sinusoidal cell types (Van Berkel, 1986; Van Berkel, 1987). After degradation of the carrier, such drugs could locally act in these cells or be slowly released from the cells to reach the other cell types in the liver. Reuptake of the parent drugs in these cells, however, may not be efficient enough to reach therapeutic concentrations, and at the same time prevent toxic levels in extrahepatic tissues. The choice of such carriers therefore is determined by a number of considerations (Meijer et al, 1992; Meijer et al, 1990; Poznansky and Juliano, 1984; Meijer and Van der Sluijs, 1989; Wu, 1988; (see Fig. 1).

In the drug targeting field often macromolecular carrier devices are employed: single macromolecules such as (neo-)glycoproteins, and immunoglobulins but also more complex type of carriers such as liposomes and nanoparticles. Uptake into cells of such macromolecular carriers occurs by endocytosis or phagocytosis.

Interestingly, in two recent studies (Kramer et al, 1992; Stephan et al, 1992), a much smaller homing moiety was employed: the bile acid molecule. The particular bile acid "prodrugs" directly deliver the coupled drugs to the cytoplasmic compartment of the cells via carrier-mediated uptake of the bile acids instead of taking the more indirect endocytic (lysosomotropic) pathway followed by the macromolecular carriers. Kramer et al, 1992, used taurocholic acid as a carrier to deliver the alkylating anti-cancer drug chlorambucil as well as to target a prolylhydroxylase inhibitor to hepatocytes. Stephan et al employed cholic acid in an attempt to improve the hepatic cholesterol lowering activity of the thyroid hormone L-T_3 and at the same time reduce its cardiovascular and central side effects (see for structures Fig. 4). The chlorambucil and the prolyl hydroxylase inhibitor were linked via ester or amide bonds to the 3α-hydroxy group, conserving the anionic bile acid side chain, whereas the thyroid hormone was coupled via an amide linkage between the C_{25} carboxylic group of cholic acid and the α-amino group of the alanine side chain of L-triiodothyronine. The nature of such chemical linkages is important in relation to the questions a) does coupling of the drug drastically change the normal properties of the (particular) carrier and b) does the chosen covalent linkage allow a proper cellular release of the active form of the drug?. Interestingly, release of the targeted drugs from the chosen bile acid carriers may not be required for the desired intracellular effects of the tested drugs. For the chlorambucil derivative an alkylating activity of the intact conjugate was demonstrated by covalent binding of bile acid carrier proteins to the plasma membranes and also to intracellular proteins. Biliary excretion of the chlorambucil bile acid derivative after in vivo administration to rats was much higher than that of chlorambucil itself. Evidently, hepatic targeting was obtained but the question whether an optimal therapeutic concentration and a proper cellular residence time of the targeted drug was reached, remained unanswered in these preliminary studies. Relevant in this respect is that hepatocytes in primary liver carcinomas have a low expression of bile acid uptake carriers if present at all (Buscher et al, 1988) and this is certainly true for secondary forms of liver cancer (for instance metastatic colon carcinomas). Coupling of such drugs to bile acids may increase distribution to the normal hepatocytes but less to primary or secondary cancer cells. The low expression of targeted receptors in hepatic carcinomas was earlier used for so called "reversed targeting" of rescue factors to the healthy cells (Wu, 1988). In this sense biodegradable bile acid conjugates of rescue factors targeted to normal hepatocytes would be attractive candidates for anti-cancer therapy. On the other hand, if at least some primary uptake would occur, cancer cells may excrete such bile acid prodrugs much slower and

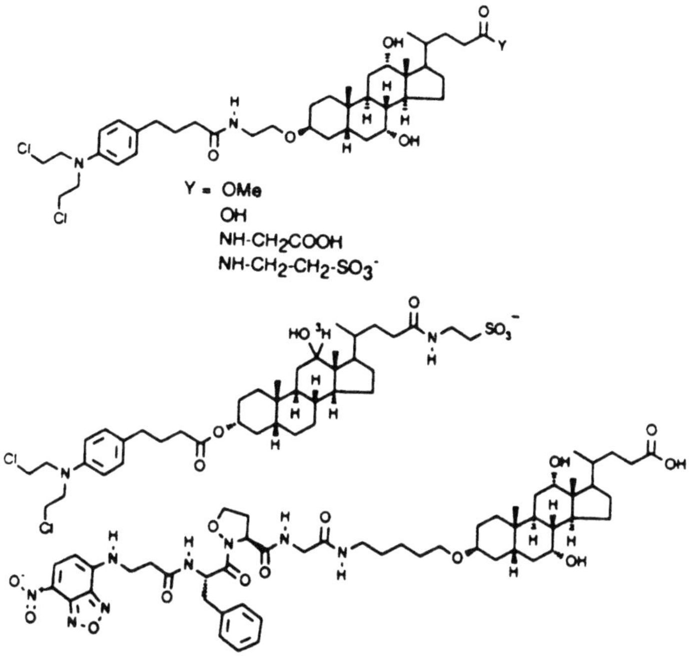

Fig. 4 Liver-specific Drug Targeting by Coupling to Bile Acids. Structural formulas of two taurocholate-chlorambucil prodrugs linked via aramide (A) or ester bonds (B) for the therapy of liver tumors and a taurocholate-oxalyl prolyl peptide prodrug (C) targeted for inhibition of collagen synthesis

therefore may still accumulate some of the cytostatic agents. In addition, biliary excretion of the bile acid conjugate will expose the alkylating drug to cells in the biliary system and may be of value for treatment of cancers of the biliary epithelia.

Interesting data were also obtained with regard to the peptidic prolyl-hydroxylase inhibitor, meant to influence fibrogenesis. It was demonstrated that improved targeting to the hepatobiliary system is obtained in the form of the unchanged peptide-bile acid derivative. In addition, through coupling to bile acid moiety, the peptide structure of the prolylhydroxylase inhibitor was protected against extra- or intracellular metabolism. However, assuming that in certain stages of liver fibrosis hepatocytes contribute to excessive collagen formation, the antifibrotic drug should be efficiently delivered into this cell and not be rapidly excreted in bile. Although certainly a step forward was made in showing that coupling to bile acids diverts drugs from urinary excretion and improves delivery of drugs to the biliary system, the efficacy of such preparations in combatting various hepatic diseases remains to be demonstrated.

In the paper of Stephen et al, 1992, an improved therapeutic profile of the bile acid-coupled thyroid hormone in lipemic rats was observed. It was shown that by coupling L-T_3 to cholic acid, at least part of the cholesterol-lowering activity is retained in vitro and in vivo. Although in vivo the conjugate was 6 times less potent than L-T_3 itself, the cardiostimulant and thyroxine suppressing side effects were 50-60 times decreased compared with the parent drug, and consequently lead to an improved therapeutic safety of the hypocholesterolemic drug. This dissociation of hepatic and cardiovascular effects can be explained by a decreased affinity to receptors in cardiac tissue and pituitary gland and at the same time to an improved delivery to hepatocytes. The anticipated higher hepatocyte concentrations compared with L-T_3 itself may compensate for the much lower intrinsic activity of the cholic acid-L-T_3 conjugate that was demonstrated by the authors in vitro. In addition, slow intracellular degradation of the amide bond and generation of the free L-T_3 may contribute to the final interaction with the L-T_3-receptor in the liver cell nuclei, an activity that finally leads to upregulation of the LDL receptors.

Further elaboration of this technology would certainly benefit from the design of suitable biodegradable bonds between the bile acid carrier and the targeted drug in order to obtain a programmed release of the drug in relation to the required biological or therapeutic response.

The final goal of any drug targeting attempt is to make the active drug available in the cell type aimed at. Drugs such as doxorubicin that are covalently linked via their NH_2-groups to COOH-groups of macromolecular carrier proteins were only slowly released in their active form. Introduction of amino-acid spacers between the drug and the protein can improve release rate and such spacers can be used to manipulate the cellular concentrations (Shen, 1989; Trouet et al, 1982; Trouet et al, 1982; Schneider et al, 1984). Drugs with functional COOH-groups or isothiocyanate moieties can be linked to lysine-NH_2-groups of proteins but the final degradation product will contain at least one amino acid (lysine) since this linkage cannot be split from the drug by the lysosomal enzymes (Van der Sluijs et al, 1985). Fluorescein isothiocyanate (FITC) covalently coupled to asialoorosomucoid is quantitatively taken up by the parenchymal cells of the liver. After degradation the dye is released but still bound to lysine. After conjugation with glucuronic acid it is excreted into bile (Van der Sluijs et al, 1985). The excretion rate of the fluorescent material in bile reflects the rate limiting lysosomal degradation of the glycoprotein carrier (Van der Sluijs et al, 1985). Uncoupling of the drug can also happen too fast: trifluorothymidine covalently linked to the hepatotropic carrier asialofetuin, is so rapidly released from lysosomes in the hepatocytes that substantial amounts escape to the general circulation and reach non-target tissues such as bone marrow (Fiume et al, 1982). Depression of bone marrow by the redistributed drug was shown in otherwise healthy animals but not in animals with experimental hepatic necrosis. This phenomenon was explained by a reduced lysosomal release rate of the drug in the liver of the latter animals (Fiume et al, 1982).

An important factor determining the rate of release from lysosomes is probably the extent of protonation of the delivered drug. Basic drugs will be largely protonated at the relatively low internal pH of this organelle (pH 5-6) and consequently will undergo persistent storage and relatively slow release. In contrast, acidic drugs will mainly be present in the undissociated form and therefore will easily leave the lysosomal compartment. This was shown for the protease inhibitor pepstatin that was administered covalently coupled to asialofetuin (Furuno et al, 1983). The carboxylic tripeptide was properly delivered to hepatocyte lysosomes but after carrier degradation it was rapidly excreted into bile. However, in spite of this rapid release, intralysosomal levels remained high enough for inhibition of cathepsin during 6 hours after injection, demonstrating the extreme potency of the pepstatin molecule.

Glycoproteins are attractive objects for drug targeting to the liver: the structure can be modified with regard to the protein backbone as well as the functional sugar groups. The relatively small molecular mass does not exclude passage of the vascular endothelium (Poznansky and Juliano, 1984; Scheider et al, 1984; Bodmer and Dean, 1988) and immunogenicity after chronic administration is limited (Fiume et al, 1987). Naturally occurring plasma-proteins such as disialylated orosomucoid (α_1-acid glycoprotein) and fetuin were often used. They display a relatively high affinity for the asialoglycoprotein receptor probably due to the clustered arrangement of the antennary oligosaccharide side chains (Monsigny et al, 1994; Meijer and Groothuis, 1991). Some artificial carrier systems mimic the geometric organization of sugar groups. Plasma-proteins such as albumin and apoprotein B as well as certain enzymes can be randomly derivatized with sugar groups. Well known methods are the reductive amination using boronhydride and lactose. In this reaction the aldehyde moiety of lactose is reacted with -NH_2 groups of lysine. Up to sixty lysine molecules can be linked to lactose in this manner. It should be realized, that the nitrogen-atom in the protein-sugar linkage can still become protonated and that consequently no positive charge is lost. Such glycosylated albumin molecules may loose their normal neutral-base transition properties and loss of flexibility in the protein molecule due to sugar

Table 1. Targeting of drugs to the liver with glycoproteins and other glycoconjugates

Drugs/agents	Carrier	Species	Reference
Antiviral agents			
Ara-AMP	Lactosaminated HSA	mouse	Fiume et al 1986
Ara-AMP	Lactosaminated HSA	human	Fiume et al 1988
Ara-AMP, acyclovir	Lactosaminated HSA	woodchuck	Fiume et al 1991
Ara-AMP	Galactosyl-L-lysine	mouse	Fiume et al 1994
PMEA	Mannosylated-poly-L-lysine	mouse	Midoux et al 1990
Ara-AMP	Lactosaminated HSA	rat	Jansen et al 1993
Ara C	Galactose/Mannose Carboxyl-methyl-Dextran	mouse	Nishikawa et al 1993
Trifluorothymidine	Asialofetuin	mouse	Fiume et al 1982
N-Acetyl muramyl peptide	Mannosylated BSA	mouse	Monsigny et al 1994
Antiparasitic agents			
Primaquine	Asialofetuin	mouse	Trouet et al 1982a
Pepstatin	Asialofetuin	rat	Furuno et al 1983
Allopurinol riboside	Mannosylated-poly-L-lysine	mouse	Nègre et al 1992
Antineoplastic agents			
Daunomycin	Lactosaminated HSA	mouse	Schneider et al 1984
Methotrexate	Poly-L-lysine	mouse	Arnold 1985
Mitomycin	Asialofetuin	rat	Kaneo et al 1991
Doxorubicin	HPMA copolymers	mouse	Seymour et al 1991
Ricin	Epidermal growth factor	mouse	Herschman et al 1984
Diagnostic agents			
Fluorescein	Asialoorosomucoid	rat	Van der Sluys et al 1985
Technetium-99m	Galactosyl-HSA	rat/rabbit	Vera et al 1985
Technetium-99m	Galactosyl-HSA	human	Stadalnik et al 1985
Agents affecting lipid metabolism			
LDL	Tris-galactosyl cholesterol	rat	Van Berkel 1985
HDL	Tris-galactosyl cholesterol	rat	Bijsterbosch et al 1991
LDL	Lactosaminated anti-apolipoprotein B antibodies	rat	Bijsterbosch et al 1991
Antioxidants/cytoprotective agents			
Uridine monophosphate	Asialofetuin/poly-L-lysine	rat	Keegan-Rogers 1990
Naproxen	Galactosyl/Mannosyl HSA	rat	Franssen et al 1993
Superoxide dismutase	Mannosylated form	rat	Takakura et al 1994

derivatization can therefore occur. Other methods to connect sugars include the thioglycoside method (Stowell and Lee, 1980) and coupling via thiophosgene activation of amino-sugars (Monsigny et al, 1988; Kataoka and Tavassoli, 1985). Especially the last method may lead to negatively charged proteins since the nitrogen-atom in the linking moiety cannot be protonized in contrast with the lysine -NH_2 groups. This charge modification is further increased by subsequent covalent coupling of drugs to the remaining -NH_2 groups. Drugs with acidic functional groups will amplify this problem (Meijer and Ziegler, 1993; Van der Sluijs and Meijer, 1991).

Alternatives for acid sensitive linkages are biodegradable spacer moieties composed of sugars (Molteni, 1979) or aminoacids (Trouet et al, 1982). However, the multi-step synthesis of such preparations is expensive and cumbersome.

It follows that caution is warranted in the synthesis of drug-protein complexes. Mild methods should be used that conserve the spatial conformation of the protein as much as possible. Deviation from the normal charge distribution in the protein may lead to immunogenicity and loss of cell specificity. Prominent hydrophobic features of the connected drugs can lead to aggregation of the carrier molecules and capturing by phagocytotic systems. In general it can be stated that the higher the drug loading, the more artefacts will be introduced with regard to cell specificity. Consequently practical compromises between optimal drug loading and relative changes in the chosen carrier protein due to derivatization should be looked for.

CURRENT STATE OF ART IN DRUG TARGETING TO THE LIVER WITH PROTEIN CARRIERS

Introduction

The liver probably is the most systematically studied organ in the field of drug targeting. Quite a few examples can be given now in which successful and selective delivery of drug to this organ was achieved in vivo (Meijer and Ziegler, 1993; Meijer et al, 1990; Wu, 1988; Fallon and Schwartz, 1989; Stahl, 1990) (see Table 1). It is of importance to note that the idea of drug targeting in fact turned around the usual rules in drug design and development: very toxic agents that seemed to be definitely put aside by pharmacologists became serious candidates again for therapeutic application.

Targeting of antiviral agents

Fiume and coworkers coupled 5-15 molecules of adenine arabinoside mono-phosphate (Ara-AMP) to lactosaminated albumin containing 20-50 terminal galactose groups per albumin molecule. This particular drug formulation reached the clinical stage and a preliminary study showed promising results in wood chuck hepatitis (Ponzetto et al, 1991) and HBV-infected patients (Fiume et al, 1988). More definite studies are awaited. In contrast to the parent drug, the Ara-AMP conjugate with lactosaminated albumin is almost devoid of side effects in bone marrow and intestinal mucosa (Fiume et al, 1982; Fiume et al, 1986). Another positive aspect is that the conjugate does not seem to evoke major immunogenic reactions after chronic administration, that is if homologous albumin is used (Fiume et al, 1987).

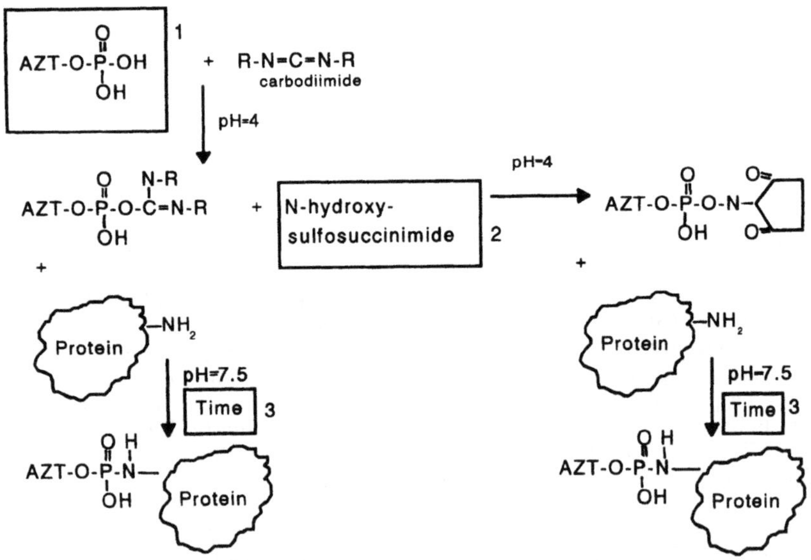

Fig. 5 Coupling of tha AZT-monophosphate to (neo)-glycoproteins. After activation of the phosphate groups at pH 4.0, the pH is raised to 7.5 in order to prevent activation of protein COOH groups and polymerization during the coupling of AZT to the protein carrier. Multivariate analysis in an optimization procedure was used to study the influence the amount of starting material (1), the time of the coupling reaction to protein (3) as well as the influence of addition of N-hydroxy sulfo succinimide (2) that is used to stabilize the carbodiimide-activated phosphate group

Chemical moieties used to link drugs to carrier proteins should be stable in the blood-stream to prevent premature degradation but they should be labile in acidic compartments within the target cells for proper intracellular release. Antiviral nucleoside analogues in their mono- or diphosphate forms can be linked to poly-L-lysine (Fiume et al, 1994) through lysine ϵ-NH_2 and histidine nitrogen of proteins via a pH-sensitive phosphoramide group (Fiume et al, 1980). The initial one-step coupling reaction (Fiume et al, 1982), in which the nucleoside monophosphate is activated with carbodiimide in the presence of the neoglycoprotein, was replaced by a two step procedure in which the phosphate group is activated at pH 4.5 and after raising the pH to 7.5 the albumin is added (see Fig. 5). This prevents simultaneous activation of -COOH groups in the protein leading to formation of dimers and trimers (Jansen et al, 1993; Molema, 1992). Such multimeric conjugates will tend to be taken up in the liver endothelial and Kupffer cells. This may corrupt the specificity of the carrier. Our studies showed indeed that monomeric AraAMP-lactosaminated albumin was solely taken up in the hepatocytes (Jansen et al, 1993) and not in the sinusoidal cell types as was earlier shown for the preparations of Fiume et al, 1987.

Crucial for keeping the drug inside the cells may be the release of the polar, monophosphated form of e.g. AZT or araA (Fiume et al, 1982). We have shown that AraAMP is remarkably stable in lysosomal fractions of liver homogenates. In this preparation predominantly AraAMP was released after degradation of the carrier rather than AraA itself (Jansen et al, 1993). In the latter study it was also shown that human hepatocytes exhibit similar uptake kinetics of AraAMP-lactosaminated albumin compared with rat hepatocytes. Another advantage of introducing phosphorylated nucleoside derivatives in the cell via covalent binding to carrier proteins may be an improved rate of cellular activation into the antivirally active triphosphate form. For example, coupling of phosphorylated derivatives may circumvent resistance at the level of thymidilate kinase, which is a serious drawback in the case of AZT.

Interesting alternatives to the glycoprotein carriers might be glycosylated carboxyl-methyl-dextran (Nishikawa et al, 1993) and HPMA polymers (Seymour et al, 1991) as well as antibodies directed to HBV surface antigens (Shouval et al, 1982) or albumin treated with glutaminase or malondialdehyde. The latter proteins display a high affinity for receptors on the virus particle as well as the hepatocytes (Michalak and Bolger, 1989; Thung et al, 1989).

Targeting of antineoplastic agents and anti-toxicants

These include cytostatic agents such as daunorubicin, vindesine, methotrexate, cyclophosphamide, treminon and toxins such as ricin, abrin, gelonin as well as diphteria toxins (see Meijer and Van der Sluijs, 1989; Wu, 1988; Fallon and Schwartz, 1989; Blakey and Thorpe, 1988). As carriers natural glycoproteins (Fiume et al, 1980; Herschman et al, 1984; Kaneo et al, 1991), (neo-)glycoproteins (Monsigny et al, 1988; Trouet et al, 1982; Bodmer and Dean, 1988), tumour specific antibodies (Ghose et al, 1982), poly-L-lysine (Shen, 1989; Arnold, 1985; Ryser and Shen, 1986; Ponpipom et al, 1984) and various other synthetic polymers (Duncan, 1992; Seymour et al, 1991) were used (see Table 1). It should be realized here that antibodies carriers per se (for instance against HBVs Ag (Shouval et al, 1982) and alpha-fetoprotein) (Hirai et al, 1983), can be effective in the in vitro and in vivo lysis of hepatoma cells. Conjugates of antibodies and lactosaminated albumin with daunorubicin increased survival of animals with implanted tumours (Trouet et al, 1982; Schneider et al, 1984) and in some cases of cancer patients (Wu, 1988; Schneider et al, 1984). Asialoglycoproteins were linked to bacterial toxins yielding extremely potent drugs that inhibit protein synthesis in the hepatoma cells at a concentration of 10^{-11}M (Cawley et al, 1981; Chang and Kullberg, 1982; Simpson et al, 1982). Since in some but not all hepatomas the density of the particular receptor is considerably decreased, this approach is apt to failure in such cases (Wright, 1989). This receptor heterogeneity can be used to rescue non-cancer cells from the toxication of high doses of methotrexate (Wu, 1988). This was accomplished by coupling folinic acid to asialoglycoproteins and combining this with antineoplastic drug. Only the receptor negative cells are killed by the combination. It can be concluded from these data that the rescue factor, after release from the carrier, does not redistribute rapidly to the cancer cells. This targeting concept was later on expanded to prevent hepatoxicity of acetaminophen (Wu et al, 1985) and galactosamine (Wu et al, 1988) through covalent coupling of N-acetylcysteine and uridinemonophosphate, respectively, to asialoglycoproteins.

It is assumed that in vivo, the particular antitoxicant preparations are equally delivered to al the hepatocytes. However, studies on the acinar distribution of injected asialoglycoproteins exhibited zonal heterogeneity (Van der Sluijs et al, 1988; Burwen and Jones, 1990) with relatively low concentrations in the pericentral cells that is if the portal concentration of the carrier exceeds the K_m for receptor mediated uptake. Since the affinity of most galactose-terminated neoglycoproteins for the receptor as well as the hepatic extraction is considerably lower than for the naturally occurring asialoglycoproteins, a major heterogeneity in distribution of the neoglycoproteins is not anticipated (Van der Sluijs and Meijer, 1991). Although galactosylated carriers also may display some affinity for Kupffer cells (van Berkel et al, 1986; Bijsterbosch et al, 1991; Bijsterbosch et al, 1992; van Berkel et al, 1985), mannosylated (neo-)glyco-proteins and polyamino acids should be preferred for targeting of drugs to macrophages (Stahl, 1990; Monsigny et al, 1994; Negre et al, 1992; Takakura et al, 1994). This concept was used for the antileishmanial drugs (Negre et al, 1992), antimalaria agents (Trouet et al, 1982), cytotoxicity inducing agents (Monsigny et al, 1994) and recently also for a hepatic cell-protective action of superoxide dismutase that was obtained by mannosylation of the protein (Takakura et al, 1994).

Targeting of genes

The process of receptor mediated endocytosis can be in principle also be employed for delivery of small fragments of DNA to certain cell types (see Fig. 6). The option here is to correct genetic deficiencies in the production of essential proteins such as peptide

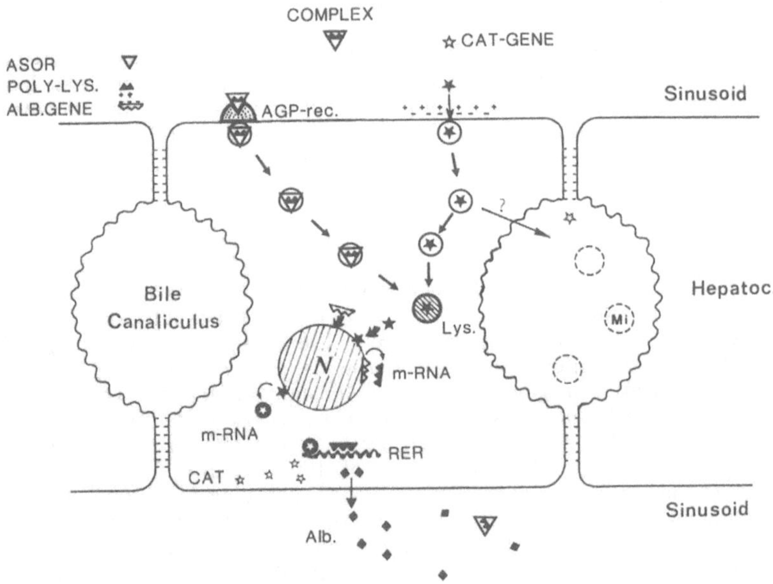

Fig. 6 Targeting of genes to the liver in vivo. In principle, two interiorization procedures can be used: cationization of the plasma membrane leading to adsorptive endocytosis of the DNA fragments [indicated for the chloramphenicol acetyl transferase (CAT) gene] or noncovalent binding of the plasmid to poly-L-lysine, covalently coupled to the asialo-glycoprotein asialoorosomucoid (ASOR) (indicated for the albumin gene). After receptor-mediated or adsorptive endocytosis, an unknown part of the endocytosed material escapes the abortive lysosomal route and part of the DNA becomes integrated in the cellular genome. Persistent expression was detected by mRNAs and synthesis of the particular proteins (See Wu et al, 1991)

hormones, plasma proteins and membrane receptors. The delivered plasmid should be taken up in the cell and integrated in the cellular genome. Transcription to m-RNA and translation into polypeptides can be monitored to detect the cellular expression of the targeted gene. Gene targeting has been attempted for insulin, using a proinsulin gene included in small liposomes (Nicolau and Cudd, 1989). In order to test the suitability of the targeted gene product, foreign "reporter" genes such as the bacterial gene coding for chloramphenicol-acetyl-transferase (CAT) are often used. In vitro, genes can be quite easily introduced in cultured cells by microinjection or Ca-phosphate-mediated permeation of the plasma-membrane. Fusion with plasmid containing liposomes (Nicolau and Cudd, 1989) and cellular penetration by viruses (Wilson, 1986) are also employed. Evidently, it is much more difficult to perform this in the intact organism. Recent developments in gene targeting to the liver in vivo indicate that persistent expression of genes might be achieved (Wu et al, 1989; Wilson et al, 1989; Wu et al, 1991; Wu et al, 1988). Receptor mediated endocytosis is used to introduce the gene, in spite of the fact that usually most of the internalized material is trafficked to the lysosomes. The crucial question here is how part of the endocytosed plasmid escapes this degradation route and reaches the cell nucleus instead. There is evidence for some interiorized ligands that part of the endocytic vesicles may become associated with the Golgi system and others, may recycle to the plasma membrane or can undergo transcytosis (Russel et al, 1983; Townsend et al, 1984). For a ligand such as epidermal growth factor, vectorial transport to the cell nucleus normally is a minor route but during regeneration (following hepatectomy) complete perturbation of cellular routing is observed. In this condition much less of the ligand is trafficked to lysosomes and much more to the cell nucleus (Marti et al, 1989). Whether this is true for other glycoproteins remains to be studied. Interestingly, integration of targeted DNA and also its expression can be greatly enhanced by prior partial hepatectomy (Wu et al, 1989; Wu et al, 1991) probably as a consequence of rapid cell replication. Agents that induce hyperplasia of the liver can also be used to improve integration of the foreign DNA. Pre-treatment of rats with the hypolipidemic agent nafenopin that induces a marked liver growth (Meijer et al, 1977), lead to persistant expression of the CAT-gene injected in vivo (Wu et al, 1990).

The prime technical item is how to design a suitable carrier for targeting DNA to organs in vivo: the complex should be sufficiently stable in the bloodstream, non-covalent binding of DNA should be preferred and DNA/carrier complex should be water soluble. Wu et al, used the basic polypeptide polylysine that strongly binds DNA. The polylysine matrix was then covalently coupled to asialooroso mucoid (ASOR) to obtain hepatocyte specificity (Wu et al, 1989). Direct galactosylation of polylysine negatively influences its binding affinity for DNA while the use of cluster glycosides with spacer arms is in principle possible though expensive (Wu, personal communication). Depending on the size of the plasmid, polylysine polymers of different length can be employed. Careful titration is necessary to keep the complex in solution. This concept was applied to target a foreign gene (CAT) combined with various promotors to the liver and clear expression of the gene was demonstrated (Wu and Wu, 1988) especially after partial hepatectomy (Wu et al, 1989). Excess asialoorosomucoid prevents DNA targeting to the liver probably through competition with the ASGP receptors. Both viral promotors and human promotor genes can be used. Recently Wu et al, succeeded in using gene targeting to correct genetic analbuminaemia in the so called Nagase rat model (Wu et al, 1991) (see Fig. 6) and cholesterolaemia due to a LDL-receptor deficiency in the rabbit (Wilson et al, 1989). A gene coding for human serum albumin came to expression in the injected rats within 48 hours after injection, raising the albumin concentration in the plasma from zero to 1.8 mg/100 ml. Persistent expression was seen until four weeks postinjection. A gene coding for the LDL receptor protein was targeted to the rabbit liver in vivo resulting in a 25-50% decrease in cholesterol plasma concentration lasting for 2 days after injection of the plasmid-polylysine-ASOR preparation. In both cases expression was controlled by detecting the particular m-RNA's as well as the proteins themselves.

These preliminary observations may mark the beginning of a new era in the manipulation of cell biology in vivo (see Meijer et al, 1992; Meijer et al, 1990; Meijer and Ziegler, 1993; Monsigny et al, 1988; Meijer and Van der Sluijs, 1989; Fallon and Schwartz, 1985). Applications in corrections of genetic disorders such as enzyme deficiencies, receptor deficiencies as well as in vivo production of pharmacologically active polypeptides

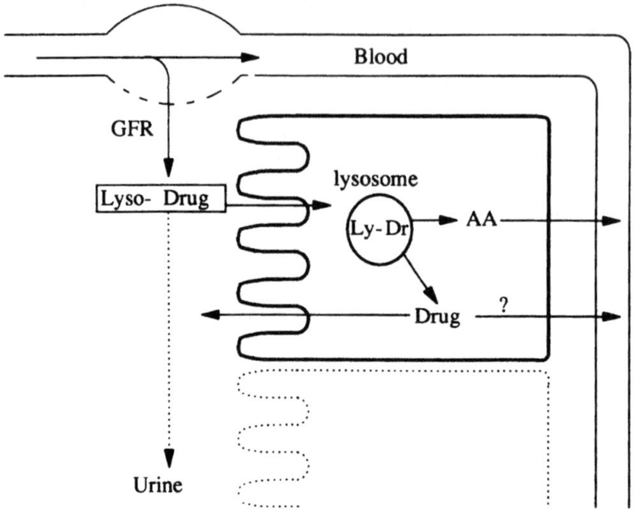

Fig. 7 The concept of renal drug targeting with low molecular weight proteins (LMWP's) such as lysozyme. After glomerular filtration and adsorptive endocytosis, the lysozyme-drug conjugate enters tubular lysosomes, at which site the protein is degraded to aminoacids (AA) and the drug is released. After local release in the tubuli, the drug may be excreted in primary urine and/or may partly re-enter the general circulation

represent novel options in the field of cell specific drug delivery.

DRUG TARGETING TO THE KIDNEY WITH PROTEIN CARRIERS

Introduction

For some years now we have worked on a drug targeting strategy for renal specific delivery and endorenal release of drugs using Low Molecular Weight Proteins (LMWPs). In general, LMWPs are known to be filtered and subsequently reabsorbed by the proximal tubular cells of the kidneys (see Fig. 7). Within these cells LMWPs are catabolized in lysosomes due to prevalence of proteolytic enzymes and a pH of 4-5. As such, these LMWPs might serve as drug carriers that release drugs site-specifically in the kidneys.

The capability of the kidney to release the parent drug from the drug-LMWP conjugates and drug-spacer derivatives by enzymatic or chemical hydrolysis of the connecting bond was tested by incubation experiments in renal cortex homogenates and lysosomal lysates. The disposition of drug-LMWP conjugates of naproxen, sulfamethoxazole and dopamine have also been studied in vivo. The pharmacokinetics of these conjugates was examined in freely moving rats and compared to those obtained after injection of equivalent doses of unreacted mixtures of drug and protein. In addition, the renal handling of the drug LMWP conjugates was studied by radioactive-imaging with a γ-camera. These studies demonstrated that the kidneys are the main site of uptake of drug conjugated LMWPs and that the drugs are renally uncoupled and finally excreted in the urine.

γ-camera experiments revealed a predominant renal uptake of radio-iodinated drug-conjugated LMWPs. Co-administration of an excess of the native LMWP resulted in a pronounced decrease of urinary (parent) drug excretion, suggesting competition for proximal tubular uptake between the LMWP and the drug-conjugated LMWP.

Table 2. Cleavability of drug-LMWP bonds in renal lysosomes[a]. Lysosomal degradability of linking moieties between various classes of drugs and the low molecular weight protein (LMWO)

Class	Drugs Structure	Therapeutic Class	Representative	Bonds Amide (direct)	Amide (peptide)	Amide (acid labile)	Ester	Spacer-type
1A	R-COOH (non-peptide)	NSAID's Diuretics	Naproxen Furosemide	-	-	n.a.	+ (E)	(poly)-α-hydroxyacids
1B	R-COOH (peptide)	ACE-inhibitors	Captopril Enalapril	n.t.	+ (E)	n.a.	n.t	none or oligopeptide
2	R-NH_2	Antibiotics Cardiovasculars Diuretics Cytostatics	Sulfa's Dopamine Aminoglycosides Adriamycin	n.t.	+/- (E)	+ (C)		oligopeptide or cis-aconityl-link

Notes: Abbreviations in the table: n.t. = not tested; n.a. not available. + = cleavable (E) denotes by enzymatic hydrolysis and (C) by chemical hydrolysis; - = not-cleavable.

Potential carrier systems for targeting of drugs to the kidney

LMWPs have attractive properties for renal targeting. (i) a selective renal handling, (ii) a local breakdown into natural constituents (amino acids), (iii) the presence of functional reactive groups in the molecule available for drug conjugation, and (iv) the immunogenicity of the carrier may be limited due to the endogenous character of these proteins. How can drugs be coupled to a LMWP, such that an adequate endorenal release of the drug is attained? In principle, non-covalent and covalent binding are possible. In the case of a non-covalent binding, an extremely high affinity is required to prevent premature dissociation of the drug-protein complex. Both physicochemical features of the drug and the properties of the chosen protein will be essential here. In the case of covalent binding, the functional groups of the drug and protein molecules should allow a cross-linking reaction in which direct binding of the drug to the protein or insertion of a spacer moiety between the drug and the protein are in principle possible.

The potentials of non-covalent association are limited, since it has been demonstrated that the pharmacokinetics of mixtures of different drugs (naproxen and sulfamethoxazole) with the LMWP lysozyme did not deviate significantly from those of equivalent single doses of these drugs (Franssen et al, 1991; Franssen et al, 1992). For the class of NSAIDs this has also been demonstrated in the case of their complexation with cyclodextrins (Frijlink et al, 1991). In this respect the only exception to this rule seems to be the quite stable binding of doxorubicin to DNA (Trouet and Jolles, 1984) and DNA to polylysine (Keegan-Rogers et al, 1991; Wu et al, 1989; Wilson et al, 1989; Wu et al, 1988; Wu and Wu, 1988).

For covalent coupling of drugs to proteins, the amino groups of the proteins can be employed for drug linkage as was shown before (Poznansky and Juliano, 1984; Trouet et al, 1982; Ryser and Shen, 1986). If such a covalent coupling is chosen, it is essential that the particular connecting bond between the drug and the carrier can be degraded in the target tissue. In this respect the metabolic make up of the target cell is of crucial importance. The kidney and particularly the lysosomes of tubular cells have been shown to exhibit a broad metabolic capacity (Haga, 1989; Barrett and McDonald, 1980; McDonald and Barrett, 1986; Barrett and Heath, 1977; Segal, 1975; Maunsbach, 1974). On this basis we have investigated various model drugs and bonds (Franssen et al, 1992). The cleavability, i.e. the extent and rate of active drug release, of these compounds was investigated in vitro. The results are summarized in Table 2. This table shows that for achieving an endorenal release of the parent drug, very different coupling modalities can be chosen depending on the chemical structures of the drugs. In most cases spacers between the drug and the carrier are required for efficient drug release. It was demonstrated that drugs with carboxyl groups (**Class 1a drugs**) can be linked via an ester bond to the protein (see Fig. 8). Potentially, this is a solution for most anti-inflammatory agents and diuretics with carboxyl groups, like

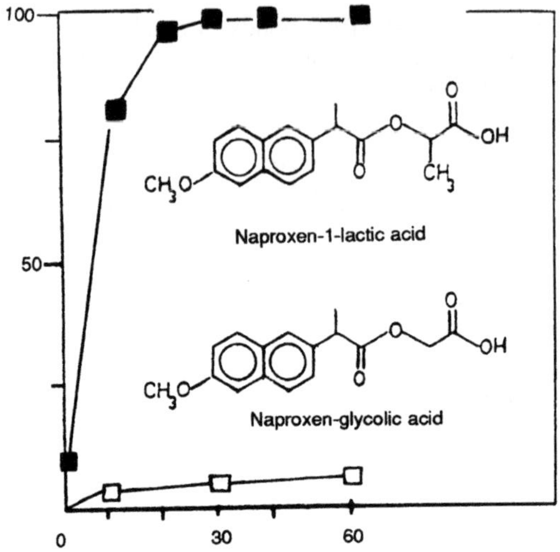

Fig. 8 Spacer technology in the preparation of drug protein conjugates. The anti-inflammatory drug naproxen can be coupled to protein carriers via biodegradable lactid acid or glycolic acid ester bonds. In a lysosomal cell fraction, naproxen is released from the two carrier moieties with very different rates. Renal generation of the parent naproxen coupled via a lactic acid spacer to lysozyme has been shown in the rat in vivo (see Franssen et al, 1992a)

furosemide and analogues. Only in the rather exceptional case that a drug has a peptide structure itself (**Class 1b drugs**), amide bonds can be used for conjugation. In this regard, a direct coupling to a LMWP may be possible for the group of angiotensin-converting-enzyme (ACE) inhibitors, although coupling via oligopeptides is an alternative. Drugs that have a primary animo groups available for conjugation (**Type 2**) can in principle be coupled to macromolecular carriers via oligopeptides or alternatively via acid sensitive bonds (Franssen et al, 1992a; Franssen et al, 1992b; Franssen, 1992).

Chemical hydrolisis of the drug based on the acid conditions of endosomes and lysosomes using acid-sensitive linkages, seems to be applicable to a broader variety of drugs and, in principle, is less dependent on the nature of the drug itself (Franssen et al, 1992a; Franssen et al, 1992b; Franssen, 1992). The rate of acid hydrolysis and drug release can even be controlled by chemical variations in the acid-sensitive moieties, as was demonstrated for the aconityl-type of spacers (Shen and Ryser, 1981). The potentials of acid-sensitive drug release from the LMWP carrier has been demonstrated for two different drugs, the antibiotic sulfamethoxazole and the vasodilating agent dopamine (both **type 2** drugs) (see Fig. 9). Both drugs could be conveniently coupled to lysozyme via a cis-aconityl-amide linkage by coupling via their amino groups (Franssen et al, 1992; Franssen, 1992). In vitro these drug-LMWP conjugates released the parent drug pH-dependently. At neutral pH's (6.8-7.4) as prevalent in the bloodstream and the primary urine, these conjugates were quite stable, whereas at low pH conditions as present in endosomes and lysosomes of tubular cells, the amide link appeared to be rapidly cleaved. Due to the fact that a protein contains many reactive groups, coupling may coincide with polymerization reactions. Due to such artifacts, the carrier characteristics of the protein may be corrupted. For instance, in the case of LMWPs, the low molecular weight character can be drastically altered due to polymerization and this may impede their glomerular filtration (Maack, 1975). In addition, polymerized material tends to accumulate to a large extent in the RES and in particular the Kupffer cells of the liver (Franssen et al, 1993).

For establishing the degree of coupling, both the amount of drug and protein present in the preparation have to be determined. Covalently coupled drug can be measured quantitatively by HPLC after alkaline or acid hydrolysis of the conjugates (Franssen et al,

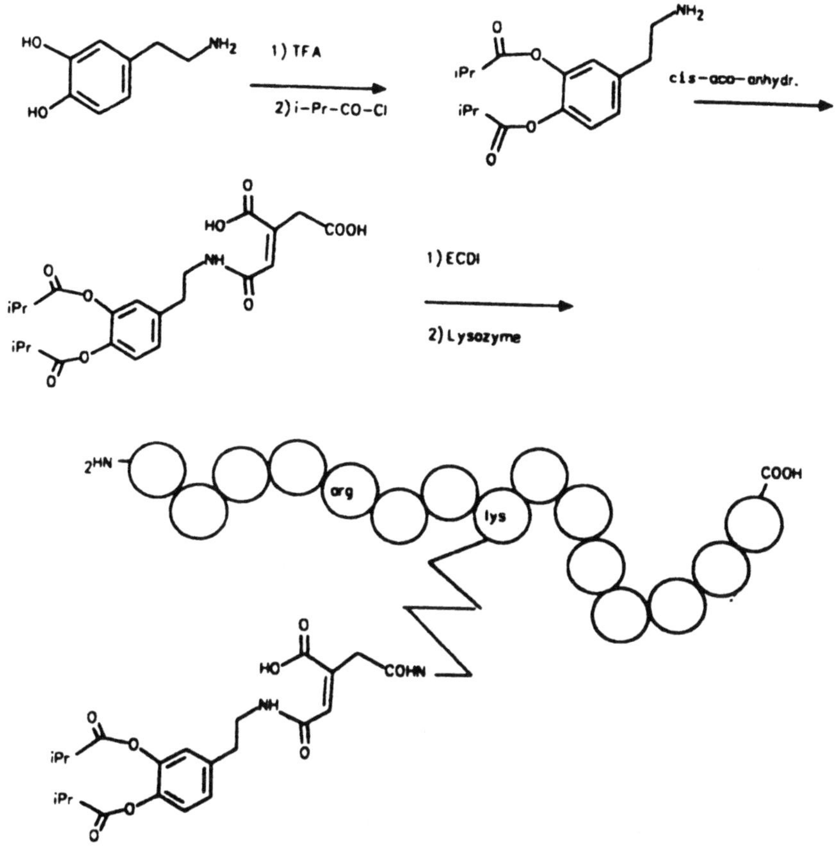

Fig. 9 Synthesis of Dibudop-aco-Lysozyme. Chemical structure of dibutyryl-dopamine coupled via an acid sensitive aconityl-spacer to lysine-NH_2 groups of low molecular weight proteins

1991; Franssen et al, 1992; Franssen et al, 1993; Franssen et al, 1993). Separation techniques, such as FPLC may provide information on charge (ion-exchange chromatography) and size (gel permeation chromatography) alterations of the protein as well as the validation of the drug conjugation procedure (Franssen et al, 1992). The identity of the drug-protein conjugate can be confirmed by ionspray mass spectrometry (Bruins et al, 1987; Covey et al, 1988). For a drug-conjugated LMWP, the expected molecular weight was corroborated by the observation of the corresponding multiple charged ion, as depicted in Figure 10. These analytical data can provide proof of covalent binding of the drug to the LMWP as well as verification of the intercalation of the spacer molecule between the drug and the LMWP (Franssen et al, 1993). Evidence for the presence of individual components in a synthetic drug-protein conjugate mixture can be obtained from separation by capillary zone electrophoresis with on-line detection by ionspray mass spectrometry (Fig. 10). It has been demonstrated that a separation of a mixture of lysozyme and its conjugates with one or more drug molecules by capillary electrophoresis is superior to that by ion-exchange chromatography using FPLC (Franssen et al, 1992; Kostiainen et al, 1993). In conclusion, a spectrum of analytical techniques is available for an appropriate characterization of suitable drug-LMWP conjugates.

State of the art on renal drug targeting with LMWP's

Thus far, only a few drug-LMWP conjugates have been tested for renal drug delivery and release in vivo. Both drug-LMWP conjugates of the **class 1** type (Franssen et

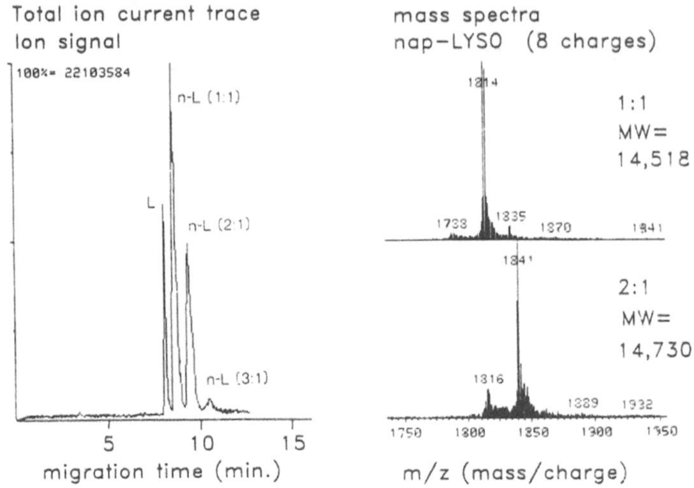

Fig. 10 Separation of naproxen-lysozyme conjugates with 1,2 and 3 naproxen molecules coupled per lysozyme molecule by capillary zone electrophoresis followed by mass spectrometry identification by mass over charge ratio's. Lysozyme exhibits 8 positive charges

al, 1991; Franssen et al, 1993) and **class 2** type (Franssen et al, 1992; Franssen, 1992) have been studied. It was demonstrated that the kidneys indeed are the main site of uptake of the drug-conjugated LMWPs (Franssen et al, 1991; Franssen et al, 1992; Franssen et al, 1992; Franssen et al, 1993).

In vivo, the **type 1** drug naproxen has been shown to be released endorenally from the LMWP conjugate, provided that coupling is performed with lactic acid via an ester bond (Franssen et al, 1993) (Fig. 8). In contrast, in the case that naproxen is directly coupled to the protein via an amide linkage (Franssen et al, 1991), no parent drugs was regenerated. Rather, renal proteolysis of the LMWP-conjugated drug resulted in formation of its lysine-derivative. Interestingly, this lysine-naproxen derivative possesses an equipotent pharmacological activity as compared to the native drug (Franssen et al, 1991).

The **type 2** drug sulfamethoxazole has been shown to be endorenally released from the drug-LMWP conjugate. Apart from the altered kinetics of the drug itself, the selective renal distribution was further demonstrated by the absence of its hepatically formed metabolite N^4-acetyl-sulfamethoxazole in urine. This metabolite was abundantly present in the control experiments after injection of the drug in its uncoupled form. The LMWP-concept was also tested for dopamine (Franssen, 1992). Within a relatively low LMWP dose range, a dose-dependent endorenal dopamine release was demonstrated. In a higher dose range the saturability of LMWP reabsorption (Maack et al, 1979; Sumpio and Maack, 1982) was demonstrated by a relatively low endorenal release and urinary excretion of dopamine. The local kinetics of the drugs are important to consider for the ultimate success of the targeting strategy. In this respect the question arises to which compartment the drugs are transported; can they reach their site of action of do they reenter the bloodstream before exerting a renal specific effect?

From the experiments with naproxen it has been inferred that free naproxen, as observed in plasma, originates from released drug in the tubular cells. Nevertheless renal selectivity is still obtained since these experiments revealed a plasma-AUC of the free drug of only 13% as compared to an equivalent dose of unconjugated drug (Franssen et al, 1993). Yet part of the released drug can escape from the lysosomes and reenter the bloodstream, either directly, or by reabsorption after initially entering the tubular lumen.

On theoretical grounds, transport across the lysosomal membrane is expected to occur rather quickly for **class 1** drugs. Within lysosomes, these acidic drugs are predominantly in their uncharged, lipophilic forms. Therefore, the passive driving force for these agents may be sufficient to allow them to enter the cytoplasmic compartment of tubular cells. It should be noted that unconjugated naproxen is also taken up in proximal tubular cells by anionic transporters present at the BLM-site (Moller and Sheikh, 1983). In its LMWP-conjugated form it is taken up at the luminal BBM-site and the lysosomally released drug may also enter the tubular cytoplasm. From this point on, it may follow similar local kinetics, including transfer to sites of action and local metabolism as its unconjugated couterpart. In this respect the released naproxen may flow from the basolateral domain into the blood, assuming that the membrane transport at that level is bidirectional. Alternatively, some of the released and unchanged drug may be excreted into the primary urine and be reabsorbed in more distal parts of the tubuli into the blood.

Interestingly, both LMWP-conjugated and unconjugated naproxen appeared to be finally excreted in the urine predominantly in the demethylated and sulphated form. This may reflect renal metabolism of NSAIDs. Renal metabolism of NSAIDs may contribute to a large extent to its body clearance, as has recently been established by us (Moolenaar et al, 1992).

For **class 2** agents no reentry of the drug from the cells into the blood could be demonstrated thus far. Apparently, these agents are predominantly transported to the tubular lumen and urinary space and are not reabsorbed significantly. In the case of LMWP-conjugated dopamine, delivery to the urinary space of the parent drug is enhanced as compared to administration in the unconjugated form. Renally formed dopamine apparently did not enter the systemic circulation.

A major portion of the natural occurring dopamine excretion in the urine originates from renal dopamine which is locally produced out of L-DOPA and excreted from the cytoplasm of tubular cells (Casagrande et al, 1989). Interestingly, LMWP-conjugated dopamine is probably delivered very closely to this site. After passing the lysosomal membrane, the regenerated dopamine may be subject to similar intrarenal transport processes as the endogeneously formed compound and may exert similar renal effects.

Passive transport across the lysosomal membrane of **class 2** drugs is expected to occur more slowly than their **class 1** counterparts. Since these agents (basic drugs are predominantly in their charged, hydrophilic forms in the acidic lysosomal environment, diffusion to the tubular cytoplasm may occur relatively slowly. In lysosomes, dibutyryl dopamine may be released from its LMWP conjugate in its (partially) esterified and in its free form. Presently, it is unknown in what form and by what kind of transport dopamine actually enters the cytoplasm. The hydrophobicity of the ester groups as applied for the LMWP-conjugated dopamine may influence the rate of passive diffusion across this membrane. In addition, various active transport systems are present in lysosomal membranes, including one for the transport of amino acids and dipeptides (Barrett and McDonald, 1980; Segal, 1975). In principle, it is possible that intralysosomally released drugs like dopamine (Franssen, 1992) of drug catabolites, like naproxen-lysine (Franssen et al, 1991) are also transported by these systems.

The LMWP-concept seems to be applicable to a relative large variety of drugs aimed to manipulate renal function or combat local infections. Drugs that can be appropriately conjugated to LMWPs include at least those with carboxyl and amino groups available for conjugation. From a therapeutic point of view, these classes of drugs cover a large number of interesting agents (see Table 1). In the future, the application of other types of spacers may allow coupling of drugs with other reactive groups (like, for instance, hydroxyl groups as present in some corticosteroids).

It is important to note that the mechanism of renal uptake and endorenal drug release by this macromolecular drug targeting concept is clearly different from that governing renally applied prodrug strategies. As has been demonstrated for gamma-glutamyl prodrugs, these preparations enter the proximal tubular cell at the basolateral site, and parent drug generation is supposed to occur by enzymes present in the cytoplasm. Limitations, such as incomplete renal uptake of conjugated drugs and major uptake in the liver have been observed in application of such prodrug strategies (Daemen et al, 1991). Therefore, the use of LMWP carriers may serve as useful complement to the other strategies of renal drug targeting.

DRUG TARGETING WITH (GLYCO-)PROTEIN CARRIERS TO VARIOUS BLOOD CELL TYPES

Introduction

The design and development of potential carriers for cell-specific delivery of therapeutics should be based on the knowledge of recognition sites on the surface of blood cells as well as insight in internalization and further cellular disposition of such macromolecules. The presence of various lectins on T-lymphocytes and monocytes/macrophages as well as immunological receptors and adhesion molecules have been demonstrated (Meijer et al, 1992; Monsigny et al, 1988). The recognition sites could in principle serve in binding to endocytosis of glyco-conjugate carriers for the delivery of drugs and genetic material. Glycoproteins with a suitable density of sugar groups or a proper geometric organization of oligosaccharide side chains may provide potential carriers that escape receptor mediated endocytic processes in other tissues such as liver, spleen and bone marrow. Other modalities are recombinant T-cell receptors (sCD_4) that bind to HIV infected cells, monoclonal antibodies raised against viral envelope proteins as well as certain cytokines. Bacterial toxins, antisense DNA, and genes that code for synthesis of antivirally active proteins as well as coupling products of nucleoside analogues such as AZT have been delivered to blood cells using this technology. Some of these drug carriers may provide an intrinsic antiviral action that may add to the effect of the coupled drug (dual targeting). Such multi-active drug targeting preparations may offer advantages of synergistic effects, counteraction of drug resistance apart form the improved body distribution aimed at.

POTENTIAL CARRIERS FOR TARGETING OF DRUGS TO BLOOD CELLS

The receptors and lectins that are present on blood cells can be directly (e.g. using neoglycoproteins) or indirectly (e.g. using monoclonal antibodies directed against the receptor) used for drug targeting purposes.

The use of monoclonal antibodies (MoAbs) to target drugs to specific cell types is a promising approach in view of the large extend of tissue specificity that is obtained. Drugs are coupled to an antibody, thereby creating a hybrid molecule with the specificity of the immunological ligand that retains the therapeutic activity of the drug (Vitetta and Uhr, 1985).

Disadvantages that can be anticipated using immunoconjugates include: toxicity of the antibodies due to their cross reactivity with non-target cells (Vitetta and Uhr, 1985), limited access of the antibody preparations to the target cell type (Hwang et al, 1984), cell heterogeneity with respect to the determinant to which the antibody is directed (Hwang et al, 1984; Olsnes et al, 1989) as well as immunogenicity of the antibody and that of the antibody-drug conjugate.

Other drawbacks may be loss of antibody specificity and/or affinity due to coupling with the drug molecule and "opsonisation" of injected antibodies and complex formation with circulating antigens (Molthoff et al, 1992).

The latter point was clearly demonstrated by the observation that tumor-specific antibodies have largely different pharmacokinetics in tumor-bearing individuals compared with normals because of a rapid clearance of the immune-complex by the liver (Shea et al, 1989). Immunotoxins may also be partly inactivated and cleared via the α_2-macroglobulin system and thereby accumulate in non-target tissue such as liver and monocytes (Ghetie et al, 1991).

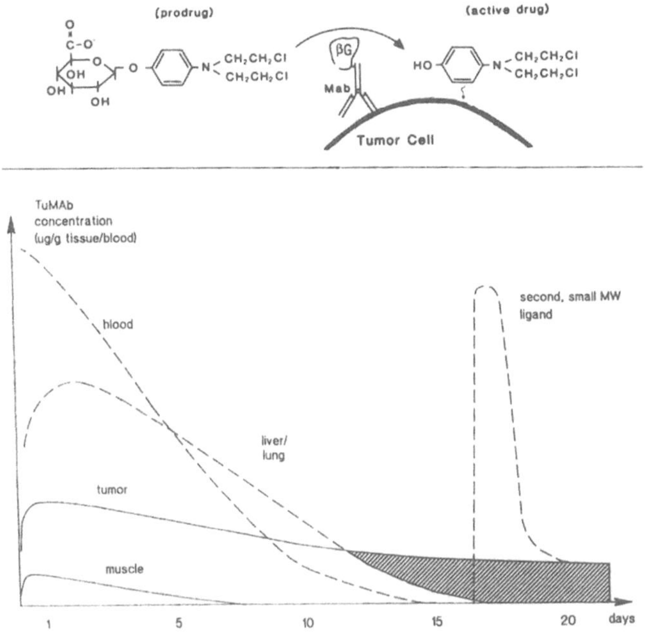

Fig. 11 Antibody directed enzyme prodrug targeting. The concept of antibody directed enzyme prodrug targeting (ADEPT) in which chlorambucil is liberated from its glucuronide prodrug at the target tumor cell that is coated by a tumor-specific antibody-β-glucuronidase conjugate (top). Biphasic treatment of the tumor is essential in which the prodrug administered at the time that the monoclonal antibody-enzyme conjugate has been degraded in normal tissue but less in the tumor

The potential problems with the MoAb carriers can be approached in various ways. Purification of the antibody, using affinity techniques, can be used to extract the cross reacting fraction (Ghose et al, 1982). Pretolerisation of patients by immunosuppressive monomethoxy-polyethyleneglycol (mPEG) conjugates prior to the treatment with the immunoconjugate itself may reduce the immunogenic reactions (Sehon, 1991). The access of MoAbs to the cells can be improved by employing immunological active fragments ($F(ab)_2$ and Fab') instead of the complete IgG protein. Transcapillary and cell wall passage of these fragments is much better than that of the intact immune products (Ghose et al, 1982; Olsnes et al, 1989). In addition, such fragments display a diminished immunogenicity compared to the entire protein. However, the blood half-lives of these fragments are significantly shorter than those of the intact IgG (Molthoff et al, 1992) and only the combination of characteristics (cross reactivity, immunogenicity, pharmacokinetic parameters) for each individual MoAb or fragment will determine its suitability in drug targeting.

An innovative strategy for the delivery of cytotoxic or antiviral agents to specific cell types is the prodrug activation by Ab-enzyme conjugates. The enzymes to be targeted are chosen for their ability to convert relatively non-toxic drug precursors into their active form. Such antibody coupled enzymes can be specifically delivered to the cell type that expresses the antigenic determinant (Bagshawe et al, 1988; Senter, 1990; Senter et al, 1993). The formation of the active derivative in the close proximity of the target cells could lead to higher cellular and lower systemic concentrations of the active drug. In fact, the need for Ab internalization, one of the major problems in the immunoconjugate application, is eliminated in this concept (see Fig. 11). Enzymes that can be used are for instance alkaline phosphatase (for the conversion of phosphated prodrugs) and carboxypeptidase (an enzyme converting inactive carboxy compounds into their active carboxylic counterparts). Haisma et al, described a monoclonal antibody-β-glucuronidase conjugate as activator of the prodrug epirubicin glucuronide for the specific treatment of cancer (Haisma et al, 1992). This strategy may not only be applicable in cancer therapy (Senter et al, 1993; Bagshawe et al, 1988), it may also be applicable in the therapy of other diseases. A potential drawback forms the presence of related (endogenous) enzymes in the general circulation of the treated individual that may lead to premature activation of the particular prodrug. Of major importance in this concept is the timing of administration of the prodrugs: only after most of the ab-enzyme conjugate is eliminated from non-target tissues and high affinity binding to the tumor prevails a real advantage of this ADEPT approach can be anticipated (see Fig. 11).

In the search for more effective anti-HIV agents, several groups reported on the use of recombinant CD4 (rCD4) as a carrier for toxins. In principle, rCD4 can exhibit the same degree of cell specificity and affinity as antibodies do. rCD4 was chosen because of its high affinity for the HIV envelope gp120. Although gp120 shows a distinct variability among different HIV strains, its CD4 binding site is highly conserved. Therefore, rCD4 may exhibit the same high affinity to gp120 as cellular CD4 and would potentially target coupled antiviral drugs to HIV infected cells.

The loss of CD4+ cells in HIV patients is partly due to the interaction of gp120 in the membrane of infected cells with CD4 molecules on non-infected cells. This leads to cell membrane fusion and syncytium formation and implies that rCD4 carriers will have an additional intrinsic therapeutic effect. A pronounced anti-HIV activity of a conjugate of rCD4 with deglycosylated ricin A chain was found in an HIV infected human T cell line. Non-infected cells, expressing MHC class II antigens (the natural ligand for CD4), were not affected (Till et al, 1988). A recombinant protein, containing the HIV-binding portion of the human CD4 molecule, was linked to active regions of Pseudomonas Exotoxin A (PE-A) and displayed selective toxicity towards infected cells expressing the gp120 (Chaudhary et al, 1988). A combination of this CD4-PE-A and reverse transcriptase inhibitors resulted in highly synergistic effects and led to a complete elimination of infectious HIV-1 from human T cell lines in vitro (Ashorn et al, 1990). In spite of these promising in vitro effects, further in vivo testing is awaited. Problems may be encountered in the immunogenicity of the rCD4 products as well as harmfull effects caused by interactions with elements of the immune system. Furthermore, destruction of HIV infected cells and the sudden release of particles may lead to massive infections of previously uninfected CD4 positive cells, although recent evidence exist that the released particles are immature and non-infectious (Vitetta, personal communication). Inactivation of CD4 based toxins by serum components present in the blood of HIV positive individuals (Ghetie et al, 1991) and the observation that most of the infected cells do not express gp120 (Schooley, 1992) may be major drawbacks in this approach. Interestingly it was recently shown that coupling of the CD4-toxin to albumin largely prevents inactivation by serum components probably by steric hindrance. Also the extremely short serum half life of CD4 linked toxins can be overcome by the use of this CD4-toxin-albumin conjugate (Ghetie et al, 1991).

Glycoproteins as drug carriers to blood cells

Glycoproteins have several favourable properties in relation to drug targeting (Poznansky and Juliano, 1984; Bodmer and Dean, 1988). Preparation of glycoprotein-drug conjugates is relatively cheap compared to e.g. that of antibody-drug conjugates. Neo-glycoproteins are generally preferred to endogenous glycoproteins. A major problem of such naturally occurring plasma proteins is the presence of the non-terminal sugars in the oligosaccharide chain that invite interactions with other receptors in the body. Furthermore, enzymatic cleavage of sugars from the chain in order to expose the required terminal sugar is often incomplete. With this procedure, glycoproteins with a mixture of terminal sugars are produced and consequently loss of cell specificity should be anticipated.

T-lymphocyte targeting of AZT with neoglycoprotein carriers

In our laboratory, we investigated the possibilities of specific delivery of the anti-HIV drug AZT to T-lymphocytes using neoglycoproteins as drug carriers. We synthesized neoglycoproteins with human serum albumin as the protein backbone and varied the type of sugar as well as the number of sugar molecules per protein molecule. AZT was, in its monophosphate form, covalently attached to the albumin using a water soluble carbodiimide as linking agent (Molema et al, 1991). The number of AZTMP molecules per protein molecule varied between 1 and 6 and the charge of protein did not alter upon conjugation (Molema et a, 1991; Molema et al, 1990; Molema, 1992). In an in vitro anti-HIV assay using the HIV-III_B strain of the virus and the human T-lymphocyte cell line MT-4 as the target cells, mannosylated AZTMP conjugates were shown to have a remarkably high activity against HIV-1 induced cytopathicity (Molema et al, 1990). The galactosylated and fucosylated AZTMP conjugates exhibited anti-HIV-1 activity in a similar concentration range. All of these conjugates were significantly more potent against HIV-1 than the nonglycosylated albumin-AZTMP conjugate and selectivity indices (the ratio of the 50% cytotoxic concentration and the 50% inhibitory concentration) were comparable to or higher than those of AZT and AZTMP freely administered.

Some of the particular neoglycoproteins with relatively low sugar density (less than 10 galactose or mannose groups per albumin molecule) exhibit a favourable pharmacokinetic profile: they are only slowly eliminated from the blood stream and are therefore adequately exposed to the target cells. The relatively low sugar substitution and minimal change in net charge will prevent recognition by the hepatic and splenic endocytic systems and may, at the same time, provide a "minimal carbohydrate structure" resembling the biantennary oligosaccharide chains that are considered to be preferentially recognized by T-cell lectins (Bezouska et al, 1985a; Bezouska et al, 1985b). Furthermore, the conjugates with high mannose substitution (more than 15 sugar groups per albumin molecule) are

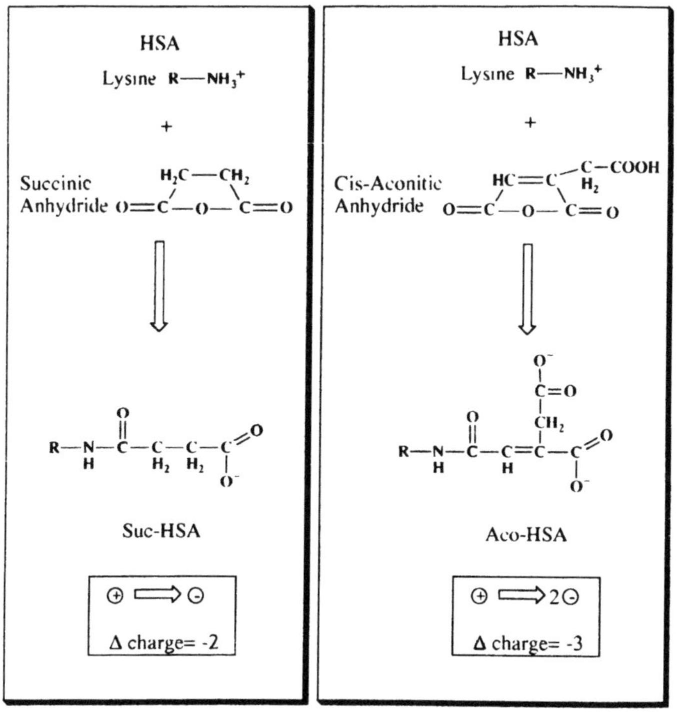

* Derivatization of HSA with anhybrides, leading to polyanionic HSA.

Fig. 12 Derivatization of HSA with anhydrides, leading to polyanionic HSA. Chemical synthesis of succinylated human serum albumin (HSA, left) and aconitylated HSA (right) using the anhydrides. Each reacted amino group in the HSA produces a net charge difference of -2 or -3 respectively. Since about 60 lysine-NH groups are available per HSA molecule, strongly negatively charged (poly-anionic) peptides are produced

attractive candidates for targeting to monocytes and macrophages (Monsigny et al, 1994).

The possible mechanisms of action of these neoglycoprotein-AZTMP conjugates that were proposed, are: 1) extracellular hydrolysis of the phosphoramide bond between AZTMP and the neoglycoprotein at the surface of the cells followed by diffusion of the AZT into the cells; 2) sugar-specific recognition and internalization of the AZTMP-conjugates by the MT-4 cells, followed by intracellular release of the drug; 3) co-endocytosis of virus and AZTMP-conjugates following sugar specific interaction between virus and neoglycoprotein-AZTMP-conjugate or 4) a combination of the above mentioned mechanisms. Extracellularly, the phosphoramide bond between AZTMP and protein appeared to be fairly stable. Experiments with ^{125}Iodine-labeled AZTMP conjugates showed that uptake of some of the conjugates was temperature dependent and could be partially inhibited by excess unlabeled conjugate, indicating a, receptor mediated uptake process (Molema, 1992).

Interestingly, the neoglycoprotein carrier containing a high ratio mannose to protein (40:1) had an own intrinsic activity against HIV-1. We considered that this intrinsic activity of the carrier might be caused by the inhibition of the interaction between HIV-1 gp120 and the MT-4 cells. The evidence provided by Curtis et al, 1992 and Haidar et al, 1992; Gattegno et al, 1991 that the gp120-binding protein can act as mannose binding lectin, at first sight supported this idea. However, another explanation for the strong anti-HIV-1 effect could be the increased negative charge of the carrier, introduced by the p-aminophenyl sugars as mentioned before.

Modification of human serum albumin with e.g. succinic anhydride (see Fig. 12) and aconitic anhydride turned the protein into highly negatively charged proteins (sucHAS and acoHSA, respectively), aco-HSA having an anti-HIV-1 activity of 10-40 times that of AZT on a molar basis (Jansen et al, 1991; Jansen et al, 1993). This high efficacy may indeed demonstrate a relationship between negative charge and anti-HIV-1 activity. The mechanism of this charge-based antiviral activity was demonstrated to be on the fusion level, both in the initial phase of the infection cycle as well as in a terminal phase, where two or more infected cells fuse with uninfected cells into giant cells (syncytia) (Jansen et al, 1991; Jansen et al, 1993). Inhibition of syncytium formation by sucHSA can occur at a concentration that neither affected primary binding of HIV to the cells nor binding of monoclonal antibodies directed against the CD4 binding domain of gp120. We postulated

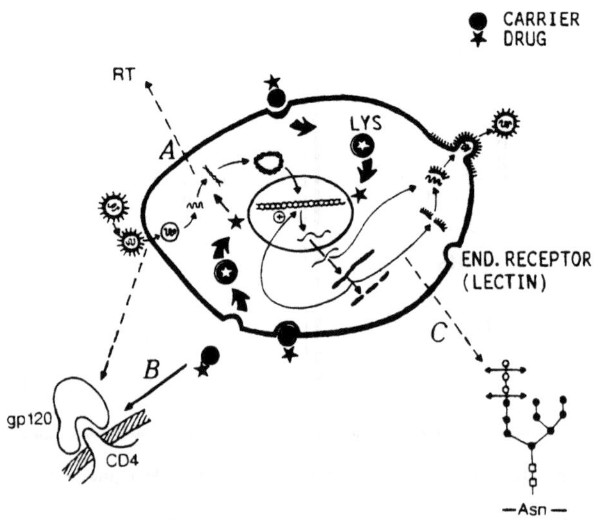

Fig. 13 The concept of dual targeting using antiviral drug conjugates. The carrier inhibits virus/cell binding and fusion extra-cellularly (B) and (after internalization) also delivers antivirally active drugs such as nucleoside analogs (A) and glycosylation inhibitors (C) for intracellular inhibition of viral replication

therefore that the negatively charged albumins interact with gp41 envelope glycoprotein or may bind to the V3 loop of gp120 that have been demonstrated to be essential for virus-cell fusion. The latter mechanism was earlier proposed for the polyanionic dextran sulfate. However, in contrast to sucHSA, dextran sulfate inhibits virus binding in the same concentration range where it inhibits syncytium formation. The peculiar mechanism of action of negatively charged albumin was recently confirmed in a study on maleylalbumin (Takami et al, 1992) and together with some plant lectins (Balzarinin et al, 1992), they constitute a novel category of anti-HIV compounds.

At present we investigate these proteins as potential carriers for AZT-like compounds potentially providing inhibition of HIV replication both at the level of virus/cell fusion and syncytium formation as well as on RNA transcription. Such dual targeting with intrinsically active carriers (see Fig. 13) could lead to synergistic effects and a preferential activity against syncytium inducing (SI) variants of HIV-1 that are predominant in the terminal phase of HIV infection. SI HIV-1 isolates have been shown to have an increased positive charge in the V3 loop of the gp120 domain due to changes in amino acid composition and may therefore represent a potential target for the modified albumins (Callahan et al, 1991; Batinic and Robey, 1992). In addition, T-lymphocytes like macrophages may express some kind of scavenger receptors that recognize negatively charged proteins (Takami et al, 1992) or may provide binding sites for polyanions via CD receptors and adhesion molecules (Meijer et al, 1992).

Another example of the direct use of the presence of a specific receptor on the target cells is the successful targeting of diphteria toxin related IL-2 fusion protein DAB_{486}-IL-2 to IL-2 receptor positive T cells. DAB_{486}-IL-2 bound to the high affinity form of the IL-2 receptor, was rapidly internalized into acidified vesicles and subsequently inhibited protein synthesis (Bacha et al, 1988).

Derivatization of a poly-lysine macromolecule with mannosyl and the antiviral drug PMEA resulted in an uncharged mannosylated polymer with a molar ratio drug to protein of 20:1. It was shown to be more potent in inhibiting herpes simplex virus type 1 replication in vitro in human macrophages than the freely administered PMEA (Midoux et al, 1990). As alternatives for retroviral vectors, neoglycoproteins and mannosylated polymers could also serve as potential carriers for genes, antisense and antigene material to virus-infected cells Monsigny et al, 1988, 1994). Antiviral therapy with genes encoding CD4, ribozymes, toxins and dominant-negative mutants of HIV-proteins were reported by various groups (see for references Bevec et al, 1992; Harrison et al, 1992). Intracellular immunization is especially attractive if gene expression can be put under control of HIV promotor sequences or trans activating proteins so that infected cells can be killed in an early stage.

CONCLUSIONS

Although so far some promising results have been achieved in vitro in the targeting of various kinds of drugs with protein carriers, much work remains to be done. More data are required to establish which fraction of the conjugated drug will finally be released in the target cells in vivo. Although this was extensively studied for lactosaminated HSA and liver (hepatocyte) targeting, further studies are required for natural- and (neo)-glycoproteins as well as for sugar derivatized polyaminoacids and polymers. In fact, receptor mapping in various tissues for such potential carriers should be performed in healthy individuals as well as in the diseased state.

Coupling of drugs to macromolecular carrier systems a priori implies that parenteral formulations have to be used. Although parenteral dosing is quite accepted for short term and even for long term clinical use (e.g. insulin and other hormone preparations) it is clear that such drug targeting preparations should have major advantages compared with the parent drug to justify their development. Such advantages could include an increased potency, an improved therapy of intracellular infections or a major reduction of the dosing frequency.

REFERENCES

Arnold, L.J., 1985, Polylysine-Drug Conjugates, in: "Methods in Enzymology", Drug and Enzyme Targeting, Part A, Academic Press, New York

Ashorn, P., Moss, B., Weinstein, J.N., Chaudhary, V.K., Fitzgerald, D.J., Pastan, I. and Berger, E.A., 1990, Elimination of infectious human immunodeficiency virus from human T-cell cultures by synergistic action of CD4-Pseudomonas exotoxin and reverse transcriptase inhibitors, Proc.Natl.Acad.Sci.USA, 87:8889

Bacha, P., Williams, D.P., Waters, C., Williams, J.M., Murphy, J.R. and Strom, T.B., 1988, Interleukin 2 receptor-targeted cytotoxicity. Interleukin 2 receptor-mediated action of a diphteria toxin-related Interleukin 2 fusion protein, J.Exp.Med., 167:612

Bagshawe, K.D., Springer, C.J., Searle, F., Antoniw, P., Sharma, S.K., Melton, R.G. and Sherwood, R.F., 1988, A cytotoxic agent can be generated selectively at cancer sites, Br.J.Cancer, 58:700

Balzarini, J., Neyts, J., Schols, D., Hosoya, M., Van Damme, E., Peumans, W. and De Clercq, E., 1992, The mannose-specific plant lectins from cymbidium hybrid and epipactis helleborine and the (N-acetylglucosamine)$_n$-specific plant lectin from urtica-Dioica are potent and selective inhibitors of human immunodeficiency virus and cytomegalovirus replication in vitro, Antiviral Res., 18:191

Barrett, A.J. and McDonald, J.K., 1980, "Mammalian Proteases: a glossary and bibliography; vol. I endopeptidases", London: Academic Press

Barrett, A.J. and Heath, M.F., "Lysosomes: a laboratory handbook", Amsterdam: Elsevier, 1977, 2nd Ed

Batinic, D. and Robey, F.A., 1992, The V3 region of the envelope glycoprotein of human immunodeficiency virus type 1 binds sulfated polysaccharides and CD4-derived synthetic peptides, J.Bio.Chem., 267:6664

Beatty, B.G., O'Conner-Tressel, M., Do, T.L., Paxton, R.J., Beatty, J.D., 1990, Mechanism of Decreasing Liver Uptake of In-111-Labeled Anti-Carcinoembryonic Antigen Monoclonal Antibody by Specific Antibody Pretreatment in Tumor Bearing Mice, Cancer Res., 50:S846

Van Berkel, T.J.C., Kruyt, J.K., Harkes, L., Nagelkerke, J.F., Spanjer, H. and Kempen, H.J.M., 1986, Receptor-dependent targeting of native and modified lipoproteins to liver cells, in: "Site Specific Drug Delivery", E. Tomlinson and S.S. Davis (Eds.), John Wiley and Sons, Ltd

Van Berkel, T.J.C., Dekker, D.J., Kruyt, J.K. and Van Eyk, H.G., 1987, The interaction in vivo of transferring and asialotransferrin with liver cells, Biochem.J., 243:715

Blakey, D.C. and Thorpe, P.E., 1988, Prevention of carbohydrate-mediated clearance of ricin containing immunotoxins by the liver, in: "Immunotoxins", A.E. Frankel, ed., Kluwer Academic Publishers, Boston/Dordrecht/Lancaster

van Berkel, T.J.C., Kruyt, J.K. and Kempen, H.J.M., 1985, Specific targeting of high density lipoproteins to liver hepatocytes by incorporation of a tris-galactoside terminated cholesterol derivative, J.Biol.Chem., 260:12203

Bevec, D., Dobronik, M., Hauber, J. and Bohnlein, E., 1992, Inhibition of human immunodeficiency virus type 1 replication in human T cells by retroviral-mediated gene transfer of a dominant-negative rev trans-activator, Proc.Natl.Acad.Sci.USA, 89:9870

Bezouska, K., Taborsky, O., Kubrycht, J., Pospisil, M. and Kocourek, J., 1985a, Carbohydrate structure-dependent recognition of desialyated serum glycoproteins in the liver and leucocytes. Two complementary systems, Biochem.J., 227:345

Bezouska, K., Karaskova, H., Taborsky, O., Kofronova, O., Vorisek, J., Kubrycht, J. and Kocourek, J., 1985b, Studies on lectins. LXVI. Binding of serum glycoprotein asialo-oligosaccharides to the porcine liver and leucocyte membrane lectins, in: "Lectins", vol. IV, T.C. Bog-Hansen and J. Breborowicz, eds., Walter de Gruyter & Co., Berlin, New York

Bijsterbosch, M.K., Bernini, F., Bakkeren, H.F., Gotto, A.M., Jr., Smith, L.C. and van Berkel, T.J.C., 1991, Enhanced Hepatic Uptake and Processing of Cholesterol Esters from Low Density Lipoprotein by Specific Lactosaminated Fab Fragments, Arteriosclerosis and Thrombosis, 11:1806

Bijsterbosch, M.K. and van Berkel, T.J.C., 1992, Lactosylated High Density Lipoprotein: A Potential Carrier for the Site-Specific Delivery of Drugs to Parenchymal Liver Cells, Mol.Pharmacol., 41:404

Bodmer, J.L. and Dean, R.T., 1988, Drug and enzyme targeting, Methods Enzymol., 112: 298
Brouwer, A., Wisse, E. and Knook, D.L., 1988, Sinusoidal endothelial cells and perisinusoidal fat-storing cells, in: "The Liver: Biology and Pathobiology", I.M. Arias, W.B. Jakoby, H. Popper, D. Schachter and D.A. Shafritz, eds., Raven Press Ltd., New York
Bruins, A.P., Covey, T.R. and Henion, J.D., 1987, Ion spray interface for combined liquid chromatography/atmospheric pressure ionization mass spectometry, Anal.Chem., 59:2642
Burwen, S.J. and Jones, A.L., 1990, Hepatocellular processing of endocytosed proteins, J.Electron Microsc.Tech., 14:140
Buscher, H.P., Gerok, W., Kurz, G., Schramm, U. and Thom, H., 1988, Hepatocyte primary culture and bile salt transport, in: "Trends in Bile Acid Research", G. Paumgartner, A. Stiehl and W. Gerok, eds., M.T.P. Press Ltd, Lancaster
Callahan, L.N., Phelan, M., Mallinson, M. and Norcross, M.A., 1991, Dextran sulfate blocks antibody binding to the principal neutralizing domain of human immunodeficiency virus type 1 without interfering with gp120-CD4 interactions, J.Virol., 65:1543
Casagrande, C., Merlo, L., Ferrini, R., Miragoli, G. and Semeraro, C., 1989, Cardiovascular and renal action of dopaminergic prodrugs, J.Cardiovasc.Pharmacol., 14 (suppl. 8), S40
Cawley, D.B., Simpson, D.L. and Herschman, H.R., 1981, Asialoglycoprotein receptor mediates the toxic effects of an asialofetuin diphteria toxin fragment A conjugate on cultured rat hepatocytes, Proc.Natl.Acad.Sci.USA, 78:3383
Chang, T.M. and Kullberg, D.W., 1982, Studies on the mechanism of cell intoxication by diphteria toxin fragment A-asialoorosomucoid hybrid toxins. Evidence for utilization of an alternative receptor-mediated transport pathway, J.Biol.Chem., 257:12563
Chaudhary, V.K., Mizukami, T., Fuerst, T.R., Fitzgerald, D.J., Moss, B., Pastan, I. and Berger, E.A., 1988, Selective killing of HIV-infected cells by recombinant human CD4-Pseudomonas exotoxin hybrid protein, Nature, 335:369
Covey, T.R., Bonner, R.F., Shushan, B.I. and Henion, J.D., 1988, The determination of protein, oligonucleotide and peptide molecular weights by oinspray mass spectrometry, Rapid Commun.Mass.Spectrom., 2:249
Curtis, B.M., Scharnowske, S. and Watson, A.J., 1992, Sequence and expression of a membrane-associated C-type lectin that exhibits CD4-independent binding of human immunodeficiency virus envelope glycoprotein-gp120, Proc.Natl.Acad.Sci.USA, 89:8356
Daemen, M.J.A.P., Thijssen, H.H.W. and Struyker-Boudier, H.A.J., 1991, Pharmacokinetic considerations in local drug delivery, Adv.Drug Deliv.Rev., 6:197
Duncan, R., 1992, Drug-polymer conjugate: potential for improved chemotherapy, Anti Cancer Drugs, 3:175
Fallon, R.J. and Schwartz, A.L., 1985, Receptor mediated endocytosis and targeted drug delivery, Hepatology, 5:899
Fallon, R.J. and Schwartz, A.L., 1989, Receptor-mediated delivery of drugs to hepatocytes, Adv.Drug Deliv.Rev., 4:49
Fiume, L., Busi, C., Mattioli, A., Balboni, P.G., Barbanti-Brodano, G. and Wieland, T., 1982, Hepatocyte targeting of antiviral drugs coupled to galactosyl-terminating glycoproteins, in: "Targeting of Drugs", G. Gregoriadis, J. Senior and A. Trouet, eds., Plenum Press, New York
Fiume, L., Busi, C., Preti, P. and Spinosa, G., 1987, Conjugates of ara-AMP with lactosaminated albumin: A study on their immunogenicity in mouse and rat, Cancer Drug Delivery, 4:145
Fiume, L., Bonino, F., Mattioli, A., Chiaberge, E., Cerenzia, M.R.T., Busi, C., Brunetto, M.R.and Verme, G., 1988, Inhibition of hepatitis B virus replication by vidarabine monophosphate conjugated with lactosaminated serum albumin, Lancet, 332:13
Fiume, L., Bassi, B., Busi, C., Mattioli, A. and Spinoza, G., 1986, Drug targeting in antiviral chemotherapy. A chemically stable conjugate of 9-β-D-arabinofuranosyl-adenine 5' -monophosphate with lactosaminated albumin accomplishes a selective delivery of the drug to liver cells, Biochem.Pharmacol, 35:967
Fiume, L., Mattioli, A., Busi, C., Balboni, P.G., Barbanti-Brodano, G., De Vries, J., Altmann, R. and Wieland, T., 1980, Selective inhibition of Ectromelia virus DNA synthesis in hepatocytes by adenine-9-β-D-arabinofuranoside (ara-A) and adenine-9-

β-D-arabinofuranosde 5'-monophosphate (ara-AMP) conjugated to asialofetuin, FEBS Lett, 116:185

Fiume, L., Mattioli, A. and Spinoza, G., 1987, Distribution of a conjugate of 9-β-D arabinofuranosyladenine 5'-monophosphate (ara-AMP) with lactosaminated albumin in parenchymal and sinusoidal cells of rat liver, Cancer Drug Delivery, 4:11

Fiume, L., Di Stefano, G., Busi, C. and Mattioli, A., 1994, A conjugate of lactosaminated poly-l-lysine with adenine arabinoside monophosphate, administered to mice by intramuscular route, accomplishes a selective delivery of the drug to the liver, Biochem.Pharmacol., 47:643

Franssen, E.J.F., Van Amsterdam, R.G.M., Visser, J., Moolenaar, F., De Zeeuw, D. and Meijer, D.K.F., 1991, Low molecular weight proteins as carriers for renal drug targeting: Naproxen-Lysozyme, Pharm.Res., 8:1223

Franssen, E.J.F., Moolenaar, F., De Zeeuw, D. and Meijer, D.K.F., 1992a, Renal specific delivery of sulfamethoxazole in the rat by coupling to the low molecular weight protein lysozyme via an acid-sensitive linker, Int.J.Pharm., 80:R15

Franssen, E.J.F., Koiter, J., Kuipers, C.A.M., Bruins, A.P., Moolenaar, F., De Zeeuw, D., Kruizinga, W.H., Kellog, R.M. and Meijer, D.K.F., 1992b, Low molecular weight proteins as carriers for renal drug targeting. Preparation of drug-protein conjugates and drug-spacer derivatives and their catabolism in renal cortex homogenates and lysosomal lysates, J.Med.Chem., 35:1246

Franssen, E.J.F., 1992, Drug targeting to the kidney with low molecular weight proteins, in: "Thesis", University of Groningen, Groningen

Franssen, E.J.F., Moolenaar, F., De Zeeuw, D. and Meijer, D.K.F., 1993a, Low molecular weight proteins for renal drug targeting: Naproxen coupled to lysozyme via the spacer L-lactic acid, Pharm.Res., 10:963

Franssen, E.J.F., Jansen, R.W., Vaalburg, M. and Meijer, D.K.F., 1993b, Hepatic and intrahepatic targeting of an anti-inflammatory agent with human serum albumin and neoglycoproteins as carrier molecules, Biochem.Pharmacol., 45:1215

Frijlink, H.W., Franssen, E.J.F., Eissens, A.C., Oosting, R., Lerk, C.F. and Meijer, D.K.F., 1991, The effects of cyclodextrins on the disposition of intravenously injected drugs in the rat, Pharm.Res., 8:380

Furuno, K., Miwa, N. and Kato, K., 1983, Receptor mediated introduction of pepstatin -asialoftuin conjugates into lysosomes of rat hepatocytes, J.Biochem., 93:249

Gattegno, L., Sadeghi, H., Saffar, L., Bladier, D., Clerget-Raslain, B., Gluckman, J.-C. and Bahraoui, E., 1991, N-acetyl-β-D-glucosaminyl-binding properties of the envelope glycoprotein of human immunodeficiency virus type 1, Carbohydr.Res., 213:79

Ghetie, M.A., Uhr, J.W. and Vitetta, E.S., 1991, Covalent binding of human α_2 -macroglobulin to deglycosylated ricin A chain and its immunotoxins, Cancer Res., 51:1482

Ghetie, V., Slaughter, C., Wheeler, H.T., Uhr, J.W. and Vitetta, E.S., 1991, CD4 peptide protein conjugates, but not recombinant human CD4, bind to recombinant gp120 from the human immunodeficiency virus in the presence of serum from AIDS patients, Proc.Natl.Acad.Sci.USA, 88:5690

Ghose, T., Blair, A.H., Vaughan, K. and Kulkarni, P., 1982, Antibody-directed drug targeting in cancer therapy, in: "Targeted drugs", E.P. Goldberg, ed., John Wiley and Sons, New York

Haga, H.J., 1989, Kidney lysosomes, Int.J.Biochem., 21:343

Haidar, M., Gluckman, J.C. and Gattegno, L., 1992, Mannosyl/N-acetyl-β-D-glucosaminyl binding properties of the envelope glycoprotein of human immunodeficiency virus type 2, Glycobiology, 2:429

Haisma, H.J., Boven, M., Van Muijen, M., De Jong, J., Van der Vijgh, W.J.F. and Pinedo, H.M., 1992, A monoclonal antibody-β-glucoronidase conjugate as activator of the prodrug epirubicin-glucuronide for specific treatment of cancer, Br.J.Cancer, 66:474

Harrison, G.S., Long, C.J., Curiel, T.J., Maxwell, F. and Maxwell, I.H., 1992, Inhibition of human immunodeficiency virus-1 production resulting from transduction with a retrovirus containing an HIV-regulated diphtheria toxin A chain gene, Hum.Gene Ther., 3:461

Herschman, H.R., Cawley, D. and Simpson, D.L., 1984, Toxic conjugates of epidermal growth factor and asialofetuin, in: "Receptor-mediated Targeting of Drugs", G. Gregoriadis, G. Poste, J. Senior and A. Trouet, eds., Plenum Press, New York

Hirai, H., Tsakada, Y., Koji, T., Ishii, N., Kaneda, H. and Kasai, Y., 1983, Attemps of

treatment of hepatoma with antibody to alpha-fetoprotein, Prot.Biol.Fluids, 31:357
Hwang, K.M., Foon, K.A., Cheung, P.H., Pearson, J.W. and Oldham, R.K., 1984, Selective antitumor effect on L10 hepatocarcinoma cells of a potent immunoconjugate composed of the A chain of abrin and a monoclonal antibody to a hepatoma-associated antigen, Cancer Res., 44:4578
Jansen, R.W., Kruijt, J., Van Berkel, T.J.C. and Meijer, D.K.F., 1993, Coupling of the antiviral drug ara-AMP to lactosaminated albumin leads to a specific uptake in rat and human hepatocytes, Hepatology, 18:146
Jansen, R.W., Schols, D., Pauwels, R., De Clercq, E. and Meijer, D.K.F., 1993, Novel, negatively charged human serum albumins display a highly potent and selective in vitro anti-human immunodeficiency virus type 1 activity, Mol.Pharmacol., 44:1003
Kaneo, Y., Tanaka, T. and Iguchi, S., 1991, Targeting of Mitomycin C to the Liver by the Use of Asialofetuin as a Carrier, Chem.Pharm.Bull., 39:999
Kataoka, H. and Tavassoli, M., 1985, Identification of lectin like substances recognizing galactosyl residues of glycoconjugates in the plasma membrane of marrow sinus endothelium, Blood, 65:1165
Keegan-Rogers, V., Wu, C.H. and Wu, G.Y., 1991, Receptor-mediated protection of normal hepatocytes during chemotherapy for hepatocellular carcinoma, in: "Liver Diseases. Targeted diagnosis and therapy using specific receptors and ligands", G.Y. Wu and C.H. Wu, eds., Marcel Dekker, Inc., New York, Basel, Hong Kong
Kostiainen, R., Franssen, E.J.F. and Bruins, A.P., 1993, Capillary zone electrophoresis-Ion spray mass spectrometry of a synthetic drug-protein conjugate mixture, J.Chromatogr., 647:361
Kramer, W., Wess, G., Schubert, G., Bickel, M., Girbig, F., Gutjahr, U., Kowalewski, S., Baringhaus, K.-H., Enhsen, A., Glombik, H., Mullner, S., Neckermann, G., Schulz, S. and Petzinger, E., 1992, Liver-specific drug targeting by coupling to bile acids, J.Biol.Chem., 267:18598
Maack, T., 1975, Renal handling of low molecular weight proteins, Am.J.Med., 58:57
Maack, T., Johson, V., Kau, S.T., Figuereido, J. and Sigulem, D., 1979, Renal filtration, transport and metabolism of low molecular weight proteins: a Review, Kidney Int., 16:251
Marti, U., Burwen, S.J. and Jones, D.L., 1989, Biological effects of epidermal growth factor, with emphasis on the gastro-intestinal tract and liver: An update, Hepatology, 9:126
Maunsbach, A.B., 1974, Isolation of kidney lysosomes, Methods Enzymol., 31:330
McDonald, J.K. and Barrett, A.J., 1986, "Mammalian proteases. Vol II: Exopeptidases", Academic Press, London
Meijer D.K.F., Jansen, R.W. and Molema, G., 1992, Drug targeting systems for antiviral agents: options and limitations, Antiviral Res., 18:215
Meijer, D.K.F. and Ziegler, K., 1993, Mechanisms for the hepatic clearance of oligopeptides and proteins: implications for rate of elimination, bioavailability and cell-specific drug delivery to the liver, in: "Biological Barriers to Protein Delivery", K.L. Audus and T.J. Raub, eds., Plenum, New York
Meijer, D.K.F., Molema, G., Jansen, R.W. and Moolenaar, F., 1990, Design of cell-specific drug targeting preparations for the liver: where cell biology and medicinal chemistry meet, in: "Trends in Drug Research, Proceedings of the seventh Noordwijkerhout-Camerino Symposium, H. Timmerman and V. Claassen, eds., Elsevier, Amsterdam
Meijer, D.K.F., and Ziegler, K., 1993, Mechanisms for the hepatic clearance of oligopeptides and proteins: implications for rate of elimination, bioavailability and cell-specific drug delivery to the liver, in: "Biological Barriers to Protein Delivery", K.L. Audus and T.J. Raub, eds., Plenum Press, New York
Meijer, D.K.F. and Van der Sluijs, P., 1989, Covalent and noncovalent protein binding of drugs: Implications for hepatic clearance, storage, and cell-specific drug delivery, Pharm.Res., 6:105
Meijer, D.K.F, and Groothuis, G.M.M., 1991, Hepatic transport of drugs and proteins, in: "Oxford Textbook of Clinical Hepatology", N. McIntyre, J.P. Benhamou, J. Bircher, M. Rizzeto and J. Rodes, eds., University Press, Oxford
Meijer, D.K.F., Vonk, R.J., Keulemans, K. and Weitering, J.G., 1977, Hepatic uptake and biliary excretion of dibromosulphthalein, albumin dependence, influence of phenobarbital and nafenopin pretreatment and the role of Y- and Z-protein, J.Pharmacol.Exp.Ther., 202:8

Michalak, T.I. and Bolger, G.T., 1989, Characterization of the binding sites for glutaraldehyde-polymerized albumin on purified woodchuk hepatocyte plasma membranes, Gastroenterology, 96:153
Midoux, P., Negre, E., Roche, A.C., Mayer, R., Monsigny, M., Balzarini, J., De Clercq, E., Mayer, E., Ghaffar, A. and Gangemi, J.D., 1990, Drug targeting: Anti-HSV-1 activity of mannosylated polymer-bound 9-(2-Phosphonylmethoxyethyl)adenine, Biochem.Biophys.Res.Commun., 167:1044
Molema, G., 1992, Optimization of the synthesis of neoglycoprotein-AZTMP conjugates: influence on stability, hepatic recognition and anti-HIV activity, thesis, University of Groningen, The Netherlands
Molema, G., Jansen, R.W., Visser, J., Herdewijn, P., Moolenaar, F. and Meijer, D.K.F., 1991, Neoglycoproteins as carriers for antiviral drugs: synthesis and analysis of protein-drug conjugates, J.Med.Chem., 34:1137
Molema, B., Jansen, R.W. Pauwels, R., De Clercq, E. and Meijer, D.K.F., 1990, Targeting of antiviral drugs to T-4 lymphocytes. Anti-HIV activity of neoglycoprotein-AZTMP conjugates in vitro, Biochem.Pharmacol., 40:2603
Molema, G., Delivery of the anti-HIV drug azidothymidine (AZT) to T-lymphocytes with neoglycoprotien carriers, thesis, University of Groningen, 1992
Moller, J.V. and Sheikh, M.I., 1983, Renal organic anion transport systems: Pharmacological, physiological and biochemical aspects, Pharmacol.Rev., 34:315
Molteni, L., 1979, Dextrans as drug carriers, in: "Drug Carriers in Biology and Medicine", G. Gregoriadis, ed., Academic Press, New York
Molthoff, C.F.M., Pinedo, H.M., Schuper, H.M.M., Nijman, H.W. and Boven, E., 1992, Comparison of the pharmacokinetics, biodistribution and dosimetry of monoclonal antibodies OC125, OV-TL3, and 139H" as IgG and $F(ab')_2$ fragments in experimental ovarian cancer, Br.J.Cancer, 65:677
Monsigny, M., Roche, A.C. and Midoux, P., 1988, Endogenous lectins and drug targeting, Ann.N.Y.Acad.Sci., 551:399
Monsigny, M., Roche, A.C., Midoux, P. and Mayer, R., 1994, Glycoconjugates as carriers for specific delivery of therapeutic drugs and genes, Adv.Drug Deliv.Rev., 14:1
Moolenaar, F., Cancrinus, S., Visser, J., De Zeeuw, D. and Meijer, D.K.F., 1992, Clearance of indomethacin occurs predominantly by renal glucuronidation, Pharm.Weedbl (Sci) 14:191
Negre, E., Chance, M.L., Hanboula, S.Y., Monsigny, M., Roche, A.-C., Mayer, R.M. and Hommel, M., 1992, Antileishmanial drug targeting through glycosylated polymers specifically internalized by macrophage membrane lectins, Antimicrob.Agents Chemother., 36:2228
Nicolau, C. and Cudd, A., 1989, Liposomes as carriers of DNA, Crit.Rev.Ther.Drug Carrier Syst., 6:239
Nishikawa, M., Kamijo, A., Fujita, T., Takakura, Y., Sezaki, H. and Hashida, M., 1993, Synthesis and Pharmacokinetics of a New Liver-Specific Carrier, Glycosylated Carboxymethyl-Dextran, and Its Application to Drug Targeting, Pharm.Res., 10:1253
Olsnes, S., Sandvig, K., Petersen, O.W. and Van Deurs, B., 1989, Immunotoxins - Entry into cells and mechanisms of action, Immunol.Today, 10:291
Ponpipom, M.M., Bugianesi, R.L., Robbins, J.C., Doebber, T.W. and Shen, T.Y., 1984, Saccharide receptor-mediated drug delivery, in: "Receptor-mediated targeting of drugs", G. Gregoriadis, G. Poste, J. Senior and A. Trouet, eds., Plenum Press, New York
Ponzetto, A., Fiume, L., Forzani, B., Song, S.Y., Busi, C., Mattioli, A., Spinelli, C., Marinelli, M., Smedile, A., Chiaberge, E., Bonino, F., Gervasi, G.B., Rapicetta, M. and Verme, G., 1991, Adenine Arabinoside Monophosphate and Acyclovir Monophosphate coupled to Lactosaminated Albumin reduce Woodchuck Hepatitis Virus viremia at lower doses that do the unconjugated drugs, Hepatology, 14:16
Poznansky, M.J. and Juliano, R.L., 1984, Biological approaches to the controlled delivery of drugs: a critical review, Pharmacol.Rev., 36:277
Roerdink, F.H., Daemen, T., Bakker-Woudenberg, I.A.J.M., Horm, G., Crommelin, D.J.A. and Scherphof, G.L., 1988, Therapeutic utility of liposomes, in: "Drug Delivery Systems, Fundamentals and Techniques", P. Johnson and J.G. Lloyd-Jones, eds., VCH
Russel, F.G.M., Weitering, J.G., Oosting, R., Groothuis, G.M.M., Hardonk, M.J. and Meijer, D.K.F., 1983, The influence of taurocholate on hepatic clearance and biliary

excretion of asialointestinal alkaline phosphatase in the rat in vivo and in isolated perfused rat liver, Gastroenterology, 85:225
Ryser, H.J.P. and Shen, W.C., 1986, Drug-poly(lysine) conjugates: Their potential for chemotherapy and for the study of endocytosis, in: "Targeting of Drugs with Synthetic Systems", G. Gregoriadis, J. Senior and G. Poste, eds., Plenum Press, New York
Schneider, Y.J., Abarca, J., Aboud-Pirak, E., Baurain, E., Ceulemans, F., Deprez-De Campaneere, D., Lesur, B., Masquelier, C., Otte-Schlachmuylder, C., Rolin-Van Swieten, D. and Trouet, A., 1984, Drug targeting in human cancer chemotherapy, in: "Receptor-mediated Targeting of Drugs", G. Gregoriadis, G. Poste, J. Senior and A. Trouet, eds., Plenum Press, New York
Schooley, R.T., 1992, Antiretroviral chemotherapy, in: "AIDS and other manifestations of HIV infection", G.P. Wormser, ed., Raven Press Ltd, New York
Segal, H.L., 1975, Lysosomes and intracellular protein turnover, in: "Lysosomes in Biology and Pathology", J.T. Dingle, ed., North Holland Publ. Comp., Amsterdam
Sehon, A.H., 1991, Suppression of antibody responses by conjugates of antigens and monomethoxypoly(ethylene glycol), Adv.Drug Deliv.Rev., 6:203
Senter, P.D., 1990, Activation of prodrugs by antibody-enzyme conjugates: A new approach to cancer therapy, FASEB J., 4:188
Senter, P.D., Wallace, P.M., Svensson, H.P., Vrudhula, V.M., Kerr, D.E., Hellstrom, I. and Hellstrom, K.E., 1993, Generation of cytotoxic agents by targeted enzymes, Bioconjugate Chemistry., 4:3
Seymour, L.W., Ulbrich, K., Wedge, S.R., Hume, I.C., Strohalm, J. and Duncan, R., 1991, N-(2-hydroxypropyl)methacrylamide copolymers targeted to the hepatocyte galactose-receptor: pharmacokinetics in DBA_2 mice, Br.J.Cancer, 63:859
Shen, T.Y., 1989, Preferential membrane permeation and receptor recognition in drug targeting, in: "Bioreversible Carriers in Drug Design. Theory and application", E.B. Roche, ed., Pergamon Press, New York
Shen, T.Y. and Ryser, J.P., 1981, Cis-aconityl spacer between daunomycin and macromolecular carriers: a model of pH-sensitive linkage releasing drug from a lysosomotropic conjugate, Biochem.Biophys.Res.Commun., 102:1048
Shouval, D., Wands, J.R., Zurawski, V.R., Jr., Isselbacher, K.J. and Shafritz, D.A., 1982, Selecting binding and complement-mediated lysis of human hepatoma cells (PLC/PRF/5) in culture by monoclonal antibodies to hepatitis B surface antigen, Proc.Natl.Acad.Sci.USA, 79:650
Van der Sluijs, P., Oosting, R., Weitering, J.G., Hardonk, M.J. and Meijer, D.K.F., 1985, Biliary excretion of FITC metabolites after administration of FITC-labeled asialoorosomucoid as a measure of lysosomal proteolysis, Biochem.Pharmacol., 34:1399
Van der Sluijs, P., Braadman, I., Meijer, D.K.F. and Groothuis, G.M.M., 1988, Heterogenous acinar localization of the asialoglycoprotein internalization system in rat hepatocytes, Hepatology, 8:1521
Van der Sluijs, P. and Meijer, D.K.F., 1991, Limitations on the specificity of targeting asialoglycoprotein-drug conjugates to hepatocytes, in: "Targeted Diagnosis and Therapy of Liver Diseases: Cell Surface Receptors and Liver-directed Agents", G.Y. Wu and C.H. Wu, eds., Marcel Dekker, New York
Simpson, D.L., Cawley, D.B. and Herschmann, H., 1982, Killing of cultured hepatocytes by conjugates of asialofetuin and epidermal growth factor linked to the A chain of ricin or diphteria toxin, Cell, 29:469
Stadalnik, R.C., Vera, R.E., Woodle, E.S., Trudeau, W.L., Porter, D.A., Ward, R.E., Krohn, K.A. and O'Grady, L.F., 1985, Technetium-99m NGA functional hepatic imaging: preliminary clinical experience, J.Nucl.Med., 26:1233
Stahl, P.D., 1990, The macrophage mannose receptor - Current status, Am.J.Respir.Cell Mol.Biol., 2:317
Stephan, Z.F., Yurachek, E.C., Sharif, R., Wasvary, J.M., Steele, R.E. and Howes, C., 1992, 'Reduction of cardiovascular and thyroxine-suppressing activities of L-T_3 by liver targeting with cholic acid, Biochem.Pharmacol., 43:1969
Stowell, C.P. and Lee, Y.C., 1980, Neoglycoproteins: the preparation and application of synthetic glycoproteins, Adv.Carbohydr.Chem.Biochem., 37:225
Sumpio, B.E. and Maack, T., 1982, Kinetics, competition and selectivity of tubular absorption of protein, Am.J.Physiol., 243:F379

Takakura, Y., Masuda, S., Tokuda, H., Nishikawa, M. and Hashida, M., 1994, Targeted delivery of superoxide dismutase to macrophages via mannose receptor-mediated mechanism, Biochem.Pharmacol., 47:853
Takami, M., Sone, T., Mizumoto, K., Kino, K. and Tsunoo, H., 1992, Maleylated human serum albumin inhibits HIV-1 infection in vitro, Biochim.Biophys.Acta, 1180:180
Till, M.A., Ghetie, V., Gregory, T., Patzer, E.J., Porter, J.P., Uhr, J.W., Capon, D.J. and Vitetta, E.S., 1988, HIV-infected cells are killed by rCD4-ricin A chain, Science, 242:1166
Thung, S.N., Wang, D., Fasy, T.M., Hood, A. and Gerber, M.A., 1989, Hepatitis B surface antigen binds to human serum albumin cross-linked by transglutaminase, Hepatology, 9:726
Townsend, R.P., Wall, D.A., Hubbard, A.L. and Lee, A.C., 1984, Rapid release of galactose-terminated ligands after endocytosis by hepatic parenchymal cells. Evidence for a role of carbohydrate structure in the release of internalized ligand from receptor, Proc.Natl.Acad.Sci.USA, 81:466
Trouet, A., Baurain, R., Deprez-De Campeneere, D., Masquelier, M. and Pirson, P., 1982, Targeting of antitumor and antiparotozoal drugs by covalent linkage to protein carriers, in: "Targeting of Drugs", G. Gregoriadis, J. Senior and A. Trouet, eds., Plenum Press, New York
Trouet, A., Masquelier, M., Baurain, R. and Deprez-De Campeneere, D., 1982, A covalent linkage between daunorubicin and proteins that is stable in serum and reversible by lysosomal hydrolysis as required for a lysosomotropic drug-carrier conjugate: in vitro and in vivo studies, Proc.Natl.Acad.Sci.USA, 79:626
Trouet, A. and Jolles, G., 1984, Targeting of daunorubicin by association with DNA or proteins: A review, Semin.Oncol., 11:64
Vera, D.R., Stadalnik, R.C. and Krohn, K.A., 1985, Technetium 99m galactosylenoglycoalbumin: preparation and preclinical studies, J.Nucl.Med., 26:1157
Vitetta, E.S. and Uhr, J.W., 1985, Immunotoxins, Annu.Rev.Immunol., 3:197
Wilson, G., 1986, Gene therapy: rationale and realisation, in: "Site-Specific Drug Delivery", E. Tomlinson and S.S. Davis, eds., John Wiley and Sons Ltd, Chichester, New York, Brisbane, Toronto, Singapore
Wilson, G., Wu, C.H., Wu, G.Y., Roy Chowdhury, N., Epstein, A., Waltman, R. and Roy Chowdhury, J., 1989, Temporary amelioration of hyperlipidemia in LDL receptor deficient rabbits by targeted gene delivery, Hepatology, 10:631
Wright, T.L., 1989, Targeted therapy for hepatic cancer: good in theory, problematic in practice, Hepatology, 9:657
Wu, C.H., Wilson, J.M. and Wu, G.Y., 1989, Targeting genes: delivery and persistent expression of a foreign gene driven by mammalian regulatory elements in vivo, J.Biol.Chem., 264:16985
Wu, G.Y., 1988, Targeting in diagnosis and therapy, in: "The Liver: Biology and Pathobiology", I.M. Arias, W.B., Jakoby, H. Popper, D. Schachter and D.A. Shafritz eds., Raven Press, New York
Wu, G.Y., Wu, C.H. and Rubin, M.I., 1985, Acetaminophen hepatotoxicity and targeted rescue: a model for specific chemotherapy of hepatocellular carcinoma, Hepatology, 5:709
Wu, G.Y., Keegan-Rogers, V., Franklin, S., Midford, S. and Wu, C.H., 1988, Targeted antagonism of galactosamine toxicity in normal hepatocytes in vitro, J.Biol.Chem., 263:4719
Wu, G.Y., Wilson, J.M., Shalaby, F., Grossman, M., Shafritz, D.A. and Wu, C.H., 1991, Receptor-mediated gene delivery in vivo. Partial correction of genetic analguminemia in nagase rats, J.Biol.Chem., 266:14338
Wu, G.Y., Wilson, J.M., and Wu, C.H., 1988, Targeting genes: Delivery and expression of a foreign gene driven by an albumin promotor in vivo, Hepatology, 8:1251
Wu, G.Y., Tangco, M.V. and Wu, C.H., 1990, Targeted gene delivery: persistence of foreign gene expression achieved by pharmacological means, Hepatology, 12:871 (Abstract)
Wu, G.Y. and Wu, C.H., 1988, Evidence for targeted gene delivery to hep G2 hepatoma cells in vitro, Biochemistry, 27:887

SUGAR SPECIFIC DELIVERY OF DRUGS, OLIGONUCLEOTIDES AND GENES

M. Monsigny, A.-C. Roche, P. Midoux and R. Mayer

Laboratoire de Biochimie des Glycoconjugués et Lectines Endogènes,
Université d'Orléans et Centre de Biophysique Moléculaire, CNRS
1, rue Haute, 45071 Orléans Cedex 2, France

INTRODUCTION

Many mammalian cells have plasma membrane proteins that are able to both selectively bind macromolecules containing a recognition signal and mediate the transfer of such macromolecules to the endosomal compartments. Amongst the various receptors known so far, there are membrane lectins that selectively recognize glycoconjugates containing complex oligosaccharides structures (Goldstein et al, 1980). In mammalian cells, lectins are either secreted water soluble proteins or proteins present in both the cytosol and the nucleus (see for reviews Hubert et al, 1989; Hart et al, 1989; Wang et al, 1991). Other lectins are type I or type II membrane proteins. The first membrane lectin was discovered 25 years ago by G. Ashwell and A. Morell and their coworkers (for a review, see Ashwell and Harford, 1982). The first evidence came from an experiment using asialoceruloplasmin: ceruloplasmin is a serum protein carrying copper ions, it is a glycoprotein which has a long lifetime in the blood; however, upon desialylation, the asialoglycoprotein is cleared off from the serum in a few minutes because it is captured in the liver by parenchymal cells (Morell et al, 1971).

The molecular mechanism has been deciphered: a membrane lectin expressed by liver parenchymal cells binds glycoproteins containing tri- or tetra-antennary glycan moieties (Lee et al, 1983) in which sugars in terminal non reducing position are β-galactosides (Table 1A). That membrane lectin mediates the transfer of its ligands into the endosome compartments. The endosome vesicles have a lumen which progressively becomes acidic: because the binding capacity of this lectin is maximum at neutral pH, and inefficient in acidic medium, the lectin is transferred back to the plasma membrane and independently the asialoglycoprotein is transferred to lysosomes; there, the asialoglycoprotein will be degraded into free sugars and amino-acids. The galactose specific lectin, also called the Ashwell/Morell receptor, is present in the plasma membrane (50,000 molecules per cell) and in the endocytic compartments; because of recycling about $1.5\ 10^6$ ligands are taken up inside a cell in one hour (Harford and Ashwell, 1982). On these bases it is obvious that this type of membrane protein is a good candidate to target drugs to any cell which expresses such a membrane lectin, by making use of glycoconjugates which contain a glycan moiety specifically acting as a recognition signal.

In addition, it has been shown, more recently, that membrane lectins are also present in a large variety of cells (see for reviews, Monsigny et al, 1988a; Sharon and Lis, 1989; Hughes, 1992; Drickamer and Taylor, 1993). Some membrane lectins are expressed on endothelial cells: LECAM 2 or E selectin (for a review, see Lasky, 1992). Such lectins are

Targeting of Drugs 4, Edited by G. Gregoriadis *et al.*
Plenum Press, New York, 1994

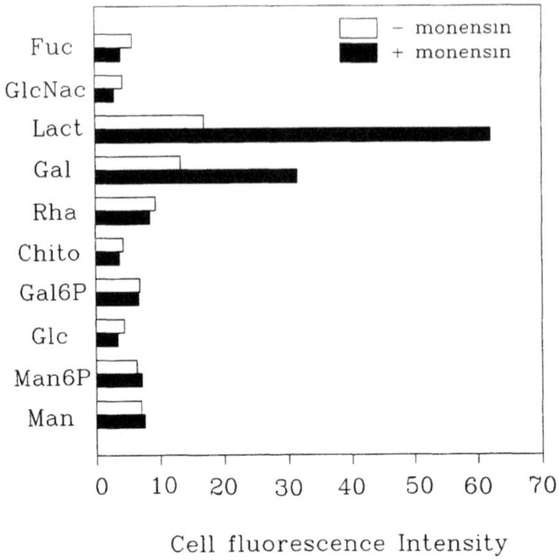

Fig. 1 Endocytosis of neoglycoproteins by HepG2 Cells. 3.10^5 plated cells were incubated at 37°C for 2 h in complete PBS containing 1% BSA in the presence of 25 µg/ml of fluorescein-labelled neoglycoproteins. The cells were harvested, washed with cold PBS containing 0.5% BSA, resuspended in cold PBS and further incubated at 4°C in the absence or in the presence of 50 µM monensin for 30 min. The cell fluorescence was measured by flow cytometry.

apparently more involved in cell recognition than in glycoconjugate uptake. These lectins recognize complex carbohydrates (Fig. 1) with quite a fine specificity. They may be ideal targets to concentrate drugs in the periphery of cells which express them, and consequently this approach may be quite relevant as putative therapeutics in the case of inflammatory or rheumathoid processes as well as in the cases of cancers or of infectious diseases. However this approach does not match with the topics of the present paper which mainly deals with drug targeting associated with endocytosis. Other membrane lectins mediating an active endocytosis of their ligands have been described. Those lectins include mannose receptor (see Stahl et al, 1978, 1984), mannose-6-phosphate receptor (see Kornfeld, 1992) and sulfo N-acetylgalactosamine receptors (for a review, see Drickamer, 1991). These membrane lectins also recognize complex oligosaccharides (Fig. 1).

A number of lectins have been cloned. Their structures have several common features. Each lectin is a protein with a short cytosolic tail, a single transmembrane domain and various numbers of exocellular domains. The carbohydrate recognition domain is either the outermost domain as in the case of selectins, the Ashwell receptor, the mannose receptor or the small mannose-6-phosphate receptor, or is located in the middle of the exocellular part of the lectin as in the case of the large mannose-6-phosphate receptor.

Macrophages and their precursors express various membrane lectins (Table 2). Amongst those present on the plasma membrane, the large mannose-6-phosphate receptor is not detectable in monocyte precursors but is present in monocytes and mediates the endocytosis of its ligand (Roche et al, 1985b); in macrophages, it is abundant and very efficiently induces the uptake of its ligands. The mannose specific receptor is not at all detectable in monocytes or in their precursors but its expression occurs along with the maturation of monocytes to macrophages (Stahl et al, 1984).

A membrane lectin of a given specificity is expressed on a limited number of cells. For instance, the galactose specific lectin is expressed in liver parenchymal cells and in a

Table 1. Structure of ligands recognized by well characterized lectins

A - Triantennary lactosamine-type asialo-oligoside : asialoglycoprotein receptor

Galβ 4GlcNAcβ 2 Manα 6 \
Manβ 4GlcNAcβ 4GlcNAcβ →
Galβ 4GlcNAcβ 4 \ /
Manα 3 /
Galβ 4GlcNAcβ 2 /

B - Tetraantennary lactosamine-type asialo oligoside : asialoglycoprotein receptor

Galβ 4GlcNAcβ 6 \
Manα 6
Galβ 4GlcNAcβ 2 / \
Manβ 4GlcNAcβ 4GlcNAcβ →
Galβ 4GlcNAcβ 4 \ /
Manα 3 /
Galβ 4GlcNAcβ 2 /

C - Lewis x : LECAM 2/3

Galβ 4 \
GlcNAcβ 3Galβ →
Fucα 3 /

D - Sialyl Lewis x : LECAM 3/2

Neu5Acα Galβ 4 \
GlcNAcβ 3Galβ →
Fucα 3 /

E - Sulfated Lewis x derivative (HNK1) : LECAM 1

(SO_3^-) 3Glc UAβ 3Galβ 4
GlcNAcβ 3Galβ 4Glc →
Fucα 3

F - Oligomannoside : mannose receptor

Manα 2Manα 6 \
Manα 6 \
Manα 3 / Manβ 4GlcNAcβ 4GlcNAcβ →
/
Manα 2Manα 2 Manα 3 /

G - Phosphorylated oligomannoside : mannose 6 phosphate receptor

(HPO_3^-) 6 \
Manα 6 \
Manα 2 / Manα 6
Manα 3 / \
Manβ 4GlcNAcβ 4GlcNAcβ →
(HPO_3^-) 6 \ /
Manα 2 — Manα 3 /
Manα 2 /

H - Sulfated lactosamine type oligosaccharide : GalNAc 4 sulfate receptor

(SO_3^-) 4GalNAcβ 4GlcNAcβ 2Manα 6 \
Manβ 4GlcNAcβ 4GlcNAcβ → →
(SO_3^-) 4GalNAcβ 4GlcNAcβ 2Manα 3 /

Table 2. Main sugar specific receptors expressed at the surface of macrophages (Adapted from Pimpaneau, 1990)

Name	Ligands	Cells
ManR	α-D-Man β-D-GlcNAc α-L-Fuc	macrophages, endothelial cells
MPR-CI MPR-CD	α-D-M6P	macrophages, precursor cells
GalR (CRP)	particle Gal	Kupffer cells, peritoneal macrophages
M-ASGP-BP	β-D-Gal α-D-GalNAc	peritoneal macrophages
Gal/FucR	α-L-Fuc β-D-Gal	Kupffer cells
β glucan	β-3-D-glucan Zymosan	monocytes, macrophages
SER	gangliosides	macrophages

R : receptor ; ManR : mannose specific lectin ; MPR-CI and MPR-CD : mannose 6 phosphate specific lectins, cation independent and cation dependent, respectively ; GalR : galactose specific lectin ; CRP : acute phase protein : M-ASGP-BP : macrophage asialoglycoprotein specific lectin ; Gal/FucR : galactose and fucose specific lectin ; SER : sialic acid specific lectin.

limited number of hepatic tumor cell. The mannose specific receptor is expressed in macrophages and other phagocytes such as Kupffer cells, but is not expressed in monocytes nor in monocyte precursors, in liver parenchymal cells, etc..

While we are still far from knowing all the existing membrane lectins on each cell type as well as the actual sugar specificity of many membrane lectins already characterized, it is already clear that glycoconjugates containing oligosaccharide structures - recognized by a membrane lectin of either liver parenchymal cells, monocytes, macrophages or endothelial cells - are suitable to act as targeted carriers of drugs.

MEMBRANE LECTINS AND ENDOCYTOSIS

Glycoproteins, isolated from the blood or from other tissues, have been used, either in their native form or after removal of one or more sugars, to determine the specificity of some lectins. For instance, Ashwell and co-workers used asialoorosomucoid to define the characteristics of mammalian liver parenchymal cell lectins, and asialoagalactoorosomucoid to define the characteristics of the avian liver lectins (see Harford and Ashwell, 1982). However, the purification of glycoproteins from biological sources may be tedious and the preparation of the modified glycoproteins is relatively expensive. Synthetic glycoproteins (Table 3) constitute another type of tool used to establish the characteristics of lectins.

Neoglycoproteins

Twenty years ago, we were working on a plant lectin, wheat germ agglutinin. Wheat germ agglutinin was known to agglutinate animal cells and specially some tumor cells; the agglutination being easily inhibited by N-acetylglucosamine, N-acetylneuraminic

Table 3. Chemical structure of common links in synthetic glycoconjugates

Linkage	
(a) azo	glycosyl - α or β - O - pC_6H_4 - N = N - pC_6H_3(OH) - CH_2 - CH(NH -)(CO -)
(b) thiocarbamyl	glycosyl - α or β - O - pC_6H_4 - NH - CS - NH - $(CH_2)_4$ - CH(NH -)(CO -)
(c) amide	glycosyl - α or β - O - pC_6H_4 - CH_2 - CO - NH - $(CH_2)_4$ - CH(NH -)(CO -)
(d) glycitylamine	glycosyl - α or β - OCH(CHOH - CH_2OH)$(CHOH)_2$ - CH_2 - NH - $(CH_2)_4$- CH(NH -)(CO -)
(e) amidine	glycosyl - α or β - S - CH_2 - CH_2 - C(NH_2)$^+$=NH - $(CH_2)_4$- CH(NH -)(CO -)

(a) Goebel & Avery, 1929 ; Privat *et al.*, 1974 ; Kieda *et al.*, 1977 ; (b) McBroom *et al.*, 1972 ; Roche *et al.*, 1983 ; (c) Derrien *et al.*, 1989 ; (d) Gray, 1974 ; (e) Krantz *et al.*, 1976 ; Lee *et al.*, 1976.

acid and derivatives. In order to study more precisely sugar lectin interactions, we prepared a neoglycoprotein by linking about 10 di-N-acetylchitobiose units on one molecule of serum albumin. This neoglycoprotein, which is easily soluble in neutral or acidic buffer, interacts with wheat germ agglutinin leading to a precipitate easily detectable (Privat et al, 1974). Furthermore, this neoglycoprotein inhibits the agglutination of red blood cells from various animals when the agglutination is induced by wheat germ agglutinin. It was found that, on the basis of the sugar concentration, the neoglycoprotein was hundred times more efficient that the corresponding free sugar. On these bases, we prepared a battery of glycosylated proteins (Monsigny et al, 1976; Kieda et al, 1977) suitable to detect cell surface glycoproteins and/or cell surface lectins by electron microscopy (Monsigny et al, 1976; Schrével et al, 1979) as well as by fluorescence microscopy (Kieda et al, 1979; Monsigny et al, 1989). The first neoglycoproteins we used were made by coupling diazophenyl-glycosides to serum albumin, ferritin or peroxidase. These neoglycoproteins were deeply colored because of the azo links and the fluorescence of such fluorescein-labelled neoglycoprotein was relatively low. Then, we prepared neoglycoproteins by using O-phenylisothiocyanate glycosides (Roche et al, 1983; Monsigny et al, 1984b). Usually, these neoglycoproteins; starting with bovine serum albumin, contain 23 ± 3 sugar moieties per protein molecule, they have a very low absorption coefficient above 350 nm. They can be easily substituted with a fluorescent tag, with a disulfide derivative, or with drugs, because serum albumin contains 57 lysine residues, and that only 40% of them need to be substituted by sugars.

Such neoglycoproteins can be easily radiolabelled with iodine because they contain tyrosine residues which are easily substituted with iodine.

Neoglycoproteins, Table 3, have been independently developed by Lee and co-workers about at the same time. These authors used (2-imino 2-methoxyethyl)-1-thioglycosides as activated glycosides which react also with the amino groups of a protein leading to an amidine keeping a positive charge (Lee et al, 1976; Krantz et al, 1976). Another way of preparing neoglycoproteins is to use the reductive alkylation proposed by Gray (1974); the sugar at the reducing position of an oligosaccharide gives an imine derivative upon reaction with an amine, this imine is selectively reduced with sodium cyanoborohydride, leading to a secondary amine which keeps a positive charge in neutral medium.

Fluorescein-labelled neoglycoproteins have been used to detect endogenous lectins in a variety of cells by using either a spectrofluorometric method (Roche et al, 1983) or a flow cytometric method (Midoux et al, 1986, 1987). In the spectrofluorometric method, cells are incubated in the presence of a given fluorescein-labelled neoglycoprotein, either at 4°C or at 37°C for 1 or 2 h. The cells are washed, dissolved in a borate buffer containing a non aromatic surfactant MAC 1gS, hexahydroxylheptanoyloctylamide CH_2OH-$(CHOH)_5$-CO-NH-$(CH_2)_7$-CH_3 which does not quench the fluorescein fluorescence (Roche et al, 1983). By this method, it is easy to rapidly evidence the presence of a cell surface lectin and to determine its sugar specificity from the fluorescence intensity of cells incubated with the neoglycoprotein at 4°C. Furthermore, upon incubation at 37°C, the fluorescence intensity of the lysate corresponds to both the neoglycoprotein linked at the surface and the neoglycoprotein present inside the cell upon endocytosis. Therefore, this method allows to select a neoglycoprotein suitable to target drugs. It is also possible to have a more precise analysis based on the fluorescence associated with single cells by using flow cytometry. Neoglycoproteins labelled with fluorescein are quite adapted to give the expected answer. Indeed, the fluorescence quantum yield of fluorescein depends on two important parameters:

1) The quantum yield is maximum at neutral pH and decreases to a very low value along with the medium acidification (Boutaric and Roy, 1939; Martin and Lindquist, 1975). Therefore, the quantum yield of fluorescein-labelled neoglycoproteins will be maximum when they are bound to the cell surface ($pH \approx 7.2$) and will be low when they are associated with either endosomal vesicules or compartments (pH range 5.5 - 6.5) or lysosomes (pH range 4.5 - 5.5).

2) The quantum yield of a fluorescein molecule bound to a protein or to a polymer is lower than that of a free fluorescein molecule or of a fluorescein molecule linked to a short peptide. Furthermore, the quantum yield of fluoresceinyl residues decreases further along with the increase of the number of fluoresceinyl residues bound to a macromolecule. Consequently, the cell associated fluorescence will be higher when the endocytosed fluorescein-labelled neoglycoprotein is degraded into short fragments by either endosomes or lysosome proteins.

In order to get rid of the variation of the quantum yield associated with the medium acidity, we neutralize the endoplasmic organelles by post-treating the cells for 30 min at 4°C in the presence of 50 µM monensin (Monsigny et al, 1984b; Midoux et al, 1986, 1987). Monensin is a sodium proton ionophore (Maxfield, 1982). In order to have more precise data about the hydrolysis extent, it is possible to use a neoglycoprotein labelled with a small number of fluorescein molecules and the same neoglycoprotein labelled with a larger number of fluorescein molecules. By comparing the fluorescence intensity of cells incubated either at 4°C or at 37°C and post-treated with monensin at 4°C in both cases, it is easy to appreciate the extent of the degradation (Midoux et al, 1986). This flow cytometric method allows an accurate appraisal of the binding capacity of the endocytosis activity and of the proteinase activity of cells under investigation and additionally allows the detection of heterogenous population with regards to these properties.

Fluorescein-labelled neoglycoprotein have been used to detect cell surface lectins on lymphocytes (Kieda et al, 1979) on tumor cells such as Lewis lung carcinoma cells (Roche et al, 1983), and L1210 lymphoma cells (Monsigny et al, 1984b), on monocytes (Roche et al, 1985a), on thyrocytes (Miquelis et al, 1987), on Plasmodium falciparum merozoites (El Moudni et al, 1993), on keratinocytes (Cerdan et al, 1991) and on melanocytes (Cerdan et al, 1992).

The specificity of the binding is clearly sugar dependent; for instance, among various neoglycoproteins used, that containing mannose-6-phosphate binds to and is endocytosed by human monocytes while the others, including that containing galactose-6-phosphate does not (Roche et al, 1985b). With HepG2 cells, HepG2 is a human hepatoma cell line, the neoglycoprotein containing β-galactoside is actively endocytosed, while the others are not (Fig 1), while with the HeLa cells, HeLa is a human carcinoma cell line, the above neoglycoprotein does not bind and is not endocytosed. Neoglycoproteins containing mannose-6-phosphate and those containing fucose are arrested by endothelial cells of rat brain while the other neoglycoproteins are not (Roche et al, unpublished results).

In some cases, neoglycoproteins may also bind to scavanger receptors which are known to recognize polyanionic globular proteins. Indeed, neoglycoproteins made by reaction of O-glycosyl-phenylisothiocyanate on serum albumin have a low isoelectric point. Serum albumin is an acidic protein and by neutralizing some 20 lysine ε-amino groups, the anion/cation ratio increases, and the isoelectric point is still lower. With cells expressing the scavanger receptor, negatively charged serum albumin is quite a good ligand (Brown et al, 1980), and neoglycoproteins may bind to both the membrane lectins and the scavenger receptor (Jansen et al, 1991).

Glycosylated Polymers

Various polymers have been developed as drug carriers (see for reviews Duncan, 1985, Vert, 1986). Some of them have been substituted by sugars as recognition signals (Duncan, 1992). Fully biodegradable glycosylated polymers have several advantages over glycosylated proteins: their synthesis can be conducted in organic solvents, leading to well-defined conjugates in very high yield, and they can be substituted by a large number of both recognition signals and drugs; we developed glycosylated polymers by substitution of all the amino groups of poly-L-lysine by three types of molecules: glycosides as recognition signals, drugs as therapeutic agents and polyhydroxyalcanoic acids (gluconoyl acid, for instance) as neutral solubilizing agents (Derrien et al, 1989); the polylysine currently used has a polymerization degree of about 190. Such glycosylated polymers are highly soluble in buffers in a large pH range, including phosphate balanced saline, pH 7.4. Those glycosylated polymers have been used to carry various drugs, including macrophage activators, and antiinfectious agents. In some cases, it can be useful to link sugar moieties to the polylysine through a peptidyl spacer arm such as glycyl-glycyl to improve the accessibility of the sugar to the binding site of the membrane lectins. A fully substituted neutral poly-L-lysine has a helicoidal structure as shown by circular dichroism studies, in contrast with the glycosylated poly-L-lysine which is random coil (Nègre, thesis 1991). Consequently, the glycosylated poly-L-lysine is more like a rod; in contrast, albumin and neoglycoproteins (glycosylated albumin) are globular. When used to carry a plasmid, we use glycosylated poly-L-lysine that is not neutralized by gluconoylation. This polycationic glycosylated poly-L-lysine is self-neutralized by interaction with the DNA which is a polyanion, leading to a roughly neutral complex.

NEOGLYCOPROTEINS AND GLYCOSYLATED POLYMERS AS CARRIERS

Neoglycoproteins and glycosylated polymers have been used successfully to carry and selectively deliver various types of drugs to several types of cells. In order to be released inside a cell, the drug is associated through a linkage which is either cleavable in acidic medium, cleavable under reducing conditions or cleavable by endosomes or lysosome hydrolases (Table 4).

Neoglycoprotein-Gelonin Conjugate

Gelonin is a single chain plant lectin which is quite efficient in blocking protein synthesis in assays using cell lysates but which has a limited activity on living cells in

Table 4 Common drug-carrier linkages

Cleavable in acidic medium

* *maleic acid derivatives* : - cis-aconitic acid (a)
- cis-ribofuranomaleic acid (b)

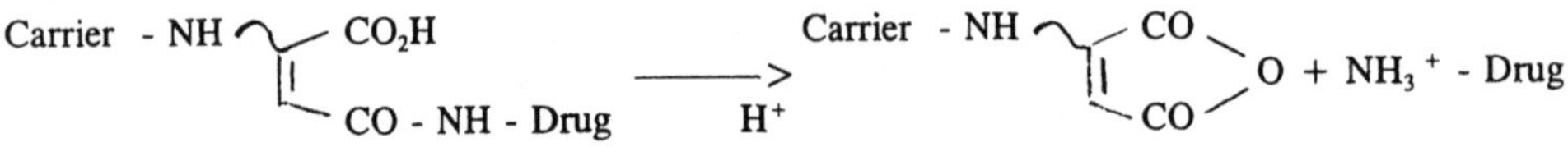

* *phenylthiocarbamoyl acidic α aminoacid derivatives* : *p*-benzylthiocarbamoyl aspartic acid (c)

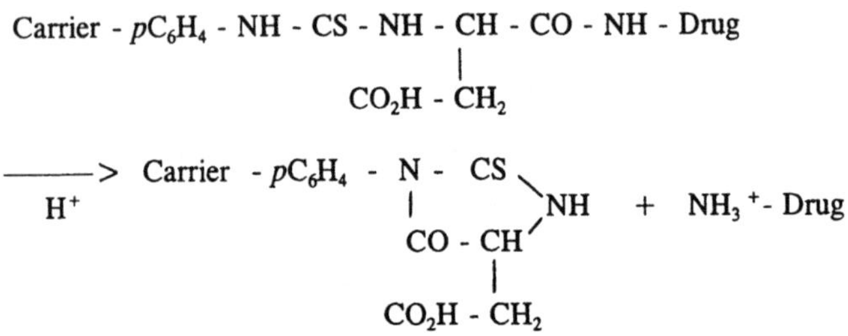

Cleavable under reducing conditions

Carrier - NH - CO - $(CH_2)_n$ - S - S - CHR - X - CO - NH - Drug

n = 2 : thiopropanoyl derivative (d)
n = 3 : thiobutanoyl derivative (e)
R = H and X= CH_2 : thiopropanoyl derivative (d,e)
R = CH_3 and X= pC_6H_4 : α-methyl-, α-thiotoluene carbonoyl derivative (f)

Cleavable by endosome or lysosome proteins

Carrier - NH - morpholinyl - thioacetyl - arginyl - leucyl - NH - Drug (g)

Carrier - NH - succinyl - alanyl - leucyl - alanyl - leucyl - NH - Drug (h)

Carrier - CO - glycyl - phenylalanyl - leucyl - glycyl - NH - Drug (i)

(a) Shen and Ryser, 1981 ; (b) Monsigny *et al.*, 1988b ; (c) Daussin *et al.*, 1988 ; (d) Roche *et al.*, 1983 ; (e)Mayer and Monsigny, unpublished data ; f) Thorpe *et al.* 1987 ; g) Monsigny *et al.*, 1980 ; Delmotte *et al.*, 1985 ; h) Trouet *et al.*, 1982 ; i) Duncan *et al.*, 1992.

contrast with other plant toxins such as ricin which contains two chains, the effector chain as in the case of gelonin, and a receptor chain which does not exist in gelonin. Gelonin was linked through a disulfide bridge to glucosylated serum albumin. This conjugate was found to be 100 times more efficient than the free gelonin in killing Lewis lung carcinoma cells (Roche et al, 1983); these cells have a membrane lectin recognizing glucosylated albumin.

N-acetylmuramyldipeptide carried by glycoconjugates

N-acetylmuramyldipeptide activates macrophages in vitro, in rendering them tumoricidal (Adam and Lederer, 1984), but is not efficient in vivo. Bound to a mannosylated serum albumin, MDP is about 100 times more efficient that free MDP in activating macrophages in vitro and is quite active in vivo in activating lungs peritoneal macrophages as well as macrophages (Monsigny et al, 1984a); the MDP-mannosylated serum albumin conjugate protects mice against the development of lung metastasis in the case of Lewis lung carcinoma, and, upon resection of the primary tumor, 70% of the mice survived after 100 days when 80% of untreated mice died within 10 days (Roche et al, 1985a, 1988). Bound to mannosylated polylysine neutralized by gluconoylation, MDP is about ten times more efficient in vitro in rendering macrophages tumoricidal than MDP bound to mannosylated serum albumin (Derrien et al, 1989); this conjugate is also efficient in vivo in improving the clearance of chromium-labelled sheep red blood cells (Monsigny et al, 1994).

MDP bound to mannosylated serum albumin as well as MDP bound to mannosylated poly-L-lysine neutralized by gluconoylation protect mice against herpes virus strains inducing hepatitis as well as strains inducing pneumotitis (Monsigny et al, 1994).

Nucleotide analogues carried by glycoconjugates

Various nucleoside and nucleotide analogues are known to have antiviral activities. Most nucleoside analogues must be phosphorylated before eliciting their activity but some cells such as macrophages are unable to add the first phosphate to a given nucleoside analogue. Therefore, it is advisable to use nucleotide analogues. However, the cell uptake of nucleotide analogues is usually not very efficient. On these bases, it appears that appropriate carriers should increase the uptake and the activity of nucleoside and nucleotide analogues. The various nucleotide analogues that have been linked to glycosylated carriers (Table 5) have a higher activity than their free counterparts. For instance, 9-(2-phosphonylmethoxyethyl)adenine (PMEA) linked to mannosylated poly-L-lysine through a glycyl-glycyl spacer arm is 50 times more efficient than free PMEA in inhibiting herpes virus multiplication in macrophages (Midoux et al, 1990) and is also efficient in vivo in inhibiting the infection of mice with HSV-1, VR3 strain (Monsigny et al, 1994). Various dideoxynucleoside analogues have been linked through a succinyl spacer arm and/or through a phosphodiester spacer arm to neoglycoproteins and glycosylated polymers (Table 5). In all cases, the phosphodiester derivatives were more efficient than the succinyl derivatives. For instance, the AZT-phosphodiester derivative bound to mannosylated polymer is more efficient than the AZT-derivative bound to mannosylated polymer, itself more efficient than free AZT in protecting human macrophages infected by HIV-1, the human acquired immunodeficiency syndrome associated virus (Monsigny et al, 1994).

Similarly, allopurinol riboside bound through a phosphodiester linkage to the mannosylated polymer is 80 times more active than the free drug in protecting macrophages infected with Leishmania, an intracellular parasite. Allopurinol riboside bound to the same polymer via a succinyl bridge has a relatively low activity (Nègre et al, 1992).

Oligonucleotide bound to glycosylated carriers

Oligonucleotides and especially antisense oligonucleotides have been shown to be active in inhibiting the expression of various proteins in several biological systems (see for a review, Hélène and Toulmé, 1990). Usually, the requested concentration to elicit a biological activity is around 10 μM (between 1 to 50 μM). Free oligonucleotides enter the cells and are mainly present in vesicles and other endosomal compartment as shown by using fluoresceinyl- or rhodaminyl-labelled oligonucleotides (Agrawal et al, 1992; Zhao et al, 1993). A small part should cross the membrane of one of the vesicular compartments to reach the cytosol, but neither the compartments where this occurs nor the molecular mechanism of the transfer is known. Conversely, it is clear that as soon as an oliogonucleotide enters the cytosol, it will reach rapidly the nucleus (Leonetti et al, 1991). Therefore several questions can be addressed. What is the nature of the molecules which allow oligonucleotides to leave the vesicle lumen to enter the cytosol by crossing the

Table 5 Structure of nucleoside analogue derivatives linked to glycosylated carriers

Phosphoramide (a)

Carrier - NH - PO_2^- - O - CH_2 - CH_2 - O - CH_2 - Adenine

Succinyl derivatives (b)

Carrier - NH - CO - CH_2 - CH_2 - CO - 5' ribavirin

Carrier - NH - CO - CH_2 - CH_2 - CO - 5' AZT

Carrier - NH - CO - CH_2 - CH_2 - CO - 5' ddI

Carrier - NH - CO - CH_2 - CH_2 - CO - 5' D4T

Carrier - NH - CO - CH_2 - CH_2 - CO - 5' allopurinol riboside

Phosphodiester derivatives (b)

Carrier - NH - CS - NH - pC_6H_4 - O - PO_2^- - 5' AZT

Carrier - NH - CS - NH - pC_6H_4 - O -PO_2^- - 5' Allopurinol riboside

Carrier - NH - CO - CH_2 - CH_2- CO-NH - $(CH_2)_6$ - O -PO_2^- - 5' AZT

(a) Midoux *et al.*, 1990 ; (b) Nègre *et al.*, 1991, 1992.

membrane? In which compartment does the oligonucleotides cross the membrane? Is it possible to increase the intracellular concentration of the oligonucleotides? Is it possible to convey the oligonucleotides to the right compartment where the membrane crossing occurs? Is it possible to help the oligonucleotides to cross the membrane? Some of these questions have already been partially solved.

A number of vehicles have been proposed as oligonucleotide carriers: they include nanoparticles, micelles, lipoproteins, liposomes, poly-L-lysine and glycoconjugates. We have shown that oligonucleotides, bound to either a neoglycoprotein or a glycosylated poly-L-lysine neutralized by gluconoylation, are more efficiently taken up by cells expressing a membrane lectin able to bind the glycoconjugate. An oligonucleotide substituted on its 3' end with either a fluoresceinyl residue or with a radioactive tag and linked through its 5' end to a glycosylated carrier is internalized up to 40 times more efficiently in cells which express a lectin able to bind the sugars borne by the glycoconjugates (Bonfils et al, 1992a). As an example, J774E cells (a macrophage cell line) internalize slowly free oligonucleotides and oligonucleotides bound to serum albumin, they internalized quite efficiently bound to mannosylated serum albumin and still more oligonucleotides bound to mannose-6-phosphate containing neoglycoproteins. The oligonucleotides are mainly present in vesicular compartments as shown by confocal microscopy. Furthermore, on the basis of quantitative data obtained by flow cytometry, this type of targeting leads to a clear increase in the quantity of oligonucleotides inside the cells as compared with that present outside the cells. Upon subcellular fractionation a larger proportion of free oligonucleotides was found in the cytosol and in the vesicular compartment when oligonucleotides were targeted than when they were used free (Monsigny et al, 1994). In addition, the biological activity was also significantly increased. For instance, an oligonucleotide specific for the sequence around the codon 12 of oncogene ras (Saison-Behmoaras et al, 1991) bound to Man6P - BSA was found to be quite active at a concentration for which the oligonucleotide protected at both ends but not carried by the appropriate neoglycoprotein was almost not active at all (Sdiqui et al, 1994). In all cases the oligonucleotides were linked to the glycoconjugate through a disulfide bridge.

Carrier - NH - CO -$(CH_2)_6$ -SS - $(CH_2)_6$ - O - PO_2^- - 5' oligonucleotide 3' - PO_2^- -O - $(CH_2)_6$ - NH - R - with R: Fluoresceinylthiocambamyl - or -CO - CH_2-pC_6H_4 - OH

Because the coupling step of the oligonucleotide derivative to the neoglycoprotein occurs in a low yield, an alternative method has been developed. The oligonucleotide is not chemically linked to the neoglycoprotein but it is substituted with biotin:

Biotinyl - NH - $(CH_2)_6$ - SS - $(CH_2)_6$ - O - PO_2^- - 5' oligonucleotide 3' - PO_2^-- O - $(CH_2)_6$) - NH - R

Independently, streptavidin is glycosylated as in the case of serum albumin. Finally solutions of biotinylated oligonucleotide and of glycosylated streptavidine are mixed, leading to a stable complex which can be isolated by gel filtration chromatography. This complex is efficiently taken up by cells expressing a membrane lectin which binds the sugar borne by the complex (Bonfils et al, 1992b).

Targeted gene transfer

Gene transfer into eukaryotic cells requires a method allowing a gene to cross a membrane to enter into the cytosol and then into the nucleus. In vitro, good results are obtained upon transient opening of the plasma membrane by an electric field (electroporation), by mechanic shear forces, by chemical shear forces, by hypoosmolarity shock, by use of calcium phosphate precipitate or of diethylaminoethyl dextran, by fusion with liposomes, as well as by various surfactants or cationic lipids (Keown et al, 1993): DOTAP or DOTMA: Dioleyloxypropyl - trimethyl ammonium, DOGS: Dioctadecyl - amidoglycylcarboxy spermine, DPPES: Dipalmitoyl-phosphatidylethanolamidocarboxy spermine, DDA or DODAC: Dimethyl - dioctadecyl ammonium, DC-Chol: dimethylaminoethylcarbamoyl cholesterol.

These techniques are relatively simple but must be optimized for each cell type because they may be noxious for the cells. Correctly adapted, they are of general use for in vitro experiments; however, they involve no recognition step and consequently they are not cell specific.

Recently, several other techniques, based on ligand-receptor recognition and ligand-induced endocytosis, have been developed: they include methods using viruses: retrovirus, adenovirus or adenoma associated virus, methods using transferrin, hormones such as insulin, glycoproteins, glycosylated polymers, etc.

Recently, poly-L-lysine substituted with various ligands of cell surface receptors have been used to carry and selectively deliver plasmids to cells. In any cases, the targeted poly-L-lysine plasmid complex is introduced inside the cell by endocytosis and migrates with the endosome vesicles to deeper intracellular compartments; eventually the plasmid reaches the cytosol. In the case of transferrin (see Hopkins, 1983, Dautry-Varsat, 1986) both the ligand and the receptor are endocytosed and recycled to the plasma membrane. Asialoglycoproteins bind their galactose specific lectin and the ligand-receptor complex is internalyzed; rapidly the luminal content of endosomes becomes acidic and the asialoglycoprotein is released. Then, the asialoglycoprotein receptor is recycled to the plasma membrane, while the ligand (asialoglycoprotein) is further transferred to deeper intracellular compartments and eventually will be delivered into lysosomes (see Ashwell and Harford, 1982, for a review). Glycoconjugates bearing β-galactoside residues act as asialoglycoproteins do.

Insulin and its receptors behave roughly as in the case of asialoglycoproteins and the galactose specific lectin respectively, but a part of insulin will be degraded inside the endosomes, another part will be recycled to the plasma membrane and released outside (Backer et al, 1990).

The intracellular traffic and the fate of the endocytosed material may be modulated by various lipophilic compounds such as chloroquine, monensin, brefeldin A, ammonia, etc.. Chloroquine is a lysosomotropic aromatic amine and as such is accumulated in acidic compartments and partly neutralyzes them. Consequently: 1) the rate of unbinding of the ligand receptor decreases, and 2) the extent of the degradation of the ligand (and of its receptor) is quite limited and 3) the transfer of the luminal content, including the ligand, to lysosomes is greatly inhibited. These properties support the large use of chloroquine in in vitro gene transfer experiments, since the pioneer work of Luthman and Magnusson (1983).

The efficiency of gene targeting estimated as the amount of the expressed reporter protein present as an active component, for instance the luciferase activity, depends on the extent of plasmid crossing a membrane of the endocytotic vesicle (or of another compartment) in order to reach the cytosol. Chloroquine, according to its main properties, may greatly enhances the gene transfer efficiency, and it has been successfully used. A more active way to help plasmid to cross the membrane is to incubate cell together with the targeted plasmid and with a fusogenic peptide, mimicking the natural path followed by the viral genome of orthomyxoviruses during the transfection step. The N-terminal moiety of the envelope glycoprotein of influenza virus is relaxed in acidic medium such as that of endosomes, and mediates the fusion between the cell membrane and the virus envelope (for reviews, see White, 1990 and 1992). Fusogenic peptides deriving from the N-terminal sequence of viral fusogenic proteins have been used to help genes to cross the endosome membrane and to enter the cytosol (see below).

Nucleic acids are polyanionic macromolecules, they cooperatively interact with polycations such as polyamines, DEAE dextrans or polylysine, etc.. Polycationic macromolecules and polycationic micelles have been used to form complex with plasmids and to transfer cells. Poly-L-lysine plasmid complexes are globular, and this feature is probably related to high efficiency in delivering DNA to cells (Wagner et al, 1991).

In 1987, Wu and Wu prepared a targeted poly-L-lysine by coupling poly-L-lysine and asialoorosomucoid. Asialoorosomucoid and polylysine were covalently conjugated through a S-S bridge. The coupling reaction was conducted in 2M guanidinium chloride to palliate the very low solubility of asialoorosomucoid and of the asialorosomucoid-poly-L-lysine conjugate. The conjugate solution was mixed with a plasmid solution. The plasmid pSV2-CAT contains the chloramphenicol acetyltransferase gene under the control of the enhancer and the early promotor of SV40 virus. Then the chaotropic agent was removed by dialysis and the complex was added to cells. HepG2 cells expressed the galactose specific lectin (Schwartz et al, 1982) and are efficiently transfected by the asialoorosomucoid poly-L-lysine conjugate complexed with pSV2-CAT plasmid. Cells, which do not express the galactose specific lectin, are not transfected with the above complex (Wu and Wu, 1987). This complex is also efficient in vivo. Upon intravenous injection of the complex, the transfected protein is transiently expressed 24 h later. When the complex is injected 30 min after a partial hepatectomy, the transfected protein is still expressed three months later (Wu et al, 1989). The first trials of gene therapy using this approach have been successful. A transient improvement of hypercholesterolemia in low density lipoprotein receptor deficient rabbits was achieved upon injection of a complex containing the gene (Wilson et al, 1992). A partial correction of genetic analbuminemia was obtained upon injection of a plasmid containing the gene of human serum albumin cDNA sequence in Nagase rats (Wu et al, 1991).

An insulin-poly-L-lysine conjugate through S-S bridge has been used to target a plasmid containing the large T antigen gene of SV40. Two hours later at 37°C, hepatoma cells which express the insulin receptor have incorporated the insulin-poly-L-lysine conjugate complexed with the plasmid and the complex has been localized around and inside the nucleus (Rosenkranz et al, 1992).

A transferrin-poly-L-lysine conjugate through S-S bridge has been used to target plasmids to growing cells (Wagner et al, 1990). It is known that growing cells must incorporate iron, that iron is carried by transferrin, and that all dividing cells express the transferrin receptor. A plasmid containing the luciferase gene complexed with a transferrin-poly-L-lysine conjugate is less efficient than when it was transfected by using DEAE dextran; however when the transfection with the complex is conducted in the presence of

100 μM chloroquine, the efficiency of this method is close to that obtained with the DEAE dextran method (Zenke et al, 1990). Chloroquine does not affect the efficiency of the gene transfer and/or expression when the DEAE dextran method is used. When transfection occurs in the presence of monensin, a H^+/Na^+ ionophore, which neutralizes all the intracellular acidic vesicles, the expression of the gene is extremely low (Cotten et al, 1990). Monensin, by neutralizing all intracellular vesicles precludes the various fusion process occuring usually between vesicles and other intracellular organelles. Various other tricks have been used to increase the transfer efficiency; among them, the use of defective adenovirus leads to a very good efficiency even better (up to 100 times) than when transfection is conducted in the presence of chloroquine (Cotten et al, 1992).

Instead of preparing a protein-poly-L-lysine conjugated, which requires the use of a chaotropic agent and which has a low solubility in physiological medium, we prepared various poly-L-lysine substituted with low molecular mass recognition signals. To target genes to cells expressing cell surface lectins, we synthesized glycosylated poly-L-lysine (Midoux et al, 1993). The synthesis of such compounds may be conducted either in water solution or preferably in organic solvent. The use of dimethylformamide as a solvent has several advantages: a large panel of reagents, which are not usable in aqueous medium, can be used; the coupling yield is usually very high as in the case of the substitution of the poly-L-lysine ε-amino groups by compounds containing a carboxylic group; the substitution is easily controlled and it is possible to precisely prepare conjugates containing a predefined number of substituents; the purification of the conjugates is simply achieved by selective precipitation upon addition of isopropanol.

A poly-L-lysine containing 190 lysine residues appeared to be suitable to prepare efficient glycosylated polymers. Conjugates containing about 60 sugar moieties (in average one lysine is substituted by one sugar moiety while two other lysines kept their cationic charge) have been found to efficiently carry a plasmid into cells expressing the related sugar specific lectin and to lead to very high expression of the transfected gene. The uptake of the glycosylated polymer/plasmid complex is sugar dependent: when sugar-free poly-L-lysine is used to prepare the polymer/plamid complex the uptake is low; when a glycosylated poly-L-lysine is used, the uptake is quite efficient into cells expressing the related sugar specific lectin. For instance, poly-L-lysine substituted with galactopyranosyl-β-4-glucopyranosyl-β-4-oxyphenylthiocarbamyl (which is a ligand of the asialoglycoprotein receptor of liver parenchymal cells and of $HepG_2$ cells) very actively transfer plasmid into these cells, but does not into other cell lines such as HeLa cells which do not express a lectin with such a specificity. When a mannosylated poly-L-lysine is used to target a plasmid into $HepG_2$ cells, which do not express a mannose specific lectin, the efficiency is as low as with the sugar-free poly-L-lysine. The uptake is also dependent on the number of sugar borne by poly-L-lysine. As it has been previously shown by us (Privat et al, 1974; Delmotte, 1976; Monsigny et al, 1983, 1988a) and others (Krantz et al, 1976; Townsend and Stahl, 1981) a high density of simple sugars bound to a globular protein leads to a high apparent affinity phenomenon related with the avidity effect as in the case of IgM type antibodies. The efficiency is also related with the nature of the bound between sugar moieties and poly-L-lysine (Table 6). Using the alkylation of poly-L-lysine residues by reducing the Schiff base formed upon condensation of galactopyranosyl-β-4-glucose(lactose) with the ε-amino groups of poly-L-lysine, the transfection was quite inefficient. Conversely, by using a cluster of galactopyranoside linked to a longer spacer arm, the transfection was efficient (Plank et al, 1992).

Glycosylated polylysines are very soluble in phosphate balanced saline or in culture media. As a consequence, the preparation of glycosylated polylysine/plasmid complex is very simple: a 300 nM glycosylated polylysine solution is simply mixed with a 3 nM solution of plasmid (pSV2-lue, 5 kbases in length), kept for 30 min at 20°C; the final solution, which contains the complex without any trace of free plasmid as shown by electrophoresis analysis, can be immediately used to transfect cells.

The presence of sugar moieties on polylysine in the complex mediates a selective recognition by the membrane lectin and an efficient uptake of the complex as shown by confocal microscopy. However, the transfected gene is expressed at a low level. To enhance the expression of the gene, it is necessary to help the plasmid to cross the

Table 6 Gene transfer: Spacer arm dependent efficiency of glycosylated poly-L-lysine bearing the same sugar in a terminal non reducing position

Type of structure	Efficiency	
Galβ - 4Glcβ - O - pC_6H_4 - NH - CS →	++	(a)
Galβ - O - pC_6H_4 - NH - CS →	++	(b)
Galβ - 4Glucityl →	-	(b,c)
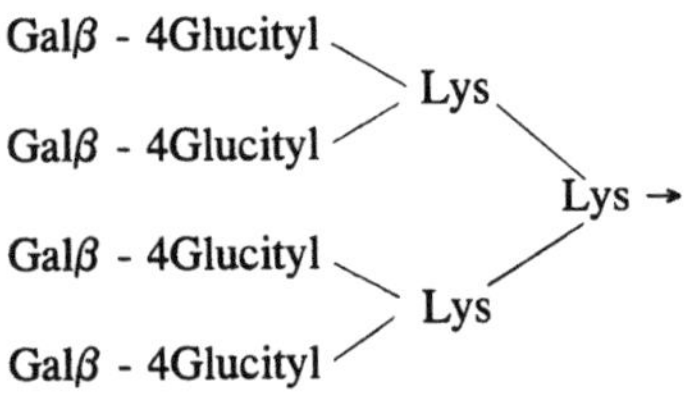	++	(c)

(a) Midoux *et al.*, 1993 ; (b) unpublished data ; (c) Plank *et al.*, 1992.

endosomal membrane in order to reach the cytosol and the nucleus. This can be achieved either by using lysosomotropic animes, such as chloroquine or by using compounds such as fusogenic peptides which destabilize the endosomal membranes.

Chloroquine increases the efficiency of gene transfer using polylysine asialoorosomucoid conjugate or lactosylated polylysine (Midoux et al, 1993) as it does with plasmid associated with either polylysine transferrin conjugate. Chloroquine is optimally efficient when it is added at the same time or 1 h after the glycosylated polylysine plasmid complex is added to cells. Optimal results are obtained when the glycosylated polylysine plasmid conjugate and 100 μM chloroquine are left in the presence of cells for 4 h. Cells are then washed, covered with fresh medium containing fetal bovine serum and grown for 48 h at 37°C.

Alternatively, the efficiency of the gene transfer can be greatly increased by using fusogenic peptides, as evidenced by the high activity of the expressed reporter protein (Wagner et al, 1992; Midoux et al, 1993). Orthomyxovirus, such as influenza virus, transfer their genetic material into the cytosol from the acidic endosomal compartments. The virus interacts with sialoglycoconjugates of the target cell thanks to a viral membrane glycoprotein, called hemagglutinin or lectin. This membrane lectin has a short intraviral C terminal sequence, a singly transmembrane sequence, a binding site which recognizes glycopeptides containing a cluster of three terminal Neu5Acα 6-Galβ 4GlcNAcβ→moieties, and a hydrophobic N-terminal sequence. This N-terminal sequence is tightly associated with the protein core in neutral medium; when the medium becomes acidic, a transconformation occurs, and the N-terminal peptide leaves the core protein to interact with the bilayer membrane, but at neutral pH it is tightly associated with the protein core. In the absence of the viral protein, a more hydrophilic peptide must be preferred. Eicosapeptides having the sequence of the N-terminal moieties of the influenza hemagglutinin or a closely related sequence have been synthesized. Peptides with five glutamyl residues at position n and n + 3 or n + 4, and with hydrophobic residues between these acidic residues are soluble in neutral or alkaline medium, and have roughly a random coil conformation. In slightly acidic medium, such peptides acquire a helix conformation as shown by circular dichroism measurements (Takahashi, 1990), and are efficient in inducing vesicle fusion (Murata et al, 1987; Rafalski et al, 1991). When the following docosapeptide

G L F E A I A E F I E G G W E G L I E G C A

was used at 5 or 10 μM concentration, together with a plasmid lactosylated poly-L-lysine complex the gene transfer was very efficient: the luciferase activity was 500-fold higher than in the absence of fusogenic peptide and about 10-fold higher than in the presence of 100 μM chloroquine. Wagner et al, 1992 used a related peptide (tricosapeptide)

G L F E A I A G F I E N G W E G M I D G G G C

in which G-8 replaces E-8, N-12 replaces G-8, M-17 replaces L-17, D-19 replaces E-19 and GGC-(21-23) replace CA-(21-22). This peptide was linked through a S-S bridge to polylysine. A complex was prepared by mixing the tricosapeptide-polylysine conjugate, the transferrin-polylysine conjugate and the plasmid. The activity of the enzyme expressed from the transfected gene was dependent on the ratio of the two polylysine conjugates. In optimal conditions, the enzyme activity was significantly higher than in the absence of the tricosapeptide-polylysine conjugate.

CONCLUDING REMARKS

Membrane lectins of various cell types actively take up glycoconjugates in a sugar specific way. On these bases, many therapeutic drugs have been rendered cell specific. In addition glycoconjugates have been shown to be suitable to tranfer oligonucleotides and genes inside cells which express a membrane lectin able to recognize the sugar moiety used as recognition signal. Several improvements are expected: they include the use of complex oligosaccharides to reach a higher cell selectivity, the use of helper peptides allowing a larger efficiency in the transfer of hydrophilic drugs, of oligonucleotides and of genes into the cytosol and into the nucleus.

Acknowledgements

We thank Philippe Bouchard for the preparation of the sugar derivatives, Emmanuelle Martin for the synthesis of peptides, and Marie-Thérèse Bousser and Suzanne Nuques for their skill in using cell biology techniques.

This work was partly supported by grants from ANRS (Agence Nationale de Recherches sur le SIDA), ARC (Association pour la Recherche sur le Cancer, WHO (World Health Organization) and CEC (Commission of the European Communities).

P.M. and A.-C.R. are Research Directors of the Institut National de la Santé et de la Recherche Médicale, R.M. is Research Director of the Centre National de la Recherche Scientifique and M.M. is Professor of Biochemistry at the University of Orléans and Head of the Department of Endogenous Lectins and Glycoconjugates, in the Centre de Biophysique Moléculaire, Orléans.

REFERENCES

Adam, A. and Lederer, E., 1984, Muramylpeptides: immunomodulators, sleep factors, and vitamins, Medicinal Res.Rev., 4:111

Agrawal, S., Sarin, P.S., Zamecnik, M. and Zamecnik, P.C., 1992, Cellular uptake and anti HIV activity of oligonucleotides and their analogs, in: "Gene Regulation and Biology of Antisense RNA and DNA" R.P. Erickson and J.G. Izant, eds., Raven Press, Ltd, New York

Ashwell, G. and Harford, J., 1982a, Carbohydrate-specific receptors of the liver, Ann.Rev.Biochem., 51:531

Backer, J.M., Kahn, C.R. and White, M.F., 1990, The dissociation and degradation of internalized insulin occur in the endosomes of rat hepatome cells, J.Biol.Chem., 265:14828

Bonfils, E., Depierreux, C., Midoux, P., Thuong, N.T., Monsigny, M. and Roche, A.C., 1992a, Drug Targeting: synthesis and endocytosis of oligonucleotide-neoglycoprotein conjugates, Nucleic Acids Res., 20:4621

Bonfils, E., Mendes, C., Roche, A.C., Monsigny, M. and Midoux, P.. 1992b, Uptake by macrophages of a biotinylated oligo-α-deoxythymidilate by using mannosylated streptavidin, Bioconjugate Chem., 3:277

Boutaric, A. and Roy, M., 1939, Pouvoir fluorescent des solutions d'uranine en fonction de leur concentration en ions H+, Compt.Rend.Acad.Sci., 209:162

Brown, M.S., Basu, S.K., Falck, J.R., Ho, Y.K. and Goldstein, J.L., 1980, The scavenger cell pathway for lipoprotein degradation: specificity of the binding site that mediate the uptake of negatively-charged LDL by macrophages, J.Supramol.Struct., 13:67

Cerdan, D., Grillon, C., Monsigy, M., Redziniak, G. and Kieda, C., 1991, Human keratinocyte membrane lectins: characterization and modulation of their expression by cytokines, Biol.Cell, 73:35

Cerdan, D., Redziniak, G., Bourgeois, C.A., Monsigny, M. and Kieda, C., 1992, C32 human melanoma cell endogenous lectins: characterization and implication in vesicle mediated melanin transfer to keratinocytes, Exp.Cell Res., 203:164

Cotten, M., Langle-Rouault, F., Kirlappos, H., Wagner, E., Mechtler, K., Zenke, M., Beug, H. and Birnstiel, M.L., 1990, Transferrin-polycation-mediated introduction of DNA into human leukemic cells: stimulation by agents that affect the survival of transfected DNA or modulate transferrin receptor levels, Proc.Natl.Acad.Sci.USA, 87:4033

Cotten, M., Wagner, E., Zatloukal, K., Phillips, S., Curiel, D.T. and Birnstiel, M.L., 1992, High-efficiency receptor-mediated delivery of small and large (48 kilobase) gene constructs using the endosome-disruption activity of defective or chemically inactivated adenovirus particles, Proc.Natl.Acad.Sci.USA, 89:6094

Dautry-Varsat, A., 1986, Receptor-mediated endocytosis: the intracellular journey of transferrin and its receptor, Biochimie, 68:375

Daussin, F., Boschetti, E. Delmotte, F. and Monsigny, M., 1988, p-benzylthiocarbamyl aspartyl-daunorubicin substituted polymer: a new drug acid labile arm-carrier conjugate, Eur.J.Biochem., 176:625

Delmotte, F., 1976, Interactions glycanne-proteine. Etude des proteines affines d'oses contenant de la N-acetylglucosamine, Thesis, Universite d'Orléans, France

Delmotte, F., Lescanne, P.J., Daussin, F., Roche, A.C., Midoux, P. and Monsigny, M., 1985, Préparation et propriétés de systèmes covalents drogue-bras-vecteur, Actualités Chimie Thérapeutique, 12:121

Derrien, D., Midoux, P., Petit, C., Nègre, E., Mayer, R., Monsigny, M. and Roche, A.C., 1989, Muramyl dipeptide bound to poly-L-lysine substituted with mannose and gluconoyl residues as macrophage activators, Glycoconjugate J., 6:241

Drickamer, K., 1991, Clearing up glycoprotein hormones, Cell, 67:1029

Drickamer, K. and Taylor, M.E., 1993, Biology of animal lectins, Annu.Rev.Cell Biol., 9:237

Duncan, R., 1985, Biological effects of soluble synthetic polymers as drug carriers, CRC Crit.Rev.Ther.Drug Carriers Syst., 1:281

Duncan, R., 1992, Drug-polymer conjugates: potential for improved chemotherapy, Anti-Cancer Drugs, 3:175

El Moudni, B., Philippe, B., Monsigny, M. and Schrével, J., 1993, Characterization of a lectin at the surface of Plasmodium falciparum merozoites, Glycobiology, 3:305

Goebel, W.F. and Avery, O.T., 1929, Chemo-immunological studies on conjugated carbohydrate proteins. IV The synthesis of p-aminophenol b-glucoside, p-aminophenol β-galactoside and their coupling with serum albumin, J.Exp.Med., 50:521

Goldstein, I.J., Hugues, R.C., Monsigny, M., Osawa, T. and Sharon, N., 1980, What should be called a lectin? Nature, 285:66

Gray, G., 1974, The direct coupling of oligosaccharides to proteins and derivatized gels, Arch.Biochem.Biphys., 163:426

Harford, J. and Ashwell, G., 1982, The hepatic receptor for asialoglycoproteins, in: "The glycoconjugates", M.I. Horowitz, ed., Academic Press, New-York

Hart, G.W., Haltwanger, R.S., Holt, G.D. and Kelly, W.G., 1989, Glycosylation in the cell nucleus and cytoplasm, Annu.Rev.Biochem., 58:841

Hélène, C. and Toulmé, J.J., 1990, Specific regulation of gene by antisense, sense and antigene nucleic acids, Biochim.Biophys.Acta, 1049:99

Hopkins, C.R., 1983, Intracellular routing of transferrin and transferrin receptors in epidermoid carcinoma A431 cells, Cell, 35:321

Hubert, J., Sève, A.P., Facy, P. and Monsigny, M., 1989, Are nuclear lectins and nuclear glycoproteins involved in the modulation of nuclear functions? Cell Differentiation and Development, 27:69

Hughes, R.C., 1992, Lectins as cell adhesion molecules, Curr.Opinion Struct.Biology, 2:687
Jansen, R.W., Molema, G., Ching, T.L., Oosting, R., Harms, G., Moolenaar, F., Hardonk, M.J. and Meijer, D.K.F., 1991, Hepatic endocytosis of various types of mannose terminated albumins. What is important, sugar recognition, net charge, or the combination of these features, J.Biol.Chem., 266:3343
Keown, W.A., Campbell, C.R. and Kucherlapatir, S., 1990, Methods for introducing DNA into mammalian cells, Methods in Enzymology, 185:527
Kieda, C., Delmotte, F. and Monsigny, M., 1977, Preparation and properties of glycosylated cytochemical markers, FEBS Letters, 76:257
Kieda, C., Roche, A.C., Delmotte, F. and Monsigny, M., 1979, Lymphocyte membrane lectins. Direct visualization by the use of fluoresceinyl glycosylated cytochemical markers, FEBS Letters, 99:329
Kornfeld, S., 1992, Structure and function of the mannose 6-phosphate/insulinlike growth factor II receptors, Annu.Rev.Biochem., 61:307
Krantz, M.J., Holtzman, N.A., Stowell, C.P. and Lee, Y.C. 1976, Attachment of thioglycosides to proteins: Enhancement of liver membrane binding, Biochemistry, 15:3963
Lasky, L.A., 1992, Selectins: interpreters of cell-specific carbohydrate information during inflammation, Science, 258:964
Lee, Y.C., Stowell, C.P. and Krantz, E.A., 1976, 2-imino-2methoxy-ethyl-1-thioglycoside: new reagents for attaching sugars to proteins, Biochemistry, 15:3956
Lee, Y.C., Towndsend, R.R., Hardy, M.R., Longren, J., Arnap, J., Haraldsson, M. and Lenn, H., 1983, Binding of synthetic oligosaccharides to hepatic Gal/GalNAc lectin. Dependent on fine structural features, J.Biol.Chem., 258:199
Leonetti, J.P., Mechti, N., Degols, G., Gagnor, C. and Lebleu, B., 1991, Intracellular distribution of microinjected antisense oligonucleotides, Proc.Natl.Acad.Sci.USA, 88:2702
Luthman, H. and Magnusson, G., 1983, High efficiency polyoma DNA transfection of chloroquine treated cells, Nucleic Acids Res., 11:1295
Martin, M.M. and Lindqvist, L., 1975, The pH dependence of fluorescein fluorescence, J.Luminesc., 10:381
Maxfield, F.R., 1982, Weak bases and ionophores rapidly and reversibly raise the pH of endocytic vesicles in cultured mouse fibroblasts, J.Cell Biol., 95:676
McBroom, C.R., Samanen, C.H. and Goldstein, I.J., 1972, Carbohydrate antigens: coupling of carbohydrates to proteins by diazonium and phenylisothiocyanate reaction, Meth.Enzymol., 28:212
Midoux, P., Mendes, C., Legrand, A., Raimond, J., Mayer, R., Monsigny, M. and Roche, A., 1993, Specific gene transfer mediated by lactosylated poly-L-lysine into hepatoma cells, Nucleic Acids Res., 21:871
Midoux, P., Nègre, E., Roche, A.C., Mayer, R., Monsigny, M., Balzarini, J., De Clercq, E., Mayer, E., Ghaffar, A. and Gangemi, J.D., 1990, Drug targeting: anti-HSV-1 activity of mannosylated polymer-bound 9-(2-phosphonylmethoxyethyl)adenine, Biochem.Biophys.Res.Commun., 167:1044
Midoux, P., Roche, A.C. and Monsigny, M., 1986, Degradation of endocytosed materials estimated by flow cytofluorometry using two neoglycoproteins containing different numbers of fluorescein molecules, Biol.Cell, 58:221
Midoux, P., Roche, A.C. and Monsigny, M., 1987, Quantitation of the binding, uptake and degradation of fluoresceinylated neoglycoproteins by flow cytometry, Cytometry, 8:327
Miquelis, R., Alquier, C. and Monsigny, M., 1987, A lectin-like N-acetylglucosamine receptor on thyrocytes: binding characteristics, J.Biol.Chem., 262:15291
Monsigny, M., Hubert, J., Obrenovitch, A. and Schrével, J., 1989, Neoglycoproteins as tools to analyze endogenous lectins and glycoconjugates, in: "Electron microscopy of subcellar dynamics" H. Plattner, ed., CRC Press, Inc., Boca Raton, FL, USA
Monsigny, M., Kieda, C., Obrenovitch, A. and Delmotte, F., 1976, Glycosylated Horse Radish Peroxydase interactions with lectins, in: "Protides of the Biological Fluids", H. Peeter, ed., Pergamon Press, Oxford
Monsigny, M., Kieda, C. and Roche, A.C., 1983, Membrane glycoproteins, glycolipids and membrane lectins as recognition signals in normal and malignant cells, Biol.Cell, 47:95

Monsigny, M., Kieda, C., Roche, A.C. and Delmotte, F., 1980, Preparation and biological properties of a covalent antitumor drug arm-carrier (D.A.C. conjugates), FEBS Letters, 119:181

Monsigny, M., Roche, A.C. and Bailly, P., 1984a, Tumoricidal activation of murine alveolar macrophages by muramyl dipeptide substituted mannosylated serum albumin, Biochem.Biophys.Res.Commun., 121:579

Monsigny, M., Roche, A.C, Kieda, C. Midoux, P. and Obrenovitch, A., 1988a, Characterization and biological implicatons of membrane lectins in tumor, lymphoid and myeloid cells, Biochimie, 70:1633

Monsigny, M., Roche, A.C. and Midoux, P., 1984b, Uptake of neoglycoproteins via membrane lectin (s) of L 1210 cells evidenced by quantitative flow cytofluorometry and drug targeting, Biol.Cell, 51:187

Monsigny, M., Roche, A.C. and Midoux, P., 1988b, Endogenous lectins and drug targeting, in:"Biological membranes in cancer cells", T.Galotti, A. Cittadini, G. Neri and A. Scarpa (eds.), Ann.N.Y.Acad.Sci., pp. 399-414

Monsigny, M., Roche, A.C., Midoux, P. and Mayer, R., 1994, Glycoconjugates as carriers for specific delivery of therapeutic drugs and genes, Adv.Drug Delivery Reviews, 4: 1

Morell, A.G., Gregoriadis, G., Scheinberg, H., Hickman, J. and Ashwell, G., 1971, The role of sialic acid in determining the survival of glycoproteins in the circulation, J.Biol.Chem., 246:1461

Murata, M., Sugahara, Y., Takahashi, S. and Ohnishi, S.I., 1987, pH-dependent membrane fusion activity of a synthetic twenty amino acid peptide with the same sequence as that of the hydrophobic segment of Influenza Virus Hemagglutinin, J.Biochem., 102:957

Nègre, E., 1991, Ciblage d'agents antiviraux et antiparasitaires par des polymères glycosylés spécifiques des lectines membranaires des macrophages, Thesis, Université d'Orléans, France

Nègre, E., Chance, M.L., Hanboula, S.Y., Monsigny, M., Roche, A.C., Mayer, R. and Hommel, M., 1992, Antileishmanial drug targeting through glycosylated polymers specifically internalized by macrophage membrane lectins, Antimicrob.Agents Chemother., 36:2228

Pimpaneau, V., 1990, Lectines endogènes de cellules myéloïde et de macrophages: caractérisation, endocytose et purification, Thesis, Université d'Orléans, France

Plank, C., Zatloukal, K., Cotten, M., Mechtler, K. and Wagner, E., 1992, Gene transfer into hepatocytes using asialoglycoprotein receptor mediated endocytosis of DNA complexes with an artificial tetra-antennary galactose ligand, Bioconjugate Chem., 3:533

Privat, J.P., Delmotte, F. and Monsigny, M., 1974, Protein-sugar interactions. Association of β-1-4 linked N - acetyl- D - glucosamine oligomer derivatives with wheat germ agglutinin (lectin), FEBS Letters, 46:224

Rafalski, M., Ortiz, A., Rockwell, A., Van Ginkel, L.C., Lear, J.D., Degrado, W.F. and Wilschut, J., 1991, Membrane fusion activity of the influenza virus hemagglutinin: interaction of HA2 N-terminal peptides with phospholipid vesicles, Biochemistry, 30:10211

Roche, A.C., Bailly, P. and Monsigny, M., 1985a, Macrophage activation by MDP bound to neoglycoproteins: metastasis eradication in mice. Invasion and Metastasis, 5:218

Roche, A.C., Barzilay, M., Midoux, P., Junqua, S., Sharon, N. and Monsigny, M., 1983, Endocytosis of glycoconjugates by Lewis lung carcinoma cells, J.Cell Biochem., 22:131

Roche, A.C., Midoux, P., Petit, C., Derrien, D., Mayer, R. and Monsigny, M., 1988, Macrophage activation by targeted biological response modifiers, in: "Immunomodulators and non specific host defense mechanisms against microbial infections, Adv.Bioscience, K.N. Masihi and W. Lange, eds., Pergamon Press, Oxford

Roche, A.C., Midoux, P., Bouchard, P. and Monsigny, M., 1985, Membrane lectins on human monocytes. Maturation-dependent modulation of 6-phohphomannose and mannose receptors, FEBS Letters, 193:63

Rosenkranz, A.A., Yachmenev, S.V., Jans, D.A., Serebryakova, N.V., Murav'ev, V.I., Peters, R. and Sobolev, A.S., 1992, Receptor-mediated endocytosis and nuclear transport of a transfecting DNA construct, Exp.Cell Res., 199:323

Saison-Behmoaras, T., Tocque, B., Rey, I., Chassignol, M., Thuong, N.T. and Hélène, C., 1991, Short modified antisense oligonucleotides directed against Ha-ras point mutation induce selective cleavage of the mRNA and inhibit T24 cells proliferation, Embo.J., 10:1111
Schrével, J., Kieda, C., Caigneaux, E., Gros, D., Delmotte, F. and Monsigny, M., 1979, Visualization of cell surface carbohydrate by a general two-steps lectin technique: lectin and glycosylated cytochemical markers, Biol.Cell, 36:259
Schwartz, A.L., Fridovich, S.E. and Lodish, H.F., 1982, Kinetics of internalization and recycling of the asialoglycoprotein receptor in a hepatoma cell line, J.Biol.Chem., 257:4230
Sdiqui, N., Arar, K., Midoux, P., Mayer, R., Monsigny, M. and Roche, A.C., 1994, Inhibition of human mammary cell line proliferation by membrane lectin-mediated uptake of Ha-ras antisense oligodeoxynucleotide, (submitted)
Sharon, N. and Lis, H., 1989, Lectins as cell recognition molecules, Science, 246:227
Shen, W.C. and Ryser, H.J.P., 1981, Cis-aconityl spacer between daunomycin and macromolecular carrier: a model of pH sensitive linkage releasing drug from a lysomotropic conjugate, Biochem.Biophys.Res.Commun., 102:1048
Stahl, P.D., Rodman, J.S., Miller, M.J. and Schlesinger, P.H., 1978, Evidence for receptor mediated binding of glycoproteins, glycoconjugates, and lysosomal glycosidases by alveolar macrophages, Proc.Natl.Acad.Sci.USA, 75:1399
Stahl, P.D., Wileman, T.E., Diment, S. and Shepherd, V.L., 1984, Mannose-specific oligosaccharide recognition by mononuclear phagocytes, Biol.Cell, 51:215
Takahashi, S., 1990, Conformation of membrane fusion-active 20-residue peptides with or without lipid bilayers. Implication of α-helix formation for membrane fusion, Biochemistry, 29:6257
Thorpe, P.E., Wallace, P.M., Knowles, P.P., Relf, M.G., Brown, A.N.F., Watson, G.J., Knyba, R.E., Warwrzynczak, E.J. and Blakey, D.D., 1987, New coupling agents for the synthesis of immunotoxins containing a hindered disulfide bond with improved stability in vivo, Cancer Res., 47:5927
Townsend, R. and Stahl, P.D., 1981, Isolation and characterization of a mannose/N-acetylglucosamine/fucose binding protein from rat liver, Biochem.J., 194:209
Trouet, A., Masquelier, M., Baurain, R. and Deprez-de-Campeneere, D., 1982, A covalent linkage between daunorubicin and proteins that is stable in serum and reversible by lysosomal hydrolases, as required for a lysosomotropic drug-carrier conjugate: In vitro and in vivo studies, Proc.Natl.Acad.Sci.USA, 79:626
Vert, M., 1986, Polyvalent polymeric drug carriers, CRC Crit.Rev.Ther.Drug carriers Syst., 2:291
Wagner, E., Cotten, M., Foisner, R. and Birnstiel, M.L., 1991, Transferrin-polylysine-DNA complexes: The effect of polycations on the structure of the complex and DNA delivery to cells, Proc.Natl.Acad.Sci.USA, 88:4255
Wagner, E., Plank, C., Zatloukal, K., Cotten, M. and Birnstiel, M.L., 1992, Influenza virus hemagglutinin HA-2 N-terminal fusogenic peptides augment gene transfer by transferrin-polylysine-DNA complexes: Towards a synthetic virus-like gene-transfer vehicle, Proc.Natl.Acad.Sci.USA., 89:7934
Wagner, E., Zenke, M., Cotten, M., Beug, H. and Birnstiel, M.L., 1990, Transferrin polycation conjugates as carriers for DNA uptake into cells, Proc.Natl.Acad.Sci.USA, 87:3410
Wang, J.L., Laing, J.G. and Anderson, R.L., 1991, Lectins in the cell nucleus, Glycobiology, 1:243
White, J.M., 1990, Viral and cellular membrane fusion proteins, Annu.Rev.Physiol., 52:675
White, J.M., 1992, Membrane fusion, Science, 258:917
Wilson, J.M., Grossman, M., Wu, C.H., Chowdhury, N.R., Wu, G.Y. and Chowdhyry, J.R., 1992, Hepatocyte-directed gene transfer in vivo leads to transient improvement of hypercholesterolemia in low density lipoprotein receptor-deficient rabbits, J.Biol.Chem., 267:963
Wu, C.H., Wilson, J.M. and Wu, G.Y., 1989, Targeting genes: delivery and persistent expression of a foreign gene driven by mammalian regulatory elements in vivo, J.Biol.Chem., 264:16985
Wu, G.Y., Wilson, J.M., Shalaby, F., Grossman, M., Shafritz, D.A. and Wu, C.H., 1991, Receptor-mediated gene delivery in vivo, J.Biol.Chem., 266:14338
Wu, G.Y. and Wu, C.H., 1987, Receptor-mediated in vitro gene transformation by a soluble

DNA carrier system, J.Biol.Chem., 262:4429
Zenke, M., Steinlein, P., Wagner, E., Cotten, M., Beug, H. and Birnstiel, M.L., 1990, Receptor-mediated endocytosis of transferrin-polycation conjugates: an efficient way to introduce DNA into hematopoietic cells, Proc.Natl.Acad.Sci.USA, 87:3655
Zhao, Q., Matson, S., Herrera, C.J., Fisher, E., Yu, H. and Krieg, A.M., 1993, Comparison of cellular binding and uptake of antisense phosphodiester, phosphorothioate, and mixed phosphorothiote and methylphosphonate oligonucleotides, Antisense Res.Develop., 3:53

THE USE OF MSH DERIVATIVES FOR TARGETING TO MELANOMAS

David R. Bard[1], C. Graham Knight[1], E. Phillip Wraight[2], J. Van Beeumen[3], B. Deveese[3] and F. Jacquemotte[4]

[1]Strangeways Research Laboratory, Worts' Causeway, Cambridge, UK; [2]Department of Nuclear Medicine, Addenbrooke's Hospital, Hills Road, Cambridge, UK; [3]Vakgroep Biochemie, Fysiologie en Microbiologie, Ledegankstraat 35, Ghent, Belgium and [4]Institut Meurice, Ave Emile Gryzon 1, Brussels, Belgium

INTRODUCTION

Malignant melanoma is one of the few human neoplasms to have shown an increased incidence throughout the developed world over the last twenty years, (Magnus, 1987; Cox et al, 1987; Little et al, 1980; Cooke et al, 1983; Schreiber et al, 1981). Due to its tendency to early metastasis and its resistance to conventional chemotherapy and radiotherapy, the only effective treatment remains excision and provided this is complete, prognosis is highly favourable, (Davis, 1991). An accurate, targeted imaging agent would assist the surgeon in reaching this ideal and to detect recurrence promptly. If the principle of targeting could be extended to the delivery of chemotherapeutic or radiotherapeutic compounds, the disappointing therapeutic ratios of existing agents might be improved sufficiently to make therapy viable.

Normal and malignant melanocytes possess cell-surface receptors for the tridecapeptide α-melanocyte stimulating hormone (MSH), (Tatro et al, 1990; Siegrist et al, 1992) and provided that this peptide can be linked to an appropriate γ-emitting isotope, the resulting complex should be capable of imaging melanomas in situ by radioscintigraphy. We have previously reported the synthesis, in vitro biological activity and in vivo biodistribution (Bard et al, 1990a; Bard et al, 1992) of an MSH derivative, *bis*MSH-DTPA. In this compound, two complete MSH sequences are linked through their *N*'-termini to a single molecule of the chelator, diethylenetriamine pentaacetic acid (DTPA). *Bis*MSH-DTPA could bind the γ-emitter, indium-111(^{111}In) stably and was as active as native MSH in stimulating the production of tyrosinase, the rate-limiting enzyme in melanin biosynthesis, in cultures of Cloudman S91 murine melanoma cells. When injected into DBA2 mice bearing Cloudman S91 melanomas, ^{111}In-*bis*MSH-DTPA rapidly localised in tumour tissue in concentrations significantly higher than in blood, skin, skeletal muscle, brain, eye, spleen, heart and lung ($P<0.001$), although uptake by liver and kidney was as high or higher than by tumour. Uptake by tumour, but not other tissues could be inhibited by an excess of native MSH co-injected with *bis*MSH-DTPA, indicating receptor mediated specific uptake (Bard et al, 1990a).

CLINICAL TRIAL OF *bis*MSH-DTPA

*Bis*MSH-DTPA was entered into a clinical trial in the Department of Nuclear

Targeting of Drugs 4, Edited by G. Gregoriadis *et al.*
Plenum Press, New York, 1994

Medicine, Addenbrooke's Hospital, Cambridge and initial reports have already been published (Wraight et al, 1992; Bard et al, 1993). Altogether, forty patients were referred for investigation of known or suspected metastatic melanoma between December 1988 and June, 1992. Three patients' scans were excluded because of a technical error in labelling *bis*MSH-DTPA. In 10 cases it was impossible to obtain an adequate clinical follow-up. Of the remaining 27 patients, there were 16 true positives, 5 true negatives and 6 false negatives. There were no false positives. The sensitivity, specificity and accuracy were 73%, 100% and 78% respectively.

As in the animal studies, high unspecific uptake was noted in the liver and kidney which presented a potential problem for imaging in these areas. It was, however, possible in the clinical studies to image liver metastases by ^{99m}Tc colloid subtraction. This method proved highly accurate and two patients referred with suspected liver metastases on the basis of ultrasound and computerised x-ray tomography (CT) scans, but who failed to image with *bis*MSH-DTPA were shown to be clear of disease by their subsequent clinical progress (Wraight et al, 1992; Bard et al, 1992; Bard et al, 1993).

COMPOUNDS BASED ON ANALOGUES OF MSH

DTPA-^{111}In Compounds

Both the animal and clinical data obtained with *bis*MSH-DTPA suggested that although this compound provided a highly accurate method of imaging melanoma, its potential use for delivering a therapeutic β-emitting isotope, such as ^{90}Y was limited by high kidney and liver uptake.

Several studies suggest that the biological activity of MSH may be enhanced by replacing Met^4 with its chiral analogue norleucine and by substituting Phe^7 with its D enantiomer, (Sawyer et al, 1980; Wilkes et al, 1983; Hadley et al, 1985). Further enhancement of activity can be obtained by replacing Glu^5 with Asp and Gly^{10} with Lys and by shortening the peptide to the 4-10 fragment, (Al-Obeidi et al, 1989b; Al-Obeidi et al, 1989a; Hadley et al, 1989). These modifications simultaneously confer greater resistance to proteolytic degradation than that of native MSH, (Hadley et al, 1989). Thus short chain analogues of MSH offer the potential of improving tumour uptake by enhanced affinity for the MSH receptor and, by their greater biological stability, of reducing unspecific uptake.

We have synthesised several analogues of MSH based on the 4-10 sequence (Fig. 1). In the first two (Fig. 1 a, b), either one (*mono*MSH$_{(4\text{-}10)}$-DTPA) or two (*bis*MSH$_{(4\text{-}10)}$-DTPA) peptide sequences were coupled to a single molecule of DTPA. The synthesis of these

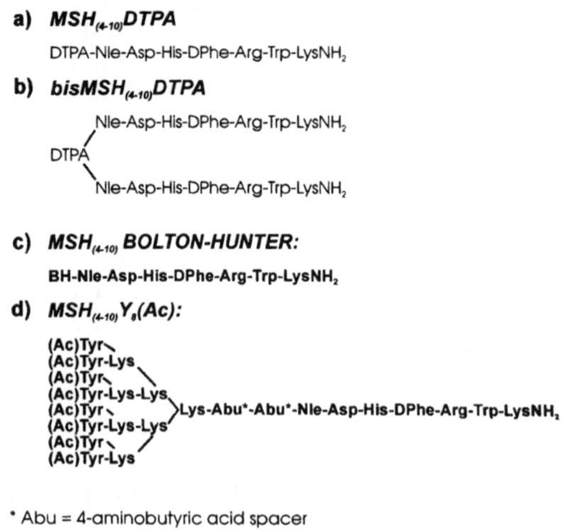

Fig. 1. Structure of MSH analogues

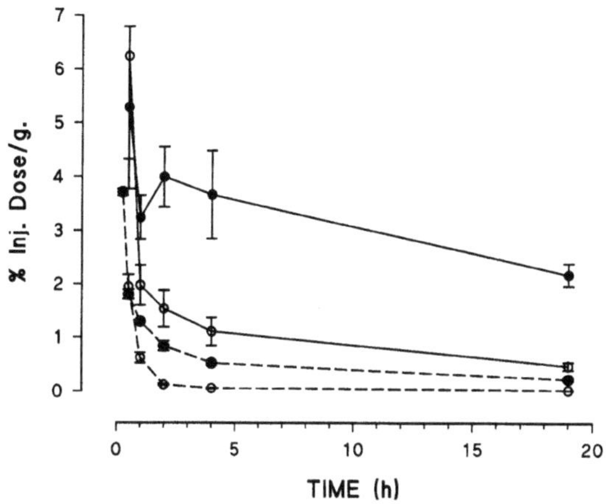

Fig. 2. Clearance of monoMSH$_{(4-10)}$-DTPA and bisMSH$_{(4-10)}$-DTPA from DBA2 mice bearing Cloudman S91 melanomas. Data expressed as % total injected dose/g. tissue (wet weight) ±SEM:

monoMSH$_{(4-10)}$-DTPA; concentration of ^{111}In in:
Tumour: open circles, solid line
Blood: open circles, dashed line

bisMSH$_{(4-10)}$-DTPA; concentration of ^{111}In in:
Tumour: closed circles, solid line
Blood: closed circles, dashed line

compounds has already been described (Bard et al, 1990b; Bard et al, 1992). Both bound indium stably for at least 24 h in 50% saline/foetal bovine serum. *Mono*MSH$_{(4-10)}$-DTPA showed activity equivalent to native MSH in the tyrosinase assay whilst *bis*MSH$_{(4-10)}$-DTPA was about twice as active. In in vivo competition experiments, both compounds showed specific uptake into Cloudman S91 tumours, but not into liver, spleen, skin, brain, eye, heart, lung or kidney. Tumour uptake was greater with *bis*MSH$_{(4-10)}$-DTPA as compared with *mono*MSH$_{(4-10)}$-DTPA and was cleared less rapidly (Fig. 2). Within 4 h of injection, tumour radioactivity, expressed as a percentage of the total radioactivity injected per gram of tissue (wet weight) was 3.66 ± 0.81 (N=11, ± SEM) with *bis*MSH$_{(4-10)}$-DTPA and 1.01 ± 0.26 (N=13, ± SEM) with *mono*MSH$_{(4-10)}$-DTPA. Radioactivity also cleared more rapidly from the circulation with *mono* as opposed to *bis*MSH$_{(4-10)}$-DTPA so that at 4 h. tumour/blood ratios were 20.6 for *mono*MSH$_{(4-10)}$-DTPA and 7.0 for *bis*MSH$_{(4-10)}$-DTPA. Unspecific liver and kidney uptake was also higher with *bis*MSH$_{(4-10)}$-DTPA, but, at 19 h, uptake of *bis*MSH$_{(4-10)}$-DTPA was higher in tumour and lower in kidney than with *bis*MSH-DTPA (P<0.001) (Fig. 3).

These data clearly indicate that tumour uptake is enhanced by using *bis* compounds, and that whilst some benefit can be gained by using the short chain analogue, a price must be paid in terms of higher liver and kidney uptake. *Mono*MSH$_{(4-10)}$-DTPA offers better tissue/blood ratios and provides the better prospect for imaging. The relatively low tumour uptake, however, suggests that high initial levels of radioactivity would be needed to obtain usable radioscinitigraphic images. The rapid clearance of this compound is an additional advantage and may enable good images to be obtained even earlier than the 4 h post-injection time-point which appears to be the optimum with *bis*MSH-DTPA.

RADIOIODINATED COMPOUNDS

Data obtained with *mono*MSH$_{(4-10)}$-DTPA and *bis*MSH$_{(4-10)}$-DTPA suggested that an

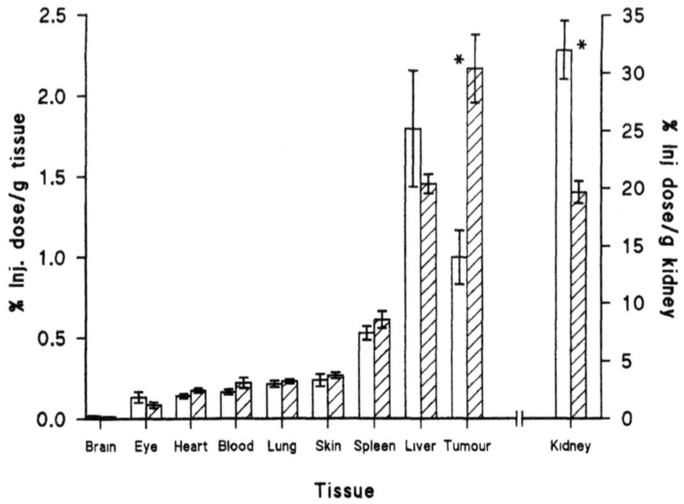

Fig. 3. Tissue distribution of $MSH_{(4\text{-}10)}$-DTPA (open bars) and bis$MSH_{(4\text{-}10)}$-DTPA (hatched bars) in DBA2 mice, 19 h post injection. Concentration of 111In expressed as % total injected dose/g tissue (wet weight)±SEM

alternative approach might be to design a compound in which a single MSH sequence was linked to a structure which permitted multiple labelling. We therefore designed two compounds in which the same $MSH_{(4\text{-}10)}$ sequence was linked through its *N*' terminus to a branched lysyl construct substituted with eight tyrosines (Fig. 1d). Because of its bulk, the polylysyl group was separated from the main peptide sequence by two γ-aminobutyric acids.

These peptides were produced by a standard Fmoc solid phase synthesis, (Atherton and Sheppard, 1989) with the polylysyl group being built up by three successive additions of Fmoc-Lys(Fmoc). On deprotection, both the α and ε amino groups of each lysine became available for substitution and a branched array was generated. As a final stage, the eight available *N*'-terminal amino groups were substituted with tyrosine. In one compound, $MSH_{(4\text{-}10)}Y_8$Ac, these tyrosines were *N*'acetylated whereas in the other, $MSH_{(4\text{-}10)}Y_8$, the *N*'-terminal amino groups were left unsubstituted. A control compound ($MSH_{(4\text{-}10)}$-BH) in which $MSH_{(4\text{-}10)}$ was substituted at its *N*'-terminus with 3-(*p*-hydroxyphenyl)propionic acid (Bolton-Hunter reagent), without the double γ-aminobutyric acid spacer, was also synthesised (Fig. 1, c.).

All these compounds were active in the tyrosinase assay. $MSH_{(4\text{-}10)}Y_8$ and $MSH_{(4\text{-}10)}Y_8$Ac were both as active as native MSH, whereas $MSH_{(4\text{-}10)}$-BH was nearly three times as active (Table 1). They were labelled with ^{125}I using Iodobeads ® (Pierce, Rockford, II, USA). Free ^{125}I was removed by reverse-phase chromatography on Tech-Sep ® cartridges (HLPC Technology, Macclesfield, UK) and solvent removed by evaporation. The resultant preparations were injected into DBA2 mice bearing Cloudman S91 melanomas and tissue localisation measured after 1 h, 2 h, 4 h, and 24 h. At 4 h, both $MSH_{(4\text{-}10)}Y_8$ and $MSH_{(4\text{-}10)}Y_8$Ac showed about twice as much tumour uptake as $MSH_{(4\text{-}10)}$-BH ($P<0.01$). Unspecific uptake by liver, kidney and spleen was much more pronounced, however, with $MSH_{(4\text{-}10)}Y_8$Ac than with either of the other two compounds. Indeed, with $MSH_{(4\text{-}103)}Y_8$. and $MSH_{(4\text{-}10)}$-BH, liver uptake was significantly lower than tumour ($P<0.01$) and kidney uptake slightly lower that tumour (Fig. 4). Kidney uptake with $MSH_{(4\text{-}10)}Y_8$, and $MSH_{(4\text{-}10)}$-BH was lower than with any of the DTPA-^{111}In compounds. These data suggest that high liver and kidney uptakes observed with the ^{111}In compounds may be partially due to transchelation of ^{111}In. Thin-layer chromatography of urine from animals injected with either *bis*MSH-DTPA or *mono*$MSH_{(4\text{-}10)}$-DTPA suggested that extensive degradation of the peptide occurred between injection and excretion, and that a significant proportion of the radioactivity was

Table 1. Tyrosinase activities of MSH derivatives

Compound	EC_{50}* ± SEM (nM)
MSH	2.83 ± 0.28
$MSH_{(4-10)}$-BH	1.05 ± 0.38
$MSH_{(4-10)}Y_8$	2.91 ± 0.08
$MSH_{(4-10)}Y_8Ac$	2.78 ± 0.96

*EC_{50} = Nanomolar concentration of peptide required to elicit half-maximum activity.

secreted as free indium. It may be that accumulation of insoluble hydroxides of indium within the kidney accounts for the high kidney uptake observed with these compounds.

This conclusion was supported by the generally lower kidney uptake observed with $MSH_{(4-10)}Y_8$ and with $MSH_{(4-10)}$-BH. The difference, however, in both kidney and liver uptake between the acetylated and non-acetylated polylysyl$MSH_{(4-10)}$ derivatives suggest that hydrophilicity may also play a part. Bagutti et al, 1993, have synthesised an $MSH_{(4-10)}$-DTPA peptide substitutes with a 2,3-dihydroxy-2S-propyl(DHP) on the ε-amino group of Lys^{10}. When this compound was used, ^{111}In uptake in the liver and kidneys was dramatically lower than with the non-DHP substituted parent compound. Similar results were seen in our experiments when $MSH_{(4-10)}Y_8$ was compared with its more hydrophobic acetylated analogue, $MSH_{(4-10)}Y_8Ac$.

STRUCTURAL STUDIES:

The conformation of MSH-DTPA has been studied by Professor Van Beeumen, Dr. B. Deveese and Dr. Francoise Jacquemotte by measuring deuterium exchange by

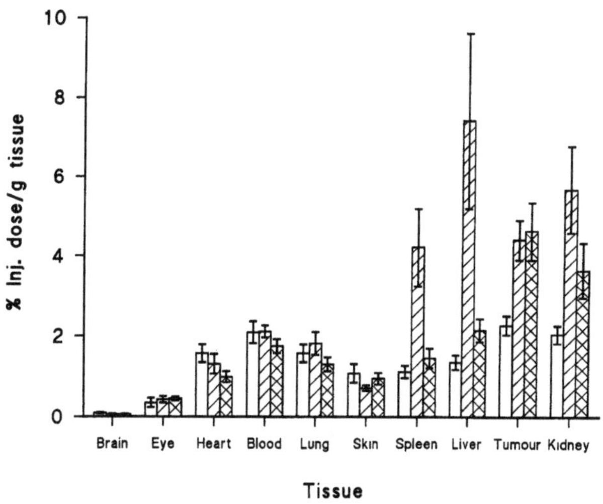

Fig. 4. Tissue distribution of $MSH_{(4-10)}BH$ (open bars), $MSH_{(4-10)}Y_8Ac$ (hatched bars) and $MSH_{(4-10)}Y_8$ (cross-hatched bars), 4 h post injection. Concentration of ^{125}I expressed as % total injected dose/g. tissue (wet weight)±SEM

Table 2. Deuterium exchange in bisMSH-DTPA

Solvent	M_r Uncomplexed	M_r Fe^{2+} Complex
$CH_2OH/H_2O/CH_3COOH$*	3602	3656
$CD_2OD/D_2O/CD_3COOD$*	3642	3698
Exchanged hydrogens	40	42

*Solvent systems consisted of: $CH_2OH/H_2O/CH_3COOH$; 1/1/0.01 (v/v) or deuterated equivalent

electrospray ionisation mass spectrometry, (Katta and Chait, 1991). This technique enables the mass of proteins and peptides to be determined with a precision of 0.01% and if the peptide is exposed to anhydrous deuterated solvents, any increase in mass due to the replacement of exchangeable hydrogen by deuterium can be readily measured. Since hydrogen exchange is inhibited by hydrogen bonding, the greater the amount of exchange in these experiments, the fewer hydrogen bonds are present and the more random the structure. Whilst the technique cannot identify which hydrogen atoms are involved in hydrogen bonding, it can give useful clues as to the degree of tertiary structure.

Deuterium exchange was measured in both chelated and unchelated forms of the peptide. The results indicated that in an Fe^{2+} complex, 42 protons (72%) were exchanged whereas in the uncomplexed peptide 40 protons (79%) were exchanged (Table 2). The difference can probably be explained by internal hydrogen bonding within the DTPA moiety.

CONCLUSION

In conclusion, the results indicate that MSH compounds provide an accurate method of targeting to melanomas. By using more hydrophillic compounds, improving the stability and number of sites of labelling and using more active analogues of MSH, unspecific uptake in liver and kidney can be significantly reduced making such compounds excellent candidates for the delivery of cytotoxic preparations.

ACKNOWLEDGEMENTS

This work was supported by grants from the Cancer Research Campaign and from CRC Technology. The attendance of DRB at the ASI was made possible by a travel grant from the Cancer Research Campaign.

REFERENCES

Al-Obeidi, F., Castrucci, A.M., Hadley, M.E. and Hruby, V.J., 1989a, Potent and prolonged acting cyclic lactam analogues of alpha-melanotropin: design based on molecular dynamics, J.Med.Chem., 32:2555

Al-Obeidi, F., Hruby, V.J., Castrucci, A.M. and Hadley, M.E., 1989b, Design of potent linear alpha-melanotropin 4-10 analogues modified in positions 5 and 10, J.Med.Chem., 32:174

Atherton, E. and Sheppard, R.C., 1989, Solid phase peptide synthesis, a practical approach, Oxford: IRL Press

Bagutti, C., Stolz, B., Albert, R., Bruns, C., Pless, J. and Eberle, A.N., 1993, [^{111}In]DTPA labelled analogues of α-MSH for the detection of MSH receptors in vitro and in vivo. Ann.NY Acad.Sci., 680:445

Bard, D.R., Knight, C.G. and Page-Thomas, D.P., 1990a, A chelating derivative of α melanocyte stimulating hormone as a potential imaging agent for malignant melanoma, Br.J.Cancer, 62:919

Bard, D.R., Knight, C.G. and Page-Thomas, D.P., 1990b, Targeting of a chelating derivative of a short-chain analogue of α-melanocyte stimulating hormone to Cloudman S91 melanomas, Biochem.Soc.Trans., 18:882

Bard, D.R., Knight, C.G., Page-Thomas, D.P., Wraight, E.P. and Maughan, T.S., 1992, The use of derivatives of α-MSH for targeting to melanomas in vivo, in: G. Gregoriadis, A.T. Florence and G. Poste (Eds.), "Targeting of drugs 3: The challenge of peptides and proteins", G. Gregoriadis, A.T. Florence and G. Poste, eds., Plenum, New York

Bard, D.R., Wraight, E.P. and Knight, C.G., 1993, BisMSH-DTPA: A potential imaging agent for malignant melanoma, Ann.NY Acad.Sci., 680:451

Cooke, K.R., Skegg, D.C.G. and Fraser, J., 1983, Trends in malignant melanoma of the skin in New Zealand, Int.J.Cancer, 31:715

Cox, N.H., Jones, S.K. and MacKie, R.M., 1987, Malignant melanoma of the head and neck in Scotland: An eight year analysis of trends in prevalence, distribution and prognosis, Quart.J.Med.New Ser., 244:661

Davis, N.C., 1991, Malignant melanoma, in: "Malignant skin tumours", A.J.J. Emmett and M.G.E. O'Rourke, eds., 2nd ed., Churchill Livingstone, Edinburgh, London, N.Y.

Hadley, M.E., Abdel-Malek, Z.A., Marwan, M.M., Kreutzfeld, K.L. and Hruby, V.J., 1985, [Nle^4, D-Phe^7]-α-MSH: A superpotent melanotropin that "irreversibly" activates melanoma tyrosinase, Endocr.Res., 11:157

Hadley, M.E., Marwan, M.M., Al-Obeidi, F., Hruby, V.J. and Castrucci, A.M., 1989, Linear and cyclic alpha-melanotropin [4-10]-fragment analogues that exhibit superpotency and residual activity, Pigment Cell Research, 2:478

Katta, V. and Chait, B.T., 1991, Conformational changes in proteins probed by hydrogen -exchange electrospray-ionisation mass spectrometry, Rapid Commun.Mass Spectrometry, 5:214

Little, J.H., Holt, J. and Davis, N., 1980, Changing epidemiology of malignant melanoma in Queensland, Med.J.Aust., 1:66

Magnus, K., 1987, Epidemiology of malignant melanoma of the skin, in: "Cutaneous melanoma: Status of knowledge and future perspective", U. Veronesi, N. Cascinelli and M. Santinami, eds., Academic Press, London, New York

Sawyer, T.K., Sanfilippo, P.J., Hruby, V.J., Engel, M.H., Heward, C.B., Burnett, J.B. and Hadley M.E., 1980, 4-Norleucine, 7-D-phenylalanine-α-melanocyte-stimulating hormone: A highly potent α-melanotropin with ultralong biological activity, Proc.Natl.Acad.Sci.USA, 77:5754

Schreiber, M.M., Bozzo, P.D. and Moon, T.E., 1981, Malignant melanoma in Southern Arizona, Arch.Dermatol., 117:6

Siegrist, W., Bagutti, C., Solca, F., Girard, J. and Eberle, A.N., 1992, MSH receptors on mouse and human melanoma cells: receptor identification, analysis and quantification, Progress in Histochemistry & Cytochemistry, 26:110

Tatro, J.B., Atkins, M., Mier, J.W., Hardarson, S., Wolfe, H., Smith, T., Entwistle, M.L. and Reichlin, S., 1990, Melanotropin receptors demonstrated in situ in human melanoma, J.Clin.Invest., 85:1825

Wilkes, B.C., Sawyer, T.K., Hruby, V.J. and Hadley, M.E., 1983, Differentiation of the structural features of melanotropins important for biological potency and prolonged activity in vitro, Int.J.Peptide Protein Res., 22:313

Wraight, E.P., Bard, D.R., Maughan, T.S., Knight, C.G. and Page-Thomas, D.P., 1992, The use of a chelating derivative of alpha melanocyte stimulating hormone for the clinical imaging of malignant melanoma, Br.J.Radiol., 65:112

THE IMMUNOGENICITY OF RECOMBINANT PROTEINS AND OF "MAGIC BULLETS" REPRESENTS THE MAIN OBSTACLE FOR THE EFFECTIVE IN VIVO TARGETING OF BIOLOGIC RESPONSE MODIFIERS (BRMs)

A.H. Sehon

Department of Immunology, The University of Manitoba
Winnipeg, Canada R3E 0W3

INTRODUCTION

This article will be limited primarily to a review of work done in my laboratory in relation to the chemical conversion of allergens and antigens to tolerogenic derivatives capable of inducing long-term specific suppression of the corresponding antibody responses. Because of space limitation no reference will be made to details of chemical and immunological methodologies developed in the course of these studies, which may be found in the appropriate references.

Use of modified allergens for immunotherapy of allergic patients

Traditional hyposensitization therapy of allergic individuals consists of injections of the appropriate allergenic extract over prolonged periods; in part, this is due to the fact that the doses administered contain minimal amounts of the allergens (AL) so as to avoid the risk of systemic anaphylactic reactions. Hence, generally, the effectiveness of classical therapeutic regimens cannot be achieved unless extended over a number of years; even then its effectiveness is frequently only marginal. To circumvent these disadvantages, many attempts have been made to decrease the allergenicity of the preparations used therapeutically by chemical modification of the AL, such as urea denaturation (Ishizaka et al, 1975), reaction with formaldehyde (Marsh, 1971) and polymerization with glutaraldehyde (Patterson et al, 1978). An original type of chemically modified AL was developed in our laboratory by the coupling of an optimum number (n) of mPEG molecules onto a given protein AL. The resulting copolymeric conjugates, $AL(mPEG)_n$, were not only essentially non-allergenic, but also immunosuppressive (Sehon, 1982). It is to be noted that in our initial studies (Lee and Sehon, 1977) the conjugates were prepared with polyethylene glycol (PEG) which contains two terminal hydroxyl groups and which, therefore, may give rise to intra- and inter-molecularly cross-linked conjugates; to avoid this possible complication we have used mPEG in subsequent experiments (Jackson et al, 1987).

In the original studies (Lee and Sehon, 1977), Chester Beatty rats were used as the model experimental animals for testing the usefulness of $Al(mPEG)_n$ for the specific suppression of IgE antibodies. The conditions and results of typical experiments are illustrated in Table 1 with respect to induction of IgE antibodies to dinitrophenylated ovalbumin $[OVA)DNP)_3]$; clearly, the animals produced both anti-OVA and anti-DNP IgE

Targeting of Drugs 4, Edited by G. Gregoriadis *et al.*
Plenum Press, New York, 1994

Table 1. Induction of tolerance in rats by OVA-PEG conjugates

Groups of rats	Treatment[a]	Rat No.	PCA titers on day 14 after sensitization[b] anti-OVA	anti-DNP	Systemic Reaction
Control Group	PBS	1	80	90	Death in 24 min.
	+ $OVA(DNP)_3$	2	100	260	Death in 33 min.
	+ $Al(OH)_3$	3	100	260	Death in 36 min.
	+ *B. pertussis*	4	80	270	Death in 28 min.
Test Group	OVA-PEG	5	<10	<10	No Symptoms
	+ $OVA(DNP)_3$	6	20	20	No Symptoms
	+ $Al(OH)_3$	7	<10	<10	No Symptoms
	+ *B. pertussis*	8	<10	<10	No Symptoms

[a] The control and test rats were pretreated with PBS and 2.5 mg of OVA-PEG, respectively; 4 h later all rats received an intraperitoneal bolus of $OVA(DNP)_3$, $Al(OH)_3$ and *B. pertussis*.

[b] The lowest dilution of the sera used for passive cutaneous anaphylaxis (PCA) was 1:10.

antibodies which could be demonstrated by passive cutaneous anaphylaxis in random-bred Long-Evans hooded rats. As is obvious from the results listed in this Table, the IgE antibody response of the rats treated with OVA-PEG conjugates was essentially completely suppressed by comparison to the IgE levels of control rats; most importantly, whereas on intravenous challenge with OVA all the animals in the test group survived, all the animals in the control group died within 40 min. Thus, these earliest experiments demonstrated that administration of tolerogenic PEG conjugates of a highly immunogenic antigen to rodents resulted in the suppression of their capacity to mount IgE antibody responses to the unmodified antigen; most importantly, these animals were also systemically anergic to the injection of the antigen.

Surprisingly, however, clinical trials in close to 300 atopic patients revealed that whereas this treatment did not affect markedly the preexisting IgE levels, the patients' AL-specific IgG levels were significantly and rapidly increased (Dreborg and Akerblom, 1990). In spite of the fact that treatment of these patients over a relatively short period of time did not result in the anticipated drop in IgE antibodies, but led to an increase in AL-specific IgG antibodies, these mPEG conjugates may still be considered useful for immunotherapy of IgE-mediated allergies since the increase in IgG antibodies is still regarded as one of the principal goals of a successful hyposentitization regimen. Moreover, it is to be noted that these conjugates were essentillay nonallergenic; this loss in allergenicity is attributable to the masking of some of the epitopes of the allergenic molecules by the mPEG chains grafted onto the AL molecules and/or to the conformational distortion of these epitopes as a result of the chemical reaction.

This perplexing discrepancy between the enhancement of IgG Abs in allergic patients and the induction of tolerance in rodents caused by the same mPEG conjugate preparation could be attributed to (1) the much higher dose (per bodyweight) of the conjugate administered to rodents, and (2) the difference in the immunological status of these two types of recipients, i.e., whereas the rats had not been stimulated by AL prior to administration of the tolerogenic $AL(mPEG)_n$, the immunological system of the patients had been activated "spontaneously" by repeated exposure to the AL prior to initiation of the immunotherapy. In essence, the results of these clinical trials confirmed the fundamental rule of immune responsiveness, viz., that it is much more difficult to inactivate or eliminate long-lived memory cells of an established ongoing immune response, which have been expanded in response to repeated exposures to a given antigen. Indeed, the plausibility of this interpretation was supported by the results of experiments in mice, which had been sensitized to produce anti-OVA IgE antibodies prior to receiving injections of the tolerogenic $OVA(mPEG)_n$ conjugates, i.e., although the IgE response of these mice was markedly reduced, their IgG responses were enhanced after secondary immunization with OVA (Sehon and Lang, 1986). It is important to stress that the allergenic preparations used for the synthesis of mPEG conjugates for the original experiments were chemically ill-defined mixtures of allergenic components. Therefore, the chemical composition of the resulting conjugates was also undefined. However, in recent years, as a result of the spectacular advances of molecular biology, many recombinant allergens have been synthesized as chemically well-defined molecules (Kraft and Sehon, 1993); therefore, it may be anticipated that tolerogenic mPEG conjugates of these well-defined molecules may prove more effective for treatment of common allergies in man. Moreover, recent results in my laboratory have indicated that treatment of pre-immunized mice, with a harmless, nonspecific pharmacological immunosuppressant for a short period appeared to inactivate the pre-existing memory B and T cells. Hence, it may be envisaged that this strategy may be followed by treatment with tolerogenic $AL(mPEG)_n$ conjugates for overcoming the hurdle discovered in the earlier clinical trials.

Possible therapeutic applications of tolerogenic mPEG conjugates of immunogenic BRMs to diseases other than pre-existing allergies

As indicated above, tolerogenic mPEG conjugates were not effective in abrogating an established antibody response, as would be also the case in auto-immunity. Hence, mPEG conjugates do not appear to be useful therapeutic agents for patients producing antibodies to antigens, prior to initiation of the immunosuppressive regimen with tolerogenic mPEG conjugates of the corresponding antigens. Nevertheless, these studies led us to anticipate that this "technology" may be exploited for the selective downregulation of the host's immune response to immunogenic BRMs, which appear to be otherwise promising therapeutic agents for a broad spectrum of diseases. Among these BRMs, one may cite (i) xenogeneic monoclonal or polyclonal antibodies (collectively referred to here as xIg) against different epitopes of the patients' $CD4^+$ cells (Lang et al, 1992) for the treatment of rheumatoid arthritis and other autoimmune diseases, or for suppression of the immune rejection of organ transplants (Ortho Multicenter Transplant Study Group, 1985), and (ii) "magic bullets" for the destruction of tumor cells (Vitetta et al, 1987), which consist of anti-tumor xIg to which are coupled toxins (Tx), or radionuclides, or chemotherapeutic drugs. However, in most cases the patients produce antibodies to the injected xIg and to the even more immunogenic immunotoxins (xIg-Tx); consequently, the therapeutic effectiveness of these immunological strategies is undermined by the patients' antibodies which prevent these bullets from reaching their target cells. In addition, the repeated administration of these agents may result in serious complications, viz. serum sickness, anaphylactic symptoms (i.e., bronchopasm, dyspnea and hypotension) and/or the deposition in the liver of toxic immune complexes leading frequently to hepatotoxicity (Dillman, 1985). Similar limitations would also apply to the use of (i) repeated passive immunization, (ii) anti-viral antibodies or ITX, e.g., ITX to HIV (Sehon, 1989), and (iii) other regulatory factors, which are often immunogenic (Katre, 1990) probably because of small differences in their conformational characteristics or in their glycosidic constituents in relation to their natural, human couterparts.

Although some encouraging clinical results have been obtained by the use of xIg or xIg-Tx, as stated earlier, one of the main complications which remains to be overcome is the inherent immunogenicity of xIg and Tx, which leads to the production by the patient of antibody responses to the epitopes of Tx, as well as to both murine isotypic and idiotypic determinants, the predominant response being anti-idiotypic (Chatenoud et al, 1986). This complication cannot be overemphasized since multiple injections of the xIg or of their immunoconjugates are required for achieving the desired clinical effects (Pimm, 1988). Moreover, the patients' antibody responses lead to changes in the pharmacokinetics of the immunoconjugates (Khazaeli et al, 1988), e.g., xIg-Tx are rapidly cleared in the form of immune complexes from the circulation. Attempts to render the xIg less immunogenic have been made by different methods, e.g., by preparing chimeric antibodies consisting of the variable region of a mouse monoclonal antibody coupled to the constant region of a human IgG molecule (Shaw et al, 1987) or by constructing humanized or "human" Abs by genetic engineering (Riechmann et al, 1988). It is expected, however, that patients would still produce antibodies at least to the idiotypic determinants of these molecules, particularly when the idiotypic regions are attached to the patients' cells (Waldmann, 1989).

Mechanism of suppression by tolerogenic $Ag(mPEG)_n$ conjugates

As stated in the abstract, for best results, the unmodified protein Px has to be injected at least 7 days after administration of the tolerogenic $Px(mPEG)_n$ conjugate. This interval has been shown to be essential for the expansion of Px-specific Ts cells generated in response to $Px(mPEG)_n$; additional injections given at various intervals maintain the proliferation of these cells and their downregulating effect on Px-specific Th cells (Chen et al, 1994). Attempts to produce nonimmunogenic $mAb(mPEG)_n$ conjugates without loss of their biological activities (i.e., antigen-binding and effector functions) have failed, probably because, as alluded to earlier, coupling of the optimal number of mPEG molecules per mAb molecule is expected to lead to (i) significant conformational distortion of the mAb, and (ii) blocking of its functional sites as a result of steric hindrance by the long molecules of mPEG. Parenthetically, it is important to note that the presence of immunogenic impurities in tolerogenic Px(mPEG) preparations may interfere with the induction of Ts cells, i.e., under these conditions, the homeostatic balance between elicitation of an immune response and its suppression is more easily tilted in favour of the former (Chatenoud, 1986).

As documented in other publications from our laboratory, we succeeded in generating non-hybridized antigen-specific Ts clones from single cells of spleens of animals immuno-suppressed by $Ag(mPEG)_n$ conjugates (Takata et al, 1990; Chen et al, 1992). These cells were shown to be $Thyl^+$, $CD3^+$, $CD4^-$, $CD5^+$, $CD8^+$ and to express the heterodimer of T cell receptors (TCR). Moreover, these cells and their soluble suppressor factor (TsF) suppressed antibody responses in an antigen-specific and MHC class I restricted manner. In recent experiments, it was also established that (i) the TsF was a disulfide linked heterodimer having properties of the soluble form of TCR (Chen et al, 1994).

We discovered recently that mice pretolerized by injection of $Ag_A(mPEG)_n$ were also unresponsive to an unrelated antigen $B(Ag_B)$, or to its haptenated derivative Ag_B-Hp_n, when Ag_B or Ag_B-Hp_n was injected in the form of a covalent adduct with Ag_A, i.e., Ag_A-Ag_B or Ag_A-Ag_B-Hp_n, but not as a mixture with Ag_A; in this study human (myeloma) IgG (HIgG) served as Ag_A, and OVA or OVA-DNP_3 served as Ag_B or Ag_B-Hp_n (OVA = ovalbumin) (Bitoh et al, 1993). Moreover, this phenomenon was reproduced in vitro, i.e., Ts cells of mice tolerized with $HIgG(mPEG)_{30}$, or the soluble monoclonal TsF of cloned Ts cells, exerted their associative suppressive effector function -- in the obligatory presence of $CD8^+$ T cells of syngeneic naive mice (T_n cells) -- on antibody formation to an Hp (DNP), when the Hp was present as a covalent adduct linked either directly to HIgG (e.g., HIgG-DNP_7), or even indirectly via OVA (as in HIgG-OVA-DNP_3); however, no suppression of the anti-DNP antibody response was observed when OVA-DNP_3 was present as a mixture with HIgG. Furthermore, it was established that the accessory (Ac) cells involved in processing the specific Ag in the presence of the Ts cells were also downregulated, as reflected by their reduced capacity for presentation of the Ag to HIgG-specific Th cells in proliferation assays. All these results demonstrate that (i) the "phenomenon of linked immunological suppression" involves the downregulation of Th cells which -- in all probability -- recognize, concomitantly with the Ts cell, the appropriate epitopes of Ag_A and Ag_B on the same Ac cell, (ii) the downregulation of these Th cells may also be a consequence of the downregulation of Ac cells by Ts cells interacting with the appropriate epitope(s) present on the Ac cells, and (iii) most remarkably the $CD8^+$ Ts cells could be substituted by $CD8^+$ T cells of naive mice, which cells had been "armed" with the specific monoclonal TsF.

Suppression of human anti-murine antibody (HAMA) responses, including anti-idiotypic responses, in hu-PBL-SCID mice

In view of the genetic diversity of the human population and since the immune response to it is under the control of the immune response (Ir) and the immune suppressor (Is) genes, the responses to immunogenic BRMs and the effect(s) of a given BRM on the immune response would differ from person to person (Bitoh, 1993). Therefore, it is difficult to establish the therapeutic efficacy of a given BRM even if one were to use extensive clinical trials involving a large number of individuals; even then, the range of the effective doses might be too wide to meet the needs of individual patients. Moreover, because of possible intoward effects, for ethical reasons, the true effectiveness of a given BRM is not being established in healthy individuals; hence it is essential to generate an appropriate system(s) for the determination of the immunogenicity and the effect(s) of a given BRM on the immune response of each patient.

To circumvent these complications for testing the tolerogenicity of mPEG conjugates of murine mAbs in relation to the human lymphoid system, we used the hu-PBL-SCID mouse model described by Mosier (Bitoh et al, 1993) with some modifications. This model involves the use of mice having the condition of severe combined immunodeficiency (SCID), which lack functional T and B lymphocytes and are, therefore, unable to reject xenogeneic cells and may be engrafted with human peripheral blood leucocytes (hu-PBL) of a healthy individual. Remarkably, very few of the hu-PBL-SCID mice exhibit a transient graft-versus-host reaction. The hu-PBL-SCID mice produce on immunization Ag-specific human Ab responses; hence the hu-PBL-SCID mouse system was deemed to represent an appropriate model for testing the immune responses of individual patients to a given BRM and the possibility of suppressing this response with tolerogenic BRM-mPEG conjugates.

Specifically, in this study we used ovalbumin (OVA) and the murine mAb directed to OVA, referred to as Mab-2, as immunogens and the corresponding Mab-2(mPEG)$_{36}$ conjugate as the tolerogen; thus, Mab-2(mPEG)$_{36}$ served in this study as an Ag-specific immunosuppressive model BRM. Utilizing this system, we demonstrated that (i) the primary and secondary human antibody responses to mouse mAb (HAMA) and to OVA were induced in hu-PBL-SCID mice, which had been engrafted with cell fractions consisting of appropriate numbers of T cells and of B plus mononuclear (B+MN) cells, (ii) treatment of the hu-PBL-SCID mice with Mab-2(mPEG)$_{36}$ 7 days before immunization with the highly immunogenic heat-treated form of Mab-2 (ha-Mab-2) induced tolerance of the human anti-murine Ig Ab (HAMA) response to Mab-2, but not to OVA, and (iii) the mechanism underlying the induction of immunosuppression by tolerogenic Mab-2(mPEG)$_{36}$ conjugates in hu-PBL-SCID mice appeared to be identical to that responsible for suppression of intact normal mice which are subjected to the same type of tolerogenic protocol, i.e., the suppression was found to be due to the generation of human suppressor CD8^{+} T (Ts) cells, which downregulated the CD4^{+} helper T (Th) cells in an Ag- and HLA class I-specific manner.

Although the suppression of HAMA in the above experiments was pronounced, we investigated also the extent of suppression of anti-idiotypic (a-Id) Abs, since it has been generally accepted that a-Id Abs, produced by patients receiving mureine mAbs, interfered with the binding of the mAbs to their cell targets (Chatenoud, 1986; Traub et al, 1988). Moreover, the a-Id response is generally more pronounced when the a-Id Abs are directed to tissue Ags; this effect is probably due to the increased immunogenicity of the corresponding anti-idiotopes as a result of their acquiring enhanced rigidity after binding to tissue antigens.

Clearly, for optimizing the tolerogenicity of the mPEG conjugates of a given mAb, it would be advisable to use a conjugate comprising the particular mAb. However, in accordance with the phenomenon of linked immunological suppression we used in this study only the tolerogenic mPEG conjugate of Mab-2 (Bitoh, 1993), in conjunction with four mAbs directed to different Ags, but possessing the same heavy and light chains. The four mAbs used were: (i) Mab-2 (γ_1,κ) directed to one of the epitopes of OVA, (ii) H_1-DNPγ-109.3 (γ_1,κ) directed to the 2,4-dinitrophenyl (DNP) residue and (iii) OKT4 (γ_{2b},κ) and Leu-3a (γ_1,κ) which are mAbs directed to different epitopes of the human CD4 marker of helper T (Th) cells. It was gratifying to find that treatment of hu-PBL-SCID mice with the tolerogenic Mab-2$(mPEG)_{36}$ suppressed the anti-Id responses with respect to not only the idiotopes of Mab-2, but also to the idiotopes of the other mAbs consisting of the same heavy and light chains (Bitoh, 1993).

CONCLUSION

This survey of studies performed in recent years by the applicant and his colleagues, which are documented in the listed references, demonstrates that (i) appropriately constructed conjugates of protein antigens and mPEG act as specific tolerogens for the protein antigens in questions, and (ii) pretreatment of mice and outbred rats with these tolerogens results in the suppression of the recipients not only to the unmodified antigens, but also to unrelated antigens presented in the form of covalent adducts with the former antigens. It remains to establish if these promising results obtained in mice and rats can be reproduced in nonhuman primates, and ultimately in man.

REFERENCES

Bitoh, S., Lang, G.M., Kierek-Jaszczuk, D., Fujimoto, S. and Sehon, A.H., 1993, Specific immunosuppression of human anti-murine antibody (HAMA) responses in hu-PBL-SCID mice, Hum.Antibod.Hybridomas, 4:134

Bitoh, S., Lang, G.M. and Sehon, A.H., 1993, Suppression of human anti-mouse idiotypic antibody responses in hu-PBL-SCID mice, Hum.Antibod.Hybridomas, 4:144

Bitoh, S., Takata, M., Maiti, P.K., Holford-Strevens, V., Kierek-Jaszczuk, D. and Sehon, A.H., 1993, Antigen-specific suppressor factors of noncytotoxic $CD8^+$ suppressor T cells downregulate antibody responses also to unrelated antigens when the latter are presented as covalently linked adducts with the specific antigen, Cell.Immunol., 150:168

Chatenoud, L., Baudrihaye, M.F., Chkoff, N., Kreis, H., Goldstein, G. and Bach, J.F., 1986, Restriction of the human in vivo response against the mouse monoclonal antibody OKT3, J.Immunol., 138:830

Chatenoud, L., 1986, The immune response against therapeutic monoclonal antibodies, Immunol. Today, 7:367

Chen, Y., Takata, M., Maiti, P.K., Rector, E.S. and Sehon, A.H., 1992, Characterization of suppressor T cell clones derived from a mouse tolerized with conjugates of ovalbumin and monomethoxypolyethylene glycol, Cell.Immunol., 142:16

Chen, Y., Takata, M., Maiti, P.K., Mohapatra, S., Mohapatra, S.S. and Sehon, A.H., 1994, The suppressor factor of Ts cells induced by tolerogenic conjugates of OVA and mPEG is serologicaly and physicochemically related to the heterodimer of the TCR, J.Immunol., 152:3

Dillman, R.O., 1985, Monoclonal antibodies in the treatment of cancer, C.R.C. Crit.Rev.Oncol.Hematol., 1:357

Dreborg, S. and Akerblom, E., 1990, Immunotherapy with monomethoxypolyethylene glycol modified allergens, CRC Crit.Rev.Ther.Drug Carrier Syst., 6:315

Ishizaka, K., Okudaira, H. and King, T.P., 1975, Immunogenic properties of modified antigen-E.II.Ability of urea-denatured antigen and alpha-polypeptide chain to prime T cells specific for antigen-E., J.Immunol., 114:110

Jackson, C.-J.C., Charlton, J.L., Kuzminski, K., Lang, G.M. and Sehon, A.H., 1987, Synthesis, isolation and characterization of conjugates of ovalbumin with monomethoxypolyethylene glycol using cyanuric chloride as the coupling agent, Anal.Biochem., 165:114

Khazaeli, M.B., Saleh, M.N., Wheeler, R.H., Huster, W.J., Holden, H., Carrano, R. and LoBuglio, A.F., 1988, Pharmacokinetics and immune response, J.Natl.Cancer Inst., 80:937

Katre, N.V., 1990, Immunogenicity of recombinant IL-2 modified by covalent attachment of polyethylene glycol, J.Immunol., 144:209

Kraft, D. and Sehon, A.H., 1993, eds., "Molecular Biology and Immunology of Allergens", CRC Press, Boca Raton, FL

Lang, G.M., Kierek-Jaszczuk, D., Rector, E.S., Milton, A.D., Emmrich, F. and Sehon, A.H.,

1992, Suppression of antibody responses in rats to murine anti-CD4 antibodies (MoAb) by conjugates of MoAb and monomethoxypolyethylene glycol (mPEG), Immunol.Letters, 32:247

Lee, W.Y. and Sehon, A.H., 1977, Abrogation of reaginic antibodies with modified allergens, Nature, 267:618

Marsh, D.G., 1971, Preparation and properties of "allergoids" derived from native pollen allergens by mild formalin treatment, Int.Archs Allergy appl.Immun., 41:199

Ortho Multicenter Transplant Study Group. A randomized clinical trial of OKT3 monoclonal antibody for acute rejection of cadaveric renal transplanst, New Engl.J.Med., 313:337

Patterson, R., Sussko, I.M., Zeiss, C.R., Pruzansky, J.J. and Bacal, E., 1978, Comparison of immune reactivity to polyvalent monomeric and polymeric ragweed antigens, J.Allergy Clin.Immunol., 61:28

Pimm, M.V., 1988, Drug-monoclonal antibody conjugates for cancer therapy: potentials and limitations, in: S.D. Bruck, ed., CRC Crit.Rev.Ther.Drug Carrier Syst.,5:189

Reichmann, L., Clark, M., Waldmann, H. and Winter, G., 1988, Reshaping human antibodies for therapy, Nature, 332:323

Sehon, A.H., 1992, Suppression of antibody responses by conjugates of antigens and monomethoxypoly(ethylene glycol) in: "Polyethylene Glycol Chemistry: Biotechnical and Biomedical Applications" J. Milton Harris, ed.

Sehon, A.H., 1982, Suppression of IgE antibody responses with tolerogenic conjugates of allergens and haptens, in: Progress in Allergy, K. Ishizaka, ed., S. Karger, Basel

Sehon, A.H. and Lang, G.M., 1986, The use of nonionic, water soluble polymers for the synthesis of tolerogenic conjugates of antigens, in: "Mediators of Immune Regulation and Immunotherapy", S.K. Singal and T.L. Delovitch, eds., Elsevier Science Publishing Inc., New York

Sehon, A.H., 1989, "Immunological strategies for therapeutic destruction of HIV and HIV -infected cells in asymptomatic patients", in: Progress in Allergy and Clinical Immunology., W.J. Pichler, B.M. Stadler, C. Dahinden, A.R. Pecoud, P.C. Frei, C. Schneider, A.L. de Weck, eds., Hogrefe & Huber Publishers, Toronto

Shaw, D.R., Khazaeli, M.B., Sun, L.K., Ghrayeb, J., Daddona, P.E., McKinney, S. and LoBuglio, A.F., 1987, Characterization of a mouse/human chimeric monoclonal antibody (17-1A) to a colon cancer tumor-associated antigen, J.Immunol., 138:4534

Takata, M., Maiti, P.K., Kubo, R.T. , Chen, Y.-H., Holford-Strevens, V., Rector, E.S. and Sehon, A.H., 1990, Cloned suppressor T cells derived from mice tolerized with conjugates of antigen and monomethoxypolyethylene glycol, J.Immunol., 145:2846

Traub, U.C., DeJager, R.L., Primus, F.J., Losman, M. and Goldenberg, D.M., 1988, Anti -idiotype antibodies in cancer patients receiving monoclonal antibody to carcinoembryonic antigen, Cancer Res., 48:4002

Vitteta, E.S., Fulton, R.J., May, R.D., Till, M. and Uhr, J.W., 1987, Redesigning nature's poisons to create anti-tumor reagents, Science, 238:1098

Waldmann, H., 1989, Manipulation of T-cell responses with monoclonal antibodies, Annu.Rev.Immunol., 7:407

TARGETING OF DRUGS TO TUMORS: THE USE OF THE PLASMINOGEN ACTIVATOR INHIBITOR AS A LIGAND

Jerzy Jankun

Department of Chemistry, The University of Toledo
Toledo, Ohio 43606, USA.

INTRODUCTION

Malignant and benign tumors are differentiated by the ability of cancer to metastasize and invade. Tumors depend on the production of proteases capable to degrade extracellular matrix and, in this way, the proteases function is essential in providing the opportunity for tumor expansion and dissemination (Liotta et al, 1987). To utilize these properties therapeutically useful markers of invasion and metastasis should be found.

Biochemical markers of malignant tumors (i.e. tumors able to invade and metastasize) are found among the proteolytic enzymes. The most commonly reported include: urokinase plasminogen activator (uPA), collagenase, gelatinize, cathepsin B and D (Duffy et al., 1990, Foekens et al, 1992; Jänicke et al., 1991; Mayer et al, 1990). The most prominent among them is uPA. Urokinase is a weak proteolytic enzyme, but it activates plasminogen to plasmin, an enzyme with a broad spectrum of proteins attacked (Mayer et al, 1990). It has been reported that uPA is expressed in high amounts in breast, prostate, colorectal, bladder and others cancers. A direct association has been found between the amount of this enzyme in cancer cells and the metastatic potential of the cancer (Duffy et al, 1990; Foekens et al, 1992; Gaylis et al, 1989; Greaff et al, 1992; Jänicke et al, 1991; Mayer et al, 1990; Nishino et al, 1988). Additionally, it has been reported that when uPA is inactivated, the metastatic potential is greatly reduced (Ossowski, 1988). Further, a high expression of uPA is associated with a poor prognosis and shortened overall survival time for breast and bladder cancer patients (Duffy et al, 1990; Foekens et al, 1992; Hausui et al, 1992). An overwhelming majority of normal cells under normal physiological conditions do not express this enzyme or express it in a very limited amount (i.e. fibroblasts) (Mayer et al, 1990). All of these facts make uPA very attractive in tumor targeting therapy.

The target therapy developed at the turn of century by Paul Erhilch is based on the idea that a drug will attack its target without damaging any other tissues. This "magic bullet" must contain cancer localizing and medicament moiety. It is, therefore, not surprise that antibody-drug conjugates are considered as a way to selectively deliver the medicament. However, in most cases, drug or toxin conjugates act only within the cell. Internalization of cell surface-attached antibodies does not necessarily happen in every case. In addition, most monoclonal antibodies applied in immunoconjugate research and clinical trials are of murine origin. This leads to negative side effects: allergic reactions, or anaphylactic shock (Raso, 1990).

Targeting of Drugs 4, Edited by G. Gregoriadis *et al.*
Plenum Press, New York, 1994

These drawbacks of antibody treatments forced us to look for an alternative way to localize the uPA-bearing tumor cells. Urokinase has properties that can be useful in cancer therapy.

PLASMINOGEN ACTIVATION SYSTEM

Components of the plasminogen activation system. Two types of plasminogen activates, uPA and tPA (tissue plasminogen activator) are known to convert nonactive plasminogen to plasmin. Both catalyze the specific reaction, converting the plasminogen to plasmin by cleavaging a single peptide bond. Both enzymes also form inactive complexes with specific inhibitors: the plasminogen activator inhibitors type 1 and 2.

Urokinase plasminogen activator. uPA has its major function in tissue-related proteolysis, and is important in the processes that entail dissolution of extracellular matrix and transversion of basement membranes. It is produced by cells and is present in extracellular fluid in a form of an inactive, single chain proenzyme (pro-uPA). Conversion of pro-uPA to active two-chain uPA by catalytic amounts of plasmin is a crucial regulatory step in plasminogen activation. This conversion provides active uPA and enables an autocatalytic acceleration of uPA formation (Mayer, 1990). uPA is overexpressed on the surface of the cancer cells if compared with its normal counterparts (Mayer, 1990; Ng, et al, 1983).

Urokinase plasminogen activator receptor. Binding of uPA to its specific receptor on the cell surface is important for the localization of uPA catalyzed plasminogen activation. Binding is rapid, saturable, and with a high affinity of 10^{-10}M. The bound enzyme is not internalized or rapidly degraded. Binding of uPA to the receptor does not involve the catalytic site of uPA; it occurs at the EGF like domain of uPA and pro-uPA (Blasi, 1988). Therefore, bound uPA retains its enzymatic activity. Single chain pro-uPA binding to the cellular receptors is followed by conversion of the bound enzyme to active, two-chain uPA on the surface of the cells. Receptor bound uPA can still bind PAI-1 or PAI-2 and maintains its susceptibility to inhibition by those inhibitors (Mayer, 1990). The inhibitor triggers a series of events leading to the internalization of the PAI-1/uPA/uPAR complex. The complex would most likely be degraded in the lysosomes and the uPA receptor would probably be recycled (Blasi et al, 1988; Cubellis et al, 1990).

Tissue plasminogen activator. tPA is a 70 kDa glycoprotein which is primarily active in fibrinolysis and thrombolysis (Collen et al, 1987). The major function of tPA, which is identical to uPA function, seems to be limited to intravascular proteolytic activity. Expression of this enzyme in tumors seems to be random and not related to the malignant state of solid tumors (Jankun et al, 1993; Mayer, 1990).

Plasminogen. Plasminogen, the physiological substrate for PAs is a 90 kDa, single chain glycoprotein that is converted to plasmin. The essential step in the activation process is cleavage of a single arginine-valine bond, which links the A- and B-chain of plasminogen. While plasminogen is activated to plasmin, it is able to attack a broad spectrum of matrix and basement membrane proteins. Additionally, the plasmin converts pro-collagenase to active collagenase and in this way the uPA and plasmin function as essential elements in matrix degradation.

Inhibitors. Emergence of premature and uncontrolled activity of PAs, and activity at inappropriate site(s), is potentially damaging. Inhibition of catalatic activity of PA(s) is, therefore, instrumental in the prevention of aberrant plasminogen activation. It is, therefore, no surprise that potent and specific inhibitors of PAs are present in cells and others extracellular compartments. The major function of these inhibitors is to regulate the catalytic activities, since the balance between enzymes and inhibitors determines net activity. Three inhibitors of uPA activity are known: PAI-1, PAI-2, and protease nexin (Mayer, 1990). Two of them are the subject of anticancer therapy: PAI-1 and PAI-2.

Plasminogen activator inhibitor type 1. Plasminogen activator inhibitor type 1 (PAI-1) is a 50 kDa single chain glucoprotein that rapidly and specifically forms equimolar, inactive complexes with two-chain uPA. PAI-1 has significant homology with members of the serine protease inhibitor (serpin) family. Serpins share structural features, a common functional principal, and are important regulators of physiological processes (Mayer, 1990). The PAI-1 forms 1:1 complexes between tPA and uPA inactivating them completely. Although, PAI-1 is synthesized in an active form, it is rapidly converted to an inactive (latent) form (Hekman et al., 1985; Hekman et al, 1988).

Molecular basis of the PAI-1 activity. Plasminogen activator inhibitor 1 (PAI-1) is a specific and fast acting inhibitor of both the tissue plasminogen activator (tPA) and the urokinase plasminogen activator (uPA). Like other serpins, PAI-1 has a reactive center located on a highly strained, exposed loop near the C-terminus of the molecule. The "reactive center" of PAI-1 contains the "bait" peptide bond between residues R(346) and M(347) (M - methionine, R - arginine), i.e. the P1-P'1 residues. This bond mimics the R(560)-V(561) (V - valine) bond of plasminogen, which is the bond cleaved by the plasminogen activators during the activation of plasminogen to plasmin. On the basis of the mechanism of the action of other serpins, it has been postulated that PAI-1 binds specifically to the plasminogen activators like substrates (Seetharm et al, 1992).

The PAI-1 can exist in two different forms: in the "active" or in the "latent" form. The active form of the protein is spontaneously converted into the latent form, which can be partially reactivated by treatment with a detergent or with the guanidine hydrochloride (Hekman et al., 1985; Seetharm et al, 1992). It has been postulated that the latent form of PAI-1 has a grater number of residues from the strained reactive center loop inserted between sheets 3 and 5 in the deducted structure of PAI-1, leading to a collapse of the strained loop and loss of inhibitory activity. The recently solved crystal structure of the latent form of rPAI-1 shows that this postulate is essentially true (Mottonen et al, 1992). Latent PAI-1 is inactive presumably because part of its reactive center loop is inaccessible or does not have conformation to bind to its cognate proteases. The residues expected to interact with protease: A(357) - Q(362) (A - alanine, Q - glutamine), all reside in the extended loop on the surface of the molecule.

Stabilization of PAI-1 activity. Active PAI- could be stored at -70°C for months with minimal loss in its activity and negligible conversion to the latent form. However, the PAI-1 lost about 50% of its activity after 2 days at 25°C or after 2 hours at 37°C due to conversion to the latent form; significant conversion to the latent form also occurred at 4°C (Seetharm et al., 1992, Reily et al, 1990). This conversion represented a major challenge in the process of purification conjugate and storing large quantities of active PAI-1 or its conjugates. The latent form of PAI-1 can be reactivated by employing the denaturating agents (6M guanidium chloride, 1% sodium dodecyl sulfate and others), or more conveniently through heat treatment (Chmielewska, 1988, Katagiri et al, 1988). Under these conditions the PAI-1 undergoes complete unfolding and could be refolded to its active form after removing the denaturating agents, yielding near 100% of its original activity. Unfortunately, the PAI-1 will be converted to its latent form again, as quickly as it was described above.

Plasminogen activator inhibitor type 2. Plasminogen activator inhibitor type 2 (PAI-2) is a 46 kDa serine protease inhibitor with the same function as a PAI-1, however with higher affinity to uPA than to tPA. The physiological function of PAI-2 is not clear at this moment, but since it has been found in plasma during pregnancy it has been postulated that it functions as a regulator of homeostasis during pregnancy (Mayer, 1990).

Internalization of uPA/uPAR complex by PAI-1 or PAI-2. uPA forms stable complexes with its receptor. Binding is saturable and with a high affinity. The uPA/uPAR complex is not internalized or rapidly degraded. The PAI type 1 or 2, upon a binding to the uPA/uPAR complex, initiates a series of events that force internalization of a formed complex into the cell. Internalization is preformed via endocytosis, uPA and PAI are probably degraded, while the receptor is probably recycled (Blasi, 1988; Blasi et al, 1988; Cubellis et al, 1990, Jankun, 1992).

However, the other elements are needed to internalize the PAI/uPA/uPAR complex. PAI-1 (or PAI-2) bound to uPA arrested to its receptor is encountered by a low density lipoprotein-related protein (LRP). After binding to LRP, the tetrameric complex is endocytosed through coated pits. The receptor-ligand-receptor complex dissociates in the acidic environment of the lysosome upon which the ligand (uPA-PAI(s)) is targeted for lysosomal degradation. LRP and the uPA receptors return to the cell surface in a recycling vesicle (Herz et al, 1992).

ANTICANCER TARGETED THERAPY

When targeted anticancer therapy is contemplated the following steps should be considered:

1. Finding the cancer markers (cell surface associated) present in common cancers.
2. Selecting a cancer marker and a high affinity binding component, which upon binding to cancer cell will be internalized into the cancer cells.
3. Providing a suitable crosslinker that is stable in circulation but cleavable inside the cancer cells to potentiate effect of medicament on the cancer cells.
4. Selecting a suitable medicament to kill the cancer cells upon internalization of the tumor localizing/medicament complex.
5. Selecting a method of crosslinking the cancer localizing factor and medicament that would prevent the conversion of the cancer localizing factor to its latent, inactive form.

<u>Cancer markers</u>. The malignant potential of solid tumors is related to their proliferation rate as well as to their capacity for invasion and metastasis. The potential of cancer cells for tissue invasion and hematogenous spread is related to their capacity to dissolve the structures in their vicinity. Since the structure of the penetrated tissues consists mainly of proteins, e. g., fibronectin, fibrin, protoglycans, or collagen, the primary substances used by a tumor cell for invasion and metastasis are proteases (Schmitt et al, 1992). Tissues of primary cancer and/or metastases of the breast, ovary, cervix uteri, bladder, prostate, liver, lung, and gastrointestinal tract have been reported to contain high levels of cathepsins, collagenase IV or urokinase plasminogen activator (uPA) if compared with benign tumors or normal tissues (Danø et al, 1985; Duffy et al, 1990; Evers et al, 1982; Gaylis et al, 1989; Jankun et al, 1991; Jankun et al, 1993; Schmitt et al, 1992; Tandon et al, 1990). Elevated levels of Cathepsin D and uPA are of clinical significance. It has been reported that increased concentration of those proteases in breast tumor tissues may predict a high risk of metastasis and shorten overall survival if compared to patients with smaller amounts of those factors (Gaylis et al, 1989; Schmitt et al, 1992; Tandon et al, 1990). Elevated levels of uPAR have been reported in breast cancer also, and overexpression of uPA is often accompanied by the coexpression of uPA (Jankun et al, 1993). These facts make uPA and uPAR proteins very attractive targets in anticancer therapy.

<u>Cancer markers or cancer associated antigens that could internalize a medicament upon binding.</u> uPA and/or uPAR could be targeted using a ligand (amino terminal fragment of uPA to target receptor) or polyclonal and monoclonal antibodies. However, these tumor localizing factors do not secure internalization of conjugated a medicament. Additionally, monoclonal antibodies, murine in origin, could, as a consequence produce negative side effects from allergic reaction to anaphylactic shock.

Another way of delivering the medicament to cancer cells is to conjugate it with PAI-1. PAI-1 binds to uPA/uPAR overexpressed on the surface of cancer cells with high affinity (10^{-13} M). The binding is strong and only then, after binding to LRP protein PAI-1 triggers internalization of the medicament/PAI-1/uPA/uPAR/LRP complex. The complex is

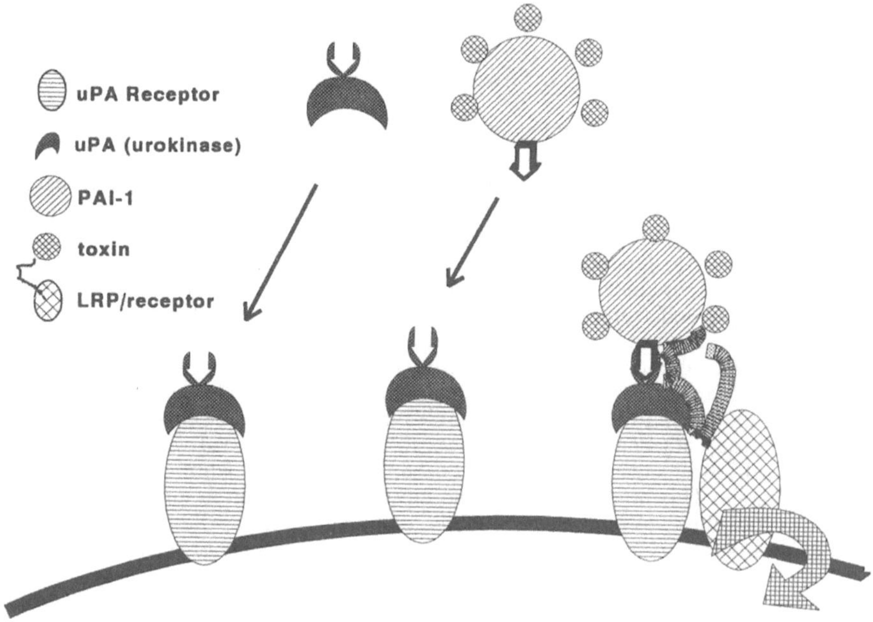

Fig. 1 A model of the delivery of the saporin (SAP) conjugated with PAI-1. uPA is bound to the GPI-linked uPA receptor at the cell surface where it encounters PAI-1/SAP conjugate. After binding to LRP, the pentameric complex is endocytosed through coated pit. The receptor-ligand-ligand-/saporin-receptor complex dissociates in the acidic enviroment of the endosome upon which ligands (uPA and PAI-1/SAP) are targeted for lysosomal degradation.

degraded, the receptors are probably recycled, and the product of proteins degradation and medicament are released to the cytosol. The idea of tumor targeting with the plasminogen activator inhibitor as a ligand is based on the overexpression of urokinase and its receptor on the surface of cancer cells when compared with normal tissue. The PAI-1 is a cancer cell localizing and internalizing molecule in a PAI-1 based medicament. As a protein of human origin, it is not immunogenic and, therefore, is superior to murine antibodies proposed for targeted cancer therapy [Fig. 1].

Additionally, new facts have been reported, making PAI-1 cancer therapy even more attractive. A direct correlation has been found between the expression of uPA and uPAR and the formation of capillary sprouts. This mechanism for new blood vessel formation (angiogenesis) is one of the essential elements of tumor formation (Peper et al, 1990; van Hinsberg et al, 1993). Destruction of capillary sprouts initially causes the blood leakage in the vicinity of the tumor mass and increases the amount of medicament diffusing into the tumor. Later, the PAI-1 based anticancer treatment could prevent the formation of new vessels and limit or prevent new growth by restricting the delivery of nutrients to the tumor mass. These findings broaden the potential use of our conjugates in anticancer therapy.

Unfortunately, PAI-1 is not stable protein, it spontaneously converts itself into a latent, inactive form. The latent form of PAI-1 does not possess the ability to bind uPA and is useless in anticancer therapy. Unexpectedly, we found that the modification of PAI-1 molecule with the heterobifunctional crosslinker SMPT [4-succinimidyloxycarbonyl-α-methyl-α-(2-pyridyldithio)-toluene] and later conjugation of modified PAI-1 with saporin prevents the conversion of an active form of PAI-1 into its inactive from. We hypothesize that saporin conjugated by SMPT with PAI-1 is creating a space hinderance in the proximity of the strained loop of PAI-1. This space hinderance is preventing the strained loop, that is responsible for PAI-1 activity, to retract into the PAI-1 molecule. As a consequence the PAI-1 molecule is "frozen" in its active state that is therapeutically desired.

Heterobifunctional crosslinker of saporin and PAI. Crosslinking or conjugating reagents are chemical compounds used to couple covalently two or more molecules by reactive functional groups in their structures. Crosslinkers contain at least two reactive groups, usually at opposite ends of the molecule, which can be homobifunctional or heterobifunctional. Homobifunctional crosslinkers are less effective than other crosslinkers and a have tendency to produce very high molecular polymers of conjugated proteins. Heterobifunctional crosslinkers involve two steps for conjugation and are often used to prevent undesirable self conjugation which occurs with the homobifunctional reagents. The type of crosslinker used to make the conjugate can affect the ability of the conjugate to kill the cell. Thiol cleavable conjugates have been shown to be more cytotoxic to tumor cells than noncleavable conjugates. SMPT [4-succinimidyloxycarbonyl-α-methyl-α-(2-pyridyldithio)-toluene], a thiol reactive and cleavable crosslinker, contains a benzene ring and a methyl group adjacent to a carbon next to the disulfide bond. These functional groups hinder the disulfide linkage and thus protect the disulfide bond from being readily reduced by thiolate anions. This hinderance prevents unwanted reduction in circulation, but allows for the release of the cytotoxic element of the conjugate in an acidic environment of lysosomes that is therapeutically desired (Lambert et al, 1988; Thorpe et al, 1985).

Medicament. PAI-1 modified by SMPT could be conjugated with virtually any anticancer drug or toxin if these substances posses a functional group (-SH) that could react with SMPT. If a substance lacks a suitable functional group, it could be modified by generally known chemical methods to introduce an -SH group. At this stage of investigation we chose to use saporin as an anticancer toxin. Saporin is a ribosome inactivating protein (RIP) type 1. Type 1 RIP contains one chain and enters cells with difficulty, since they are devoid of the binding B-chain that possess RIP type 2. As a consequence the saporin is much safer to work with, but is equally potent as type 2 RIP if delivered into the cancer or normal cells. The entry of RIP type 1 or A-chain of RIP type 2 into the cell inactivates the ribosomes; thus the protein synthesis is arrested and cell death occurs (Barbieri et al, 1990; Thorpe et al, 1985).

ANTICANCER ACTIVITY OF PAI-1/SAPORIN CONJUGATE

Materials and Methods

Conjugation of the saporin and plasminogen activator inhibitor PAI-1. An immunotoxin has been produced by crosslinking saporin and the plasminogen activator inhibitor using the heterobifunctional crosslinker SMPT [4-succinimidyloxycarbonyl-α-methyl-α-(2-pyridyldithio)-toluene]. This conjugate incorporates a disulfate bond adjacent to the benzene ring and methyl group. Such a configuration imparts plasma stability (a half life over 24 h in comparison to less than 12 h for SPDP) and an acid-labile disulfide bond which liberates saporin once it is internalized into the acidic environment of lysosomes. This action potentiates (up to 1000 times) the cytotoxic action of ribosome inactivating saporin (Trial et al, 1993).

Heterobifunctional crosslinking of saporin and PAI-1 was performed, with some modification, as described by Blakely (Blakly et al, 1987). Briefly, a PAI-1/SMPT derivative was synthesized by adding the acetonitrile solution of SMPT to the phosphate buffered saline (PBS), 1% SDS (sodium dodecyl sulfate), pH 7.4, solution of PAI-1. The reaction was performed for 2 hours at room temperature. The crosslinker was used in different concentrations that vary from 0.1 mM to 10 mM. The conjugate was used in a 0.1-4 fold excess of PAI-1. The reaction mixture was purified by gel filtration (Speedy Desalting Columns, Pierce, No.43243 X) to remove an excess of crosslinker.

It is not clear if saporin contains thiols or reducible disulfide and thus requires

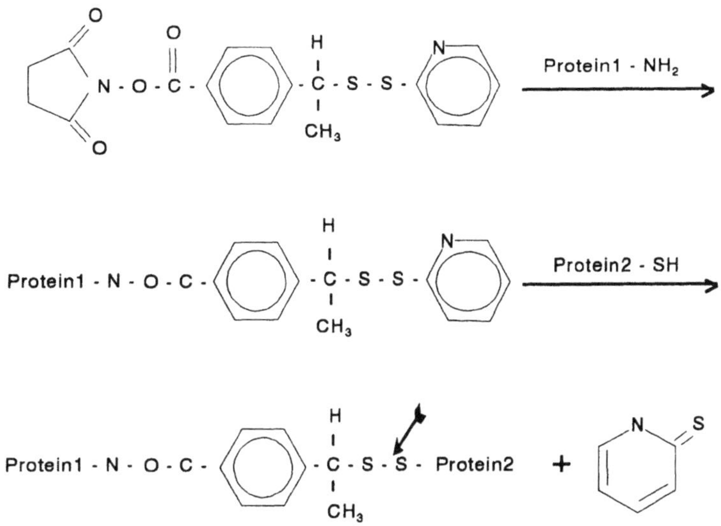

Fig. 2 Conjugation of PAI-1 and saporin by SMPT heterobifunctional crosslinker. Where: Protein1 is PAI-1, Protein2 is saporin. Arrow indicates a cleavable disulfide bond that liberates saporin in acidic enviroment.

introduction of thiols onto the surface of the toxin. For this purpose Traut's reagent (2-Iminothiolane) was used. The saporin in the TRIS-HCl buffer, pH 8.0, was incubated with Traut's reagent in a 5 X molar excess at room temperature for 45 minutes. Separation of the iminothiolated protein from the excess of 2-Iminothiolane was performed on the desalting column.

The second part of the reaction for the production of the immunotoxin via sulfhydryl group to the SMPT modified PAI-1 was done in PBS, 1 mM EDTA, 1% SDS buffer, pH 7.4. This reaction is considerably slower and was carried out for 72 hours. The efficiency of the second reaction was quantified by measuring the amount of pyridine-2-thione released as a by-product (the extension coefficient of pyridine-2-thione is $8.08 \pm 0.3 \times 10^3 M^{-1}cm^{-1}$ at 343 nm) (Blakley et al, 1987). The immunotoxin was separated from the by-products by gel filtration, then aliquoted and frozen in liquid nitrogen for future use [Fig 2].

Determination of PAI-1 activity of PAI-1/saporin conjugate. The PAI-1/saporin was incubated with an equimolar amount of uPA (1µg) in PBS, pH 7.4, for 10 minutes. Next, the solution was incubated with 0.25 mM of SPECTROZYME ® UK (American Diagnostica, Inc.) and the absorbance of generated free chromophore was determined spectroscopically at 405 nm (Harvey et al, 1988). A recombinant PAI-1 (DuPont - Experimental Station) was used as a positive control. As a reference and negative control, the non-inactivated uPA was used.

Cells and Cell culture. The sources of cells are as follows: normal fibroblast - KD and fibrosarcoma HT1080 are from American Type Culture Collection, bladder carcinoma derived cells - BL28 were obtained from Dr. Pamela Russell of the Oncology Research Center, University of New South Wales, Randwick, Australia. The cell lines were routinely cultured in αMEM medium containing penicillin (100 units/ml) and streptomycin (100 µg/ml). The medium was supplemented with 10% new-born calf serum (HyClone, Logan, UT). The cells assayed for PA and cancer cell killing activity were in exponential growth at the time they were tested. The cells assayed for these activities were cultured in a medium without antibiotics 24 h prior to and during assays.

In Vitro Survival Studies. Cells were plated in 96-well microtiter flat-bottomed cell culture plates (approximately 250 cells/well). They were treated in triplicate with different concentrations of PAI-1/SAP immunotoxin for 72 h in an antibiotics-free cell culture medium and maintained at 37°C in 5% CO_2. At the end of the incubation period, the cells were washed, fixed, and stained in 0.1% Crystal Violet solution. Next, the cells were counted under the microscope and normalized to control population.

Assay of Antitumor Activity of PAI-1/SAP *In Vivo.* SCID mice (T and B cell deficient) weighing 20-25g, were used. The animals were housed, 1/sterile cage, with access to sterile rodent chow and sterile water *ad libitum*. The photoperiod was maintained on a 12 hour light, 12 hour darkness cycle. The experiments were approved by the Animal Care and Use Committee at the University of Toledo.

Cancer cells were injected in the rear flank, 1 x 10^6 HT1080 fibrosarcoma cells. Animals were monitored for evidence of tumor growth. The mice were treated by injections of 200, and 2000 µg of PAI-1/SAP per kg of body weight into the tumors when tumors reached a volume of approximately 1000 mm^3.

At the end of the experiment the animals were sacrificed and the tumors were surgically removed, examined, weighed and fixed in formaldehyde. Next, the formaldehyde fixed tumors were paraffin embedded, sectioned, and stained with hematoxylin and eosin then examined under the microscope.

RESULTS

Effect of SMPT modification of activity of PAI-1. Heavy modification of PAI-1 by SMPT could have a negative effect on the activity of the plasminogen activator inhibitor. To check for this, the PAI-1 was modified using different concentrations of SMPT. We found that the modification of PAI-1 using a concentration of SMPT 2µM and higher reduces the activity of PAI-1 by up to 5 times. In all future experiments we used a concentration of SMPT below 2µM that does not reduce the activity of PAI-1.

Conjugation of the saporin and plasminogen activator inhibitor PAI-1. Saporin is a ribosome inactivating protein (RIP) and it was estimated that a single molecule in the cytoplasm is sufficient to kill the cell (Barbieri et al, 1990). The LD_{50} is estimated to be 2000 µg/kg of the mouse's body weight (Cavallaro et al, 1993; Barbieri et al, 1990). Although it has been reported that while saporin is conjugated with antibodies or growth factors, it can be introduced into liposomes and later into the cytosol. This greatly increases the toxicity of saporin (Barbieri et al, 1990). For these reasons saporin was chosen as the cytotoxic part of the conjugate.

Initial conjugation of PAI-1 and saporin using a covalent heterobifunctional cross linker (no SDS was used) has produced a conjugate that possesses very little PAI-1 activity (less than 1-2%). Furthermore, the activation of this conjugate by 1.0% SDS resulted in PAI-1's activity being less than 5%.

To overcome the problem of the lack of PAI-1 activity, all conjugation was preformed under a denaturating condition using a 1.0% solution of SDS. The PAI-1/SAP conjugate synthesized in this way possesses 89% of PAI-1 original activity. To minimize the loss of PAI-1 activity, the time needed to crosslik PAI-1's and saporin was analyzed by monitoring the amount of pyridine-2-thione released as a by-product during conjugation. The maximum amount of pyridine-2-thione was reached after 72 h and future conjugation was performed in the same manner.

Stabilization of PAI-1 activity after conjugation with saporin. Non-conjugated PAI-1 lost most of its activity after 72 h of incubation when saporin/PAI-1 (1:4) conjugate still possesses at least 60% of its original activity.

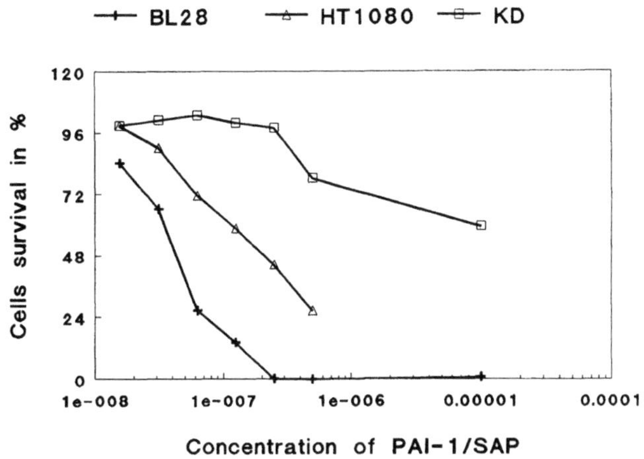

Fig. 3 Cell survival of cancer cells (HT1080 and BL28) and normal fibroblasts (KD). Cells were treated for 72 hours in different concentratiopns of PAI-1/saporin conjugates in cell culture medium. Cell survival is representd by normalized to control number of cells in the cell culture dish after the treatment.

Studies of cancer cell killing *In Vitro*. To study cell killing three different cell lines were selected: KD - normal fibroblasts (expressing less than 0.2 IU/10^6cells/24 h receptor bound uPA), HT1080 fibrosarcoma derived fibroblasts (expressing 22.7 IU/10^6cells/24 h receptor bound uPA), and BL28 bladder carcinoma epithelial cells (expressing 37.6 IU/10^6cells/24 h receptor bound uPA).

The cells were treated as described in the materials and methods section and were killed by the activated PAI-1/SAP cytotoxin. Very little killing was observed in the case of normal fibroblast expressing little receptor-bound uPA, more in the case of HT1080 expressing a moderate amount of receptor-bound uPA, and the highest rate of killing was observed in the case of BL28 possessing the highest amount of receptor-bound uPA. Moreover, the cell killing was PAI-1/SAP concentration dependent. No cell killing was observed using PAI-1/saporin conjugate that did not posses PAI-1 activity [Fig 3].

Assay of Antitumor Activity of PAI-1/saporin *In Vivo*. The PAI-1/saporin cytotoxin was injected into the tumor on the seventeenth day, after tumor cell implantation. Drastic differences were observed in the tumor growth of treated and untreated animals. In the case of animals treated with the highest dose, tumor reduction was observed (2 times in volume) while the control tumor increased its volume 5 times [Fig. 4]. At day 24, the animals were sacrificed, tumors were surgically removed, analyzed, fixed, and the histopathological morphology was analyzed. The tumor of an untreated animal weighed 421 mg and showed no evidence of necrotic tissue. The tumor of an animal treated with a dose 200 µg/kg weighed 251 mg with some necrotic tissue present. Finally, the tumor of an animal treated with the highest dose weighed 201 mg and contained the highest number of necrotic foci.

DISCUSSION

The inhibitors of the serpin family (PAI-1 is a member of this family) have a reactive center located on the exposed loop, referred to as the strained loop, situated near the carboxyl terminus of the molecule. Although the precise mechanism(s) by which the serpins inhibit their target proteases are not well understood, they are known to interact with their target proteases by providing a so called "bait" residue (P1 residue), located within the strained loop. This loop mimics the normal substrate of the target protease

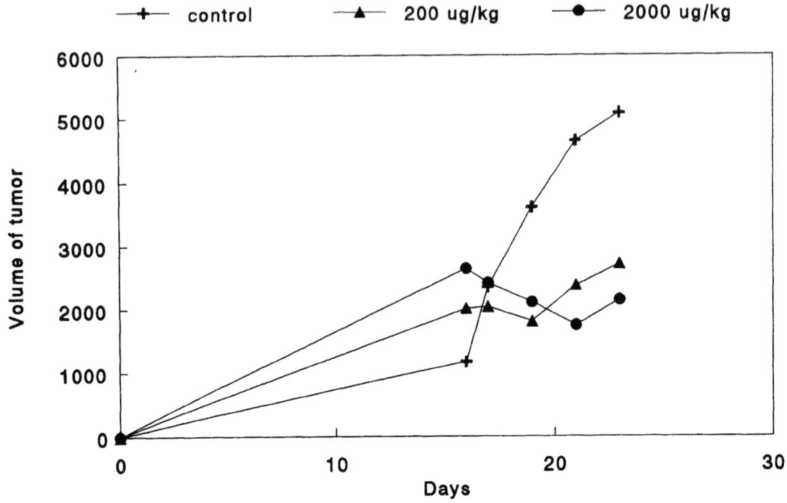

Fig. 4 SCID mice were injected at days 17 and 19 with different amounts of PAI-1 conjugated with saporin. The different tumor sizes, and tumor weights of treated and untreated tumors were observed, which depended on the amount of PAI-1/saporin conjugate injected into the animal.

(Lawrence et al, 1994). However, other residues in close proximity to the P1 also may play a role in determining inhibitor specificity.

The heterobifunctional crosslinker binds to primary amines of aminoacids; *e.g.* lysine and the aminoterminal of the protein. The strained loop does not possesses lysine and is not located on either end of the protein chain. No binding sites are available for chemical modification by SMPT and no loss of activity is expected from the direct binding of SMPT. However, if PAI-1 was to be heavily modified, it may loose uPA binding activity by creating a space hinderance in close proximity to the PAI-1 active site. This unwanted effect was avoided by modification of PAI-1 by SMPT that was below 2μM.

A very unexpected finding was the protection of PAI-1 activity in the saporin conjugated plasminogen activator inhibitor. This finding could be explained in the following way. The serine protease inhibitors all function by forming tight complexes with their target proteases. The external position of the reactive center loop makes it susceptible to proteolytic cleavage and serpins are readily inactivated in this way. Cleavage of the reactive center results in profound structural changes, with the peptide loop being inserted into the protein A-sheet to give the stable, inactive relaxed (R) form of the molecule. This change from the native stressed (S) form (often called active form) to relaxed (R) form is a characteristic feature of inhibitory serpins (Carrell et al, 1991). However, serpins can be inactivated by the insertion of the center loop into a protein without the proteolytic cleavage of the strained loop. This form is called the the locked (L) form or often called latent form. PAI-1 and other serpins can be reactivated from the L conformation by exposure to denaturating agents and refolding to a fully active S conformation. Thus, PAI-1 can be protected from conversion into the L conformation by binding to the vitronectin that holds the reactive center loop out of the A-sheet. A similar effect could be achieved by creating a space hinderance of SMPT molecules or SMPT conjugated proteins. It seems that the molecular ratio of 1:4 of PAI-1 to SMPT in conjugated proteins prevents the collapse of the reactive center of PAI-1 into the molecule, and does not reduce its activity in at least some conjugates. This effect is therapeutically desired, and makes it possible to synthesize, purify and use saporin/PAI-1 conjugates in anticancer treatment.

As we expected, the result stated above demonstrated the therapeutic value of localization and destruction of cancer cells by the proposed method. This concept of cancer therapy is based on the ability of PAI-1 to locate cancer cells. Therefore, the PAI-1 conjugated with toxins can function in the same way as antibody-drug conjugates do. If PAI-1 possesses only this property, it can eventually be used in cancer therapy. But, PAI-1 has several advantages over antibody mediated therapy. The PAI-1 upon binding to the cell surface, will initiate a series of events leading to active delivery of the cytotoxic component into the cell. The second advantage of the proposed method is the human origin of the plasminogen activator inhibitor. The PAI-1 is part of a conjugated cytotoxin will cause very little, if any, immunogenic response.

The ultimate goal of this work was to demonstrate the effectiveness of the PAI-1 in recognition of cancer cells and the proficiency of the plasminogen pathway in the delivery of the cytotoxic compound into the cell. This goal has been achieved, however, in-depth toxicological studies are needed to fully prove therapeutic value of this method. Particularly critical in the delineation of PAI-1 mediated toxicities will be the binding of PAI-1 cytotoxic conjugates within the normal tissues expressing uPA and uPAR. However, in conclusion, we believe that therapies utilizing cytotoxin designed to target cells expressing cell surface antigens will provide a viable alternative approach to conventional cancer treatments. Toxins coupled to FGF or interleukin 2 are already in different stages of preclinical or clinical trials, and some excellent clinical responses have been reported (Beitz, et al, 1992; Hesketh, et al, 1990).

Acknowledgements

We are grateful to Dr. Richard Hart, President of American Diagnostica for a generous supply of antibodies and proteins used in this study. This work was supported in part by grants from: American Diagnostica Inc., American Cancer Society, Inc., Ohio Division and NIH (BSRG 978303-SOC, BSRG 212429-SOC).

REFERENCES

Barbieri, L., Battelli, M. G., and Stripe, F., 1990, Blood clearance and organ distribution and tissue concentration of native, homopolimerized and IgG-conjugated ribosome inactivating proteins, Xenobiotica, 20:1331

Beitz, J. G., Davol, P., Clark, J. W., Kato, J., Medina, M., Frackelton, R. A. Jr., Lappi, A. D., Baird, A., and Calabresi, P., 1992, Antitumor activity of basic fibroblasts factor-saporin mitotoxin in vitro and in vivo, Cancer Res., 52:22

Blakely, D., Watson, G., Knowels, P. and Thrope, P., 1987, Effect of chemical deglycolysation of ricin A-chain on the in vivo fate and cytotoxic activity of an immunotoxin composed of ricin A-chain and anti-Thy 1.1 antibody, Cancer Res., 47:947

Blasi, F., 1988, Surface receptors for urokinase plasminogen activator receptor, Fibrinolysis, 2:73

Blasi, F., Stoppelli, P., and Cubellis, M. V., 1988, The receptor for urokinase plasminogen activator, J. Cell. Biochem., 32:179

Carrell, R.W., Evans, D.L., Stein, P.E., 1991, Mobile reactive centre of serpins and the control of thrombosis, Nature, 10:353

Cavallaro, U., del Vecchio, A., Lappi, A. D. and Soria, M., 1993, Antitumor activity of uPA/Saporin conjugate In Vitro, NATO Advanced Study Institute, Targeting of Drugs:Advances in System Constructs, 24 June - 5 July, 1993, Cape Sounion Beach, Greece

Chmielewska, J., Rauby M. and Miman, B., 1988, Kinetics of inhibition of plasminogen activators by the plasminogen activator inhibitor, Biochem, 15:327

Collen, D., 1987, Molecular mechanism of fibrinolysis and their application to fibrin-specific thrombolytic therapy, J. Cell Biochem., 33:77

Cubellis, M., V., Wun, T., C., and Blasi, F., 1990, Receptor-mediated internalization and degradation of urokinase is caused by its specific inhibitor PAI-1, EMBO J, 9:1079
Danø, K., Andreasen, P. A., Grondhal-Hansen, J., Kristensen, P., Nielsen, L. S. and Skriver, L., 1985, Plasminogen activators, tissue degradation and cancer, Adv. Cancer Res. 44:139
Duffy, M. J., Reilley, D., O'Sullivan, C., O'Higgins, N., Fennelly, J. J., and Andreasen, P., 1990, Urokinase plasminogen activator, a new and independent prognostic marker in breast cancer, Cancer Res. 50:6827
Evers, J. L., Patel, J., Madeja, J. M., Schneider, S. L., Hobika, G., H., Camiolo, S. M. and Markus, G., 1982, Plasminogen activator activity and composition in human breast cancer, Cancer Res., 42:219
Foekens, J. A., Schmitt, M., van Putten, W. L. J., Peters, H. A., Bontenbal, M., Jänicke, F., and Klijn, J. G. M., 1992, Prognostic value of urokinase plasminogen activator in 671 primary breast cancer patients, Cancer Res., 52:6101
Gaylis, F. D., Keer, H. N., Wilson, M. J., Kwaan, H. C., Sinha, A. A. and Kozlowski, J. M., 1989, Plasminogen activators in human prostate cancer cell lines and tumors; correlation with aggressive phenotype, J.Urol. 142:193
Greaff, H., Harbeck, N., Pache, L., Wilhelm, O., Jänicke, F., and Schmitt, M., 1992, Prognostic impact and clinical relevance of tumor-associated proteases in breast cancer, Fibrinolysis, Suppl. 4:45
Hasui, Y., Marutska, K. Suzumiya, J., Kitada, S., Osada, Y. and Sumyoshi, A., 1992, The content of urokinase-type plasminogen activator antigen as a prognostic factor in urinary bladder cancer, Int.J.Cancer, 50:872
Harvey, S. R., Lawrence, D. D., Madeja, J. M., Abbey, S. J., and Markus, G., 1988, Secretion of plasminogen activators by human colorectal and gastric tumor explants, Clinical Experimental Metastasis, 6:431
Hekman, C.M. and Loskutoff, D. J., 1985, Endothelial cells produce a latent inhibitor of plasminogen activators that can be activated by denaturants, J.Biol.Chem. 260:11581
Hekman, C. M., Loskutoff, D. J., 1988, Bovine plasminogen activator inhibitor 1: specificity demonstration and comparison of the active, latent, and guanidine activated forms, Biochem., 27:2911
Herz, J., Clouthier, D.E. and Hammer, R. E., 1992, LDL receptor-related protein internalize and degrades uPA-PAI-1 complexes and is essential for embryo implantation, Cell, 71:411
Hesketh, P., Caguioa, P., Bulger, K., Koh, H., Pihan, G., Carey, R., Mihm, M., Parker, K., Woodworth. T., and McCaffrey, R. Clinical response in cutaneous T-cell lymphoma to an IL-2-diphtheria hybrid toxin ($DAB_{486}IL2$). 1990, Blood, 76:352
Jänicke, F., Schmitt M., and Greaff, G., 1991, Clinical relevance of urokinase type and tissue-type plasminogen activators in and their type-1 inhibitor in breast cancer, Thrombosis Hemostasis, 17:303
Jankun, J., Maher, V. M., McCormick, J. J., 1991, Malignant transformation of human fibroblasts correlates with increased activity of receptor-bound plasminogen activator, Cancer Res., 51:1221
Jankun, J., 1992, Antitumor activity of the type 1 plasminogen activator inhibitor and cytotoxic conjugate in Vitro, Cancer Res., 52:5829
Jankun, J., Merrick, H. M., Goldblatt, P. J., 1993, Expression and localization of elements of the plasminogen activaton system in breast tumors, J. Cellular Biochemistry, 53:135
Katagiri, K., Okada, K., Hattori, H., Yano, M., 1988, Bovine endothelial cell plasminogen activator inhibitor; purification and heat activation, Eur.J.Biochem., 176:81
Lambert, J.M. and Blattler, W.A., 1988, Purification and biochemical characterization of immunotoxins, Cancer Tret.Res., 37:323
Lawrence, D.A., Olson, S.T., Palaniappan, S., Ginsburg, D., 1994, Engineering plasminogen activator inhibitor 1, mutants with increased functional stability, Biochem., 33:3643
Liotta, L. A., Rao, C. N., and Wewer, U. M., 1987, Biochemical interactions of tumor cells with basement membranes, Annu. Rev. Biochem., 55:1037
Mayer, M., 1990, Biochemical and biological aspects of plasminogen activation system, Clin.Biochem. 23:197
Mottonen, J., Strand, A., Symersky, J., Sweet, R. M., Danley, D. E., Geoghegan, K. F.,

Gerard, R. D., and Goldsmith, E. J., 1992, Structural basis of latency in plasminogen activator inhibitor-1, Nature, 355:271
Ng, R., Kellen, J. A., Wong, A. C. H., 1983, Plasminogen activators as markers of tumor colonization potential, Invas.Metast., 3:243
Nishino, N., Aoki, K., Tokura, Y., Sakaguchi, S., Takada, Y., Takada, A., 1988, The urokinase type of plasminogen activator in cancer of digestive tracts, Tromb. Res. 50:527
Ossowski, L., 1988, Plasminogen activator dependent pathways in the dissemination of human tumor cells in chick embryo, Cell, 52:321
Peper, M. S., Belin, D., Montesano, R., Orci, L., Vasselli, J. D., 1990, Transforming growth factor-beta modulates basic fibroblast growth factor-induced proteolytic and angiogenic properties of endothelial cells In Vitro, J. of Cell Biology, 111:743
Raso, V., 1990, The magic bullet - nearing the century mark, Seminars in Cancer Biology, 1:227
Reilly, T. M., Seetharam, R., Duke, J. L., Davis, G. L., Pierce, S. K., Walton, H. L., Kingsley, D, and Sisk, W. P., 1990, Purification and characterization of recombinant plasminogen activator inhibitor-1 from Escherichia Coli, J. of Biological Chemistry, 165:9570
Schmitt, M., Jänicke, F., Graeff, H., 1992, Tumor associated proteases, Fibrinolysis, 6: Supple, 4:231
Seetharm, R., Dwivedi, A. M., Duke, J. L., Hayman, A. C., Walton, H. L., Huckins, N. R., Kamerkar, S. M., Corman, J. I., Woodeshick, R. W., Wilk, R. R., Reilly, T. M., 1992, Purification and characterization of active and latent forms of recombinant plasminogen activator inhibitor 1 produced in Escherichia coli. Biochem., 31:9877
Tandon, A. T., Clark, G. M., Chamnes, G. C., Chrigwin, J. M., and McGuire, W. L., 1990, Cathepsin D and prognosis in breast cancer, N. Engl. J. Med. 322:297
Thorpe, P.E., Brown, A.N., Bremmer, J.A., Foxwell, B.M. and Stripe, F., 1985, An immunotoxin composed of monoclonal ant-Thy 1.1 antibody and a ribosome-inactivating protein from Saponaria Officialis, J.Natl.Cancer Inst., 75:151
Trail, P. A., Wilner, D., Lasch, S. J., Henderson, A. J., Hofstead, S., Casazza, A. M., Firestone R. A., Hellström, I., Hellström, K. E., 1993, Cure of xenografted human carcinomas by BR96-doxorubicin immunoconjugates, Science, 261:212
van Hinsberg, V. W. M., Kooistra, T., Koolwijk, P., 1993, The vascular fibrinolytic system: pathophysiological modulation, Fibrynolysis, 7, Supple 1:11

ADVANCES IN DIAGNOSIS AND TREATMENT OF NEOPLASIA USING RADIOACTIVE MONOCLONAL ANTIBODIES

Anthony Maraveyas and Agamemnon A. Epenetos

Tumour Targeting Laboratoy, Imperial Cancer Research Fund Oncology Unit
Dept of Clinical Oncology, Royal Postgraduate Medical School
Hammersmith Hospital, DuCane Rd., London W12 OHS

INTRODUCTION

The immunoglobulin molecule has long been the focus of investigators as a potential diagnostic and therapeutic tool for cancer (Bale et al, 1955). The first preparations of polyclonal antibodies seemed to bear out this possibility (Goldernberg, et al, 1978), however the difficulty in obtaining clinically relevant quantities through experimental animal immunisation protocols that existed and the great variability between preparations (Mach et al, 1980) made the transit from the test-tube to the clinic very difficult. Apart from limited availability, elements of unpredictability and irreproducibility hampered the research progress of cancer immunotherapy. It was the hybridoma technology (Kohler and Milstein, 1975) and the resulting monoclonal antibodies that gave new impetus to medical diagnosis and treatment. There are still however a number of obstacles to be overcome. Absolutely tumour-specific antigens are rare so crossreactivity is a potential problem. Non specific targeting of tissues rich in reticuloendothelial tissue (bone marrow, liver) is of concern. Penetration into solid tumours, due to adverse physiological or anatomical properties, is poor (Jain, 1990). Antibody dilution in the large volume of the human host circulation and shed antigen-antibody complexes may pose further pharmacokinetic problems. A sensitive host immune system is a further impediment as a part from the well characterised response to the murine antibody (HAMA), it can also recognise even relatively small chemical alterations to humanised molecules caused by linkers (Johnson et al,), thus limiting the therapeutic potential of repetitive treatments. Even the use of humanised antibodies repeatedly has been shown to result in the production of anti-idiotype antibodies (Lockwood et al, 1993). Absolute uptake in human cancer is much less than that achieved in murine xenografts (Epenetos et al, 1986). A variety of problems stemming from the actual imaging or therapeutic agents attached to the antibody have also been identified e.g. toxin immunogenicity (Pai et al, 1991), ^{111}In proclivity for the liver (Sands and Jones, 1987).

Despite the long list of impediments to immunotherapy/immunoimaging that have been recognised, one can finally say that this fledgling field has breached the "cult" status with the advent of the monoclonal antibody technology. In this review we will try to view the problems from the clinical setting, glean information from the most promising results reported in immunoscintigraphy and radioimmunotherapy, relate them to the observed limitations and finally attempt to reach a set of guidelines or else elementary principles that we feel are optimum for the technology as it stands.

Targeting of Drugs 4, Edited by G. Gregoriadis *et al.*
Plenum Press, New York, 1994

Cancer diagnosis, particularly documentation of the extent of disease, can be a notoriously difficult, time-consuming, labour-intensive exercise with a variety of invasive procudures often needed. It would be an advantage to develop a test that not only indicated the existence but also defined the exact anatomical diagnosis and extent of the tumour. Using radiolabelled monoclonal antibodies injected intravenously to localise and image the tumour was an exciting prospect, promising one such straigtforward test. It soon became apparent that the immunoglobulin is not the perfect carrier. High background in the liver due to nonspecific uptake in hepatocytes and reticuloendothelial system, increased levels of activity in the spleen, urinary retention in the bladder, frequent activity in the bowel mucosa or lumen and large central vasculature actually leave a very diminished field of low background activity where unequivocal diagnosis can be easily made.

Pharmacokinetics of most IgG mouse immunoglobulins reveal that, at 24 hours about 80% of activity is still in the body, only at about 48-72 hours does the tumour reach and overtake the levels of injected dose per gram circulating in the blood. However the best imaging isotopes available (^{99m}Tc, ^{123}I) do not offer imaging times compatible with the pharmacokinetics of the murine monoclonal antibody and ^{131}I which does, has got very poor resolution. This leaves ^{111}In, which has been the focus of intense research (Abdel Nabi and Doerr, 1993; Breitz et al, 1993) and is the isotope labelling the only currently commercially available antitumour antibody (CYT-103), marketed as Oncoscint. However this isotope is also beset not only by a high cost but also by a propensity to accumulate in the liver. Britton and Granowska have advocated the use of ^{99m}Tc pointing out that high number of counts is more essential than a low background. The simplification however of "background" to an homogeneous 81%, disregarding the fact that most of this activity is usually concentrated in liver spleen, bladder, central vasculature (these organs can hide a significant surface and volume in the abdomen) and bowel, making diagnosis of small lesions close to these organs often equivocal, is counterintuitive. The excellent results demonstrated by these researches (Granowska et al, 1990; Granowska et al, 1993) with PR1A3 and SM-3 antibodies in colorectal and ovarian cancer may also be due to a ^{99m}Tc-radioimmunoconjugate with a much faster half-life than the equivalent ^{123}I- or ^{111}In-radioimmunoconjugates (Gooden C. unpublished data). One possibility is that the modified Schwartz technique developed by Mathers et al, may be causing antibody structural alterations. There is also some experimental data showing that in some instances 20% of label is lost immediately to host cysteine residues (Thakur et al, 1991).

Combining the favourable characteristics of ^{99m}Tc with fast clearing molecules like an antibody fragment (Fab, $F(ab')_2$), a single chain antibody (ScFv) or an antibody derived peptide is logical. Already a number of clinical studies support this idea (Goldenberg et al, 1990; Lamki et al, 1990). Imaging times of 3-6 hours lead to clinical applicability i.e. a short test needing the minimum use of resources, least interference with other tests, least patient inconvenience and giving results in a short period of time and less staff exposure.

Decision making in oncology is dependant on histological diagnosis and in the competitive radioimaging world where, guided minimally invasive procedures lead to pathological diagnoses, an antibody scan based decision, will depend on a "histological" specificity of almost 100%, with sensitivity at least equivalent to that of the best conventional imaging study available. Being able to image areas of unsuspected disease process is a "bonus" that does not carry any clear benefit in many clinical settings at this point. At the level of efficacy of current treatments very often treating unsuspected asymptomatic disease is not an option at all. Upstaging high stage patients rarely changes management substantially.

Screening or primary diagnosis will need the use of pancarcinoma antibodies or a series of scans with antibodies specific for all major cancers. However pancarcinoma antibodies, useful as all-purpose primary diagnosis reagents, will be of limited value in differentiating a recurrence from a new cancer. Cues will still be needed from classical diagnostic methods. Follow-up, post-primary diagnosis, assessment of serum markers and qualification of masses demonstrated by other radiological means are all indications that need individualising and tailoring to the cancer, patient and antibody available.

A review of the latest published results of a number of studies shows that the trend is towards antibody fragments labelled with ^{99m}Tc. In colorectal cancer (IMMU-4) (Podoloff et al, 1992; Swayne et al, 1991; Serafini et al, 1991) they have been found to do at least as well as the conventional ^{111}In scans with imaging time reduced to 6 hours. When complemented by SPECT, 100% of known lesions in one report (Podoloff et al, 1992) were imaged successfully by 6 hours and confirmed on surgery. False negatives were reported for lesions smaller than 0.3 cm, small lymph nodes and anastomotic recurrences of less than 1 cm diameter. No immunogenicity or side effects have been reported in another series of 200 patients imaged with this fragment (Hansen et al, 1992).

An interesting phase III multicenter study was reported with another Fab fragment, NR-LU-10-^{99m}Tc, in patients with small cell lung carcinoma (Lamki et al, 1990). Positive predictive value was 95-100% similar to that for the whole battery of usual diagnostic tests. No other single test alone was as sensitive or accurate. Results were also good in non-small cell lung carcinoma as well (Friedman et al, 1990) with only bony metastases detection rates being inexplicably low. Adverse effects were minimal. After 148 infusions only 2 patients developed mild urticaria and 2 patients transient temperature elevation. 7.5% of SCLC patients and 25% of NSCLC developed transient human anti-mouse antibodies (HAMA). Imaging time with SPECT was at about 17 hours compared to the usual 120 hours for ^{111}In. SPECT was deemed essential in improving size and location clarification and even locating additional lesions.

Some more recent advances in imaging with antibody fragments have included ^{99m}Tc-F(ab)'$_2$ for head and neck cancer (VanDongen et al, 1992) and a ^{99m}Tc-decapenta-peptide, derived from an anti-mucin antibody and having a cysteine engineered into it for Technetium-99m labelling, in breast cancer patients (Sivolapenko et al, 1993). Preliminary findings include high detection rates with optimal scanning time of 1 hour. In the same vein two step approaches using a slow first step (bispecific antibody or streptavidin conjugated antibody) and small molecule second step. ^{111}In-DTPA (Stickney et al, 1989) or ^{111}In-biotin (Kalofonos et al, 1990) respectively) have led to short imaging times and low background levels. A three step approach (Paganelli et al, 1991) reminiscent of in vivo Elisa is another imaginative strategy in this field.

We conclude that although it has been slow progress, research is on the right track. As the technology stands, isotope decay has to be compatible with the carrier molecule half-life in the human host. Hence we believe that apart from the technological improvement of being able to incorporate ^{99m}Tc into the immunoglobulin molecule a further improvement in the carrier molecule clearance was needed so as to match the isotope. The use of small antibody based molecules, in a variety of guises, illustrates steps in the right direction. It is however doubtful that the golden standards of radioisotope detection of thyroid cancer, neuroectodermal tumour (Hoefnagel et al, 1987) and somatostatin receptor positive endocrine tumours, will be easily duplicated for the common cancers. In these cases we are dealing with gated channel, metabolite uptake and receptor-ligand binding (Kd>,10^{-12}m) and ligand internalisation, where extraction from the periphery of the imaging molecule is an active energy driven process with the potential to perform successfully against negative gradients.

The major clinical issue that remains unresolved is that of impact on the management of a patient. This is the issue that needs further clarification and unequivocal guidelines coupled to the technological simplification that we are witnessing may lead to more widespread acceptance and practice of radioimmunoscintigraphy.

TREATMENT

Radioimmunotherapy has lagged behind in development as compared to radio-immunoscintigraphy. Absolute antibody uptake in the tumour and even antibody distribution in the tumour are two factors that do not critically influence RIS but are of paramount importance in RIT (Epenetos et al, 1986). Radioimmunotherapy is founded on the principle that crossfire effects may also kill cells not expressing the antigen thus partially compensating for inefficient targeting.

While tumour detection is relatively straight-forward to appraise, assessing tumouricidal activity, either directly antibody driven or indirectly antibody transported, is a much more complicated issue. One should also appreciate that it makes sense to look primarily at the curability of radiosensitive tumours rather than treating radioresistant tumours with the low doses and dose-rates achievable with RIT (Fowler, 1990). Theoretical macro- and microdosimetry studies indicate that the choice of emitters will have to be dictated by the therapeutic objectives and intent as well as by matters of convenience, cost or availability (Humm, 1986; Wheldon et al, 1991). Howel et al for example (Howell et al, 1989) demonstrate in an artificial model, that for larger tumours (more than 1 cm diameter) high energy beta-emitters would be more effective while for smaller tumours (around 1 mm diameter) medium energy beta emitters are preferable. For micrometastases ^{193m}Pt was proposed as the best isotope. Clinical knowledge of the pathology of advanced cancer, where one cannot reasonably expect all nodules to be of uniform size, puts the proposition in perspective i.e. one has to consider using a number of emitters simultaneously, or using a medium range beta-emitter with external beam radiotherapy with the intent to "sterilise" over a wide range of cancer volumes designated as "microscopic" residual disease (Wheldon and O'Donoghue, 1990). Bulky disease will remain outside the capabilities of targeted radiotherapy, as successful treatment will depend on the mitigation of the shortcomings of the delivery system and the heterogeneous antibody distribution in the cancer.

^{131}I is the isotope initially used for radioimmunotherapy. Familiar and relatively straightforward chemistry, low cost and proven clinical efficacy in thyroid carcinoma and pheochromocytoma were some of the considerations behind this choice. A number of shortcomings though, not least the environmental hazards posed to the laboratory and nursing staff, have led to the investigation of a number of other β-emitting isotopes.

Promising isotopes exhibiting some of the positive properties of ^{131}I while lacking the problematic high energy gamma ray, seem to be rhenium-186 (Langmuir et al, 1992) and samarium-153 (Bardies et al, 1990). However obtaining a carrier free product is a problem for the former (Goldenberg, 1992) while a worrying bone seeking property evident in preclinical experiments (Boniface et al, 1989) may affect the therapeutic window in the latter. Questions of general availability and cost are other substantial drawbacks. Nonetheless the existence of a simple one step labelling method for ^{186}Re, similar to the Schwarz method for ^{99m}Tc labelling (John et al, 1993), and efficacy of well known DTPA-based labelling chemistry for ^{153}Sm make these isotopes worthwhile candidates for further investigation. The fact that they also possess low energy easy to image γ-radiation is a further advantage.

^{32}P another isotope with a successful clinical track record in the treatment of polycythemia rubra vera is also receiving attention as a radioimmunotherapeutic agent. Kemptide a biologically active peptide which binds ^{32}P, has been conjugated to antibodies (Creighton et al, 1993) and a clinical phase I study has been initiated.

Another area where improvements are needed is that of chelator technology. Inability to administer even the amounts of isotope calculated to treat minimal disease, due to inordinate in vivo chelator instability, has often proved the case e.g. with yttrium-90-DTPA-antibody conjugates (Stewart et al, 1990). The latest results published by Vriesendorp et al demonstrate the inability of a new chelating agent, GYK-DTPA, to improve ^{90}Y-antibody stability (Vriesendorp et al, 1993). CITC-DTPA and DOTA are some of the most stabe chelating agents currently available (Moi and Meares, 1988).

It has become apparent that the systemic route of administration is inefficient in targeting solid malignancies. Only haemopoietic malignancies have shown some evidence of consistent responses (Brown et al, 1989). Hence a great deal of effort has been put into evaluating intracavitary/regional/intratumoural treatments with the hope of increasing efficacy. Experimental findings however, backed by clinical studies indicate that the issue is far from clear, as intraperitoneal seedlings may display unpredictable uptake profiles (Chatal et al, 1989). The working hypothesis is that deep-seated penetrating metastases and retroperitoneal lymph nodes may be more accessible to intravenous administration, compared to more superficial seedlings and free floating cells that may be better targets for regional treatment (Ward et al, 1987).

COMMON FACTORS LINKING ENCOURAGING RIT RESULTS

Consistenely promising results have been obtained in the treatment of lymphomas and chronic lymphocytic leukemia. Monoclonal antibodies given intravenously either "unarmed" or bearing a toxic agent seem to achieve at least peripheral clearance of abnormal cells (Brown et al, 1989; Waldman et al, 1988). Increased susceptibility due to radiation induced apoptosis may be a factor, or the underlying reason may lie in better access to the tumour. The more accessible the disease (peripherial blood lymphocytes, bone marrow) the more efficient the lytic effect. The more solid the disease (lymph nodes) the more protracted and difficult it is to achieve a significant antitumour result (Hale et al, 1988).

Epithelial neoplasms

Heterogeneity of antigen expression, poor perfusion of large tumour areas, exuberant stromal reaction, dense cellularity and high interstitial pressure due to lack of lymphoid vessels, are just a sample of the known problems posed to targeted immunotherapy by the anatomic and pathological properties of the solid tumour. It is unlikely that targeting of each and every cancer cell is achievable.

Intralesional therapy. Gliomas treated with intratumour radiolabelled antibody injections have shown some gratifying responses (Riva et al, 1991). Again the probable factor that dominates in this model is that of access of the whole therapeutic dose to the treatment site. Failures are probably due to poor percolation of the injectate to only a shell of a few cell diameters around the tumour (Papanastassiou et al, 1993).

Concurrent radiation and RIT. Hepatocellular carcinoma a radioresistant tumour has shown a measure of response when attacked simultaneously with intraarterial radioimmunotherapy, external beam radiotherapy and combination chemotherapy (Order et al, 1985). However the simultaneous use of these modalities makes interpretation of data very difficult if not impossible.

Of great interest is the concept of using MIBG treatment for neuroectodermal tumours together with total body irradiation (Wheldon et al, 1993) so as to increase the likelihood of eradicating micrometastases. Judiciously, this is a radiosensitive malignancy with an energy driven process being the target of MIBG, and historically even large hepatic metastases have often been shown to recede when low dose external radiotherapy has been used. Wheldon et al have demonstrated the homogenous transition pattern of radiation achieved over a wide size variation of metastases with the combination of external beam radiotherapy and ^{131}I as the therapeutic isotope (Wheldon, 1991)

Adjuvant treatment. In a phase II study on intraperitoneal treatment for patients with ovarian cancer it was shown that patients with FIGO stage IIb and above who were negative on second look laparoscopy and treated with an ^{90}Y-labelled antimucin antibody instilled in their peritoneal cavity had improved survival, with 90% of patients being alive at 64 months post antibody treatment or 70 months post diagnosis (Hird et al, 1993). Median follow up time was 36 months post-treatment. This compared favourably with historical controls and therefore a phase III randomised study has been initiated. The question of why this treatment may be efficient is far from clear. It is worthwhile noticing that microdosimetry studies on microspheres have shown that ^{90}Y is at a disadvantage at the level of minute micrometastases and small volume disease as compared to ^{131}I, ^{153}Sm and ^{186}Re due to the fact that the β-particle delivers its energy at a path length well beyond the diameter of the targeted cell and sometimes the targeted cell cluster (Bardies et al,). It may be consequent that responses in minimal disease were not as readily noted with the ^{90}Y immunoconjugate (Stewart et al, 1990), as compared to the studies with ^{131}I labelled antibodies (Epenetos et al, 1987; Stewart et al, 1989). Increase in HAMA was reported in almost all the patients treated in this study and anti-idiotypic responses were found in many patients. It may be that an anti-idiotypic cascade is induced in these patients effectively immunising them against antigen positive tumours (Courtenay-Luck et al, 1986). On the other hand it could be due to better access to and penetration of the few existing tumour pockets by the radioimmunoconjugate. Even non specific immunotherapy regimens with BCG or Corynobacterium Parvum have shown positive results in patients with minimal intraperitoneal disease (Bast et al, 1983), cell mediated cytotoxicity being the most likely cause (Kosmas et al, 1991). However those results were not followed up and no survival benefit was yet to be comprehensively reported. There are also indications that the haemopoietic growth factor GMCSF may function as a potent immunostimulatory agent when used in conjuction with an immunising agent (Ragnhammar et al, 1993; Dranoff et al, 1993).

LESSONS FROM OTHER IMMUNOTHERAPY STRATEGIES

Magnification of the Toxic Insult

The golden standards of targeted isotope treatment are: elemental ^{131}I for thyroid cancer, a metabolite uptake (MIBG) for neuroectodermal tumours, both energy dependant amplification processes working against gradients and mediated by non immunogenic components. It is possible that the exceptional sensitivity of a few solid cancers to platinum derivatives, that have made up the success story of clinical oncology i.e. testicular cancer, may also be in part an energy dependant process (Gately et al, 1993).

ADEPT. Using an enzyme-antibody conjugate to activate a non-toxic pro-drug at the tumour site is a field of investigation of great interest to us due to its inherent logic. Enzyme mediated magnification of a toxic insult, although still at its beginning, is an exciting possibility (Senter, 1990). Although not directly relevant to radioimmunotherapy it has the vital lesson to teach, that an amplification process of the meagre uptake of antibody at the tumour site is of overriding importance, if any real progress is to be achieved.

UNIFYING GUIDELINES FOR RIT EDUCED FROM THE PRINCIPLES OF ACCESS, MULTIMODALITY TREATMENT AND AMPLIFICATION

The intrinsic problem of immunotherapy is the passivity of the whole immunotargeting affair, even a large component of antibody internalisation may be a passive membrane reconstruction (Kyriakos et al, 1992) procedure. The question is whether any other amplification recourse exists that can be resorted to, since, inducing pump or gated channel properties in cancer cells is more in the domain of gene therapy. The only logical way around this seems to come again from physiological principles, and that is to target vital structures, however disorganised, on which the tumour counts on for survival. In a cancer lacking any hormonal, neurological or immunological control, only nutrients via vascular access seem the weak link (Maeda, 1992). Tumour viability depends on this hastily constructed vasculature. Antibodies against tumour vascular endothelium, together with systemic or local normal vessel vasodilatation to induce vascular steal syndromes, is a hypothetical model. In effect there is continous access to the maximum available dose in the circulation, no need for penetration, relatively small number of cells to target and vessel occlusion leads to greater downstream nutrient starvation. Mechanisms involved based on platelet activation are operable in the tumour microenvironment as compared to a variety of cell mediated immunological responses that may not be effective due to a postulated immunosuppresive tumour microenvironment (Floutsis et al, 1989). Concurrent judicious use of GMCSF with the dual purpose of immunostimulation and haemopoietic preservation can be contemplated. Single or multiple strategies based on these principles may be the way forward.

Hepatocellular carcinoma would be of interest as a model. Local treatment based on the fact that 90% of the tumour is perfused by the hepatic artery (Lin et al, 1984) is a well practised modality. Anecdotal good results have been obtained with antibodies and starch microsphere embolisation (Wilson and Epenetos, 1987). An indwelling catheter could be a conduit for alternative anti-vascular antibody labelled with ^{186}Re and continous (Smyth et al, 1987) vasodilator infusion treatment, hence reducing tumour perfusion by dilating normal vessels. This induced vascular steal syndromr should further enhance the nutrient starvation of the tumour brought about by the endothelial destruction by the radioactive antibody. Concurrent external beam radiotherapy could kill the more viable well oxygenated tumour cells. When humanised/human antibodies (Steinitz et al, 1988) become widely available multiple fractionated treatments may further improve the prospects for treating bulky tumours.

CONCLUSION

We recognise very well that the current status of immunotherapy and immunoscintigraphy is well below what we would like it to be. There are many reasons why it is so. In this review we have endeavoured to educe some guidelines based on the most promising results published to date. We have tried to approach them from an angle of physiology and clinically based logic, despite the lack of knowledge for many of the possible facets that may contribute to each and every successful case. There is of course the oncoming surge of molecular biology and designer chemistry and the enormous variety of composite novel molecules that can be thought of and constructed (Neuberger et al, 1984). Nevertheless it is always useful to stop and take note of what has been achieved and try to extract maximum benefit from existing and practicable thechnology.

REFERENCES

Abdel Nabi, H. and Doerr, R.J., 1993, Clinical applications of Indium-111-labelled monoclonal antibody imaging in colorectal cancer patients, Seminars in Nuclear Medicine, 23:99

Bale, W.F., Sparl, F., Goodland, R.L., and Wolfe, D.E., 1955, In vivo and in vitro studies of labelled antibodies against rat kidney and Walker carcinoma, Proceedings of the Society for Experimental Biology and Medicine, 89:564

Bardies, M., Lame, J., Meyers, M.J., et al, 1990, A simplified approach to beta dosimetry for small spheres labelled on the surface, Phys.Med.Biol., 35:1039

Bardies, M., Thedrez, P., Gestin, J.F., Marcille, B.M., Guerreau, D., Faivre-Chauvet, A., Mahe, M., Sai-Maurel, C. and Chatal, J.-F., (1992) Use of multicell spheroids of ovarian carcinoma as an intraperitoneal radioimmunotherapy model: Uptake, retention kinetics and dosimetric estimations, Int.J.Cancer, 50:984

Bast, R.C. Jr., Berek, J.S., Obrist, R., Griffiths, C.T., Berkowitz, R.S., Hacker, N.F., Parker, L., Lagasse, L.D. and Knapp, R.C., 1983, Intraperitoneal immunotherapy of human ovarian carcinoma with Corynobacterium Parvum, Cancer Res., 43:1395

Berek, J.S., Knapp, R.C., Hacker, N.F. et al, 1985, Intraperitoneal immunotherapy of epithelial ovarian carcinoma with Corynobacterium Parvum, Am.J.Obstet.Gynecol., 152:1003

Boniface, G.R., Izard, M.E., Walker, K.Z., McKay, D.R., Sorby, P.J., Turner, J.H. and Morris, G.R., 1989, Labelling of monoclonal antibodies with samarium-153 for combined radioimmunoscintigraphy and radioimmunotherapy, J.Nucl.Med., 30:683

Breitz, H.B., Sullivan, K. and Nelp, W.B., 1993, Imaging lung cancer with radiolabelled antibodies, Seminars in Nuclear Medicine, 23(2):127

Britton, K.E. and Granowska, M., 1992, The role of radiolabelled antibodies in cancer diagnosis, in: "Monoclonal Antibodies: applications in clinical oncology", A.A. Epenetos, ed., Chapman & Hall, London

Brown, S.L., Miller, R.A., Horning, S.J., et al, 1989, Treatment of B-cell lymphomas with anti-idiotype antibodies alone and in combination with alpha interferon, Blood, 73:651

Chatal, J.F., Saccavini, J.C., Gestin, J.F., Thedrez, P., Curtet, C., Kremer, M., Gurreau, D., Nobile, D., Fumoleau, P. and Guillard, Y., 1989, Biodistribution of indium-111-labelled OC125 monoclonal antibody intraperitoneally injected into patients operated on for ovarian carcinomas, Cancer Res., 49:3087

Courtenay-Luck, N.S., Epenetos, A.A., Larche, M. and Ritter, M., 1986, Development of primary and secondary immune responses to mouse monoclonal antibodies used in the diagnosis and therapy of malignant neoplasms, Cancer Res., 46:6489

Creighton, A.M., Band, H.A., Jeffery, W.A., Long, J. and Welsh, J., 1993, The development of ^{32}P technology for radioimmuntotherapy in "Monoclonal Antibodies 2: Applications in clinical oncology", A.A. Epenetos, ed., Chapman, & Hall, London

Dranoff, G., Jafee, E., Lazenby, A., Golumbek, P., Levitzky, H., Brose, K., Jackson, V., Hamada, H., Pardoll, D. and Mullingan, R.C., 1993, Vaccination with irradiated tumor cells engineered to secrete murine granulocyte-macrophage colony-stimulating factor stimulates potent, specific, and long-lasting anti-tumor immunity, Proc.Natl.Acad.Sci.USA, 90:3539

Epenetos, A.A., Snook, D., Durbin, H., Johnson, P.M. and Taylor-Papadimitriou, J., 1986,

Limitations of radiolabelled monoclonal antibodies for localisation of human neoplasms, Cancer Res., 46:3183
Epenetos, A.A., Munro, A.J., Stewart, S., Rampling, R., Lambert, H.E., McKenzie, C.G., Soutter, P., Rahemtulla, A., Hooker, G., Sivolapenko, G.B., Snook, D., Courtney-Luck, N., Dhokia, B., Krausz, T., Taylor-Papadimitriou, J., Durbin, H. and Bodmer, W.F., 1987, Antibody-guided irradiation of advanced ovarian cancer with intraperitoneally administered radiolabelled monoclonal antibodies, J.Clin.Oncol., 5:1890
Floutsis, G., Ulsh, L. and Ladisch, S., 1989, Immunosuppressive activity of human neuroblastoma tumor gangliosides, Int.J.Cancer, 43:6
Fowler, J.F., 1990, Radiobiological aspects of low dose rates in radioimmunotherapy, Int.J.Radiation Oncology Biol.Phys., 18:1261
Friedman, S., Sullivan, K., Salk, D., et al, 1990, Staging of non-small cell carcinoma of the lung using technetium-99m-labelled monoclonal antibodies, Hematol.Oncol.Clin. North Am., 4:1069
Gately, D.P. and Howell, S.B., 1993, Cellular accumulation of the anticancer agent cisplatin: A review, Br.J.Cancer, 67:1171
Granowska, M., Mather, S.J., Britton, K.E., Bentley, S., Richman, P., Phillips, R.K.S., and Northover, J.M.A., 1990, ^{99m}Tc radioimmunoscintigraphy of colorectal cancer, Br.J.Cancer, 62 Suppl X:30
Granowska, M., Britton, K.E., Mather, S.J., Lowe, D.G., Ellison, D., Bomanji, J., Burchell, J., Papadimitriou-Taylor, J., Hudson, C.R. and Shepherd, J.H., 1993, Radioimmunoscintigraphy with technetium-99m-labelled monoclonal antibody, Sm3, in Gynaecological cancer, Eur.J.Nucl.Med., 20:483
Goldenberg, D.M., Deland, F., Kim, E. et al, 1978, Use of radiolabelled antibodies to carcinoembryonic antigen for the detection and localisation of diverse cancers by external photoscanning, New England J. Medicine, 298:1384
Goldenberg, D.M., Goldenberg, H., Sharkey, R.M. et al, 1990, Clinical studies of cancer antibody fragments labelled with ^{123}I or ^{99m}Tc, Cancer Res., 50:909s
Goldenberg, D.M., 1992, Radioimmunotherapy of cancer: Arming the missiles, J.Nucl.Med., 33:1110
Hale, G., Dyer, M.J.S., Clark, M.J., et al, 1988, Remission induction in non-Hodgkin lymphoma with reshaped monoclonal antibody CAMPATH-1H, Lancet, 2:1394
Hansen, H.J., Sharkey, R.M., Jones, A.L., et al, 1992, Lack of human anti-mouse antibody (HAMA) response in patients imaged with IMMU RAID-CEA(Tc-99m) and IMMU RAID NM3(Tc-99m), Antibody Immunoconj.Radiopharmaceut., 5:354
Hird, V., Maraveyas, A., Snook, D., Dhokia, B., Soutter, W.P., Meares, C., Stewart, J.S.W., Mason, P., Lambert, H.E. and Epenetos, A.A., 1993, Adjuvant therapy of ovarian cancer with radioactive monoclonal antibody, Br.J.Cancer, 68:403
Hoefnagel, C.A., Voute, P.A., deKraker, J. and Marcuse, H.R., 1987, Radionuclide diagnosis and therapy of neural crest tumours using Iodine-131 metaiodo-benzylguanidine, J.Nucl.Med., 28:308
Howell, R.W., Rao, D.V. and Sastry, K.S., 1989, Macroscopic dosimetry for radio-immunotherapy: Nonuniform activity distributions in solid tumors, Med.Phys., 16:66
Humm, J.L., 1986, Dosimetric aspects of radiolabelled antibodies for tumour therapy, J.Nucl.Med., 27:1491
Jain, R.K., 1990, Physiological barriers to the delivery of monoclonal antibodies and other macro molecules in tumours, Cancer Res., 50:814
John, E., Thakur, M.L., DeFulvio, J., McDevitt, M.R. and Damjanov, I., 1993, Rhenium-186 labelled monoclonal antibodies for radioimmunotherapy: Preparation and evaluation, J.Nucl.Med., 34:260
Johnson, D.A., Barton, R.L., Fix, D.V. et al, Induction of immunogenicity of monoclonal antibodies by conjugation with drugs, Cancer Res., 51:5774
Kalofonos, H.P., Rusckowski, M., Siebecker, D.A., et al, 1990, Imaging of tumour in patients with Indium-111-labelled biotin and streptavidin-conjugated antibodies: preliminary communication, J.Nucl.Med., 31:1791
Kohler, G. and Milstein, C., 1975, Continuous culture of fused cells secreting antibody of predefined specificity, Nature, 256:465
Kosmas, C., Epenetos, A.A. and Courtenay-Luck, N.S., 1991, Patients receiving murine monoclonal antibody therapy for malignancy develop T cells that proliferate in vitro in response to these antibodies as antigens, Br.J.Cancer., 64:494

Kyriakos, R.J., Shih, L.B., Ong, G.L., Patel, K., Goldenberg, D.M. and Mattes, M.J., 1992, The fate of antibodies bound to surface of tumour cells in vivo, Cancer Res., 52:835-842

Lamki, L.M., Zukiwski, A.A., Shanken, L.J. et al, 1990, Radioimaging of melanoma using 99mTc-labelled Fab′ fragment reactive with a high molecular weight melanoma antigen, Cancer Res., 50 (suppl):904s

Langmuir, V.K., Mendonca, H.L., Vanderheyden, J.L. and Fu-Min Su, 1992, Comparisons of the efficacy of ^{186}Re- and ^{131}I-labelled antibody in multicell spheroids, Int.J.Radiation Oncology Biol.Phys., 24:127

Lin, G., Hagerstrand, I. and Lunderquist, A., 1984, Portal blood supply of liver metastases, AJR: 143:53

Lockwood, C.M., Thiru, S., Isaacs, J.D., Hale, G. and Waldmann, H., 1993, Long term remission of intractable vasculitis with monoclonal antibody therapy, Lancet, 341:1620

Mach, J.P., Carrel, S., Forni, E., et al, 1980, Tumour localisation of radiolabelled antibodies against carcinoembryonic antigen in patients with carcinoma, New England J.Medicine, 303:5

Maeda, H., 1992, The tumor blood vessel as an ideal target for macromolecular anticancer agents, J.Control Release, 19:315

Mathers, S.J. and Ellison, D., 1990, Reduction mediated technetium-99m labelling of monoclonal antibodies, Journal of Nuclear Medicine, 31:692

Moi, K.M. and Meares, C.F., 1988, The peptide way to macrocyclic bifunctional chelating agents; synthesis of 2-p-nitrobenzyl-1.4.7,10,-tetraazacyclododecane-N',N'',N''',-tetraacetic acid and study of its yttrium(III) complex, J.Am.Chem.Soc., 110:6266

Neuberger, M.S., Williams, G.T. and Fox, R.O., 1984, Recombinant antibodies possessing novel effector functions, Nature, 312:604

Order, S.E., Stillwagon, G.B., Klein, J.L., et al, 1985, Iodine - 131 antiferritin: a new treatment modality in hepatoma: a Radiation Therapy Oncology Group Study, J.Clin.Oncol., 3:1573

Paganelli, G., Magnani, P., Zito, F., et al, 1991, Three-step monoclonal antibody tumour targeting in carcinoembryonic antigen-positive patients, Cancer Res., 51:5960

Pai, L.H., Bookman, M.A. and Ozols, R.S., 1991, Clinical evaluation of intraperitoneal Pseudomonas exotoxin immunoconjugate OVB3-PE in patients with ovarian cancer, J.Clin.Oncol., 9:2095

Papanastassiou, V., Pizer, B., Tzanis, S., Zananiri, A., Moseley, R., Ashpole, R., Bullimore, J., Coakham, H.B. and Kemshead, J.T., 1993, Use of radioimmunoconjugates in the treatment of malignant glioma in: "Monoclonal Antibodies 2: Applications in clinical oncology", A.A. Epenetos, ed., Chapman & Hall, London, 179

Podoloff, D.A., Patt, Y.Z., Curley, S., et al, 1992, Impact of surgical management by the preoperative evaluation of patients with prior colorectal cancer and rising serum CEA using a Tc-99m labelled anti-CEA monoclonal antibody, Radiology, 185:261

Ragnhammar, P., Fagerberg, J., Frodin, J.E., Hjelm, A.L., Lindemalm, C., Magnusson, I., Masucci, G. and Mellstedt, H., 1993, Effect of monoclonal antibody 17-1A and GM-CSF in patients with advanced colorectal carcinoma-long lasting, complete remissions can be induced, Int.J.Cancer, 53:751

Riva, P., Arista, A., Mariani, M., et al, 1991, Direct intratumour administration of BC-2 radiolabelled MAb: a phase I study in brain glioma, 8th International Hammersmith Meeting, Greece

Sands, H. and Jones, P.L., 1987, Methods for the study of the metabolism of radiolabelled monoclonal antibodies by liver and tumour, J.Nucl.Med., 28:390

Senter, P., 1990, Activation of prodrugs by antibody-enzyme conjugates: A new approach to cancer therapy, FASEB J., 4:188

Serafini, A.N., Vargas-Cuba, R., Benedetto, P., et al, 1991, Technetium-99m-labelled Fab fragment of anti-CEA monoclonal antibody for the radioimmunodetection of colorectal adenocarcinoma, Antibody Immunoconj.Radiopharm., 4:561

Sivolapenko, G., Douli, V., Sirmalis, G., Pectasides, D., Skarlos, D., Merkouri, E., Kapsalakis, J.Z. and Konstantinides, K., 1993, Radiolabelled tumour-associated synthetic peptides in cancer diagnosis: the tip of the iceberg, Proceedings of the Tenth International Hammersmith Conference, p.16

Smyth, M.J., Pietersz, G.A. and McKenzie, I.F.C., 1987, Use of vasoactive agents to

increase tumour perfusion and the antitumour efficacy of drug-monoclonal antibody conjugates, J.Natl.Cancer Inst., 79:1367
Steinitz, M., Tamir, S., Frodin, J.E., Lefvert, A.K. and Mellstedt, H., 1988, Human monoclonal anti-idiotypic antibodies. I. Establishment of immortalized cell lines from a tumour patient treated with mouse monoclonal antibodies, J.Immunol., 141:3516
Stewart, J.S.W., Hird, V., Snook, D., Dhokia, B., Sivolapenko, G., Hooker, G., Taylor Papadimitriou, J., Rowlinson, G., Sullivan, M., Lambert, H.E., Coulter, C., Mason, W.P., Soutter, W.P. and Epenetos, A.A., 1990, Intraperitoneal Yttrium-90-labelled monoclonal antibody in ovarian cancer, J.Clin.Oncol., 8:1941
Stewart, J.S.W., Hird, V., Snook, D., Sullivan, M., Hooker, G., Courtenay-Luck, N., Sivolapenko, G., Griffiths, M., Myers, M.J., Lambert, H.E., Munro, A.J. and Epenetos, 1989, Intraperitoneal radioimmunotherapy for ovarian cancer: pharmacokinetics, toxicity, and efficacy of I-131 labelled monoclonal antibodies, Int.J.Radiation Oncology Biol.Phys., 16:405
Stickney, D.R., Slater, J.B., Kirk, G.A., et al, 1989, Bifunctional antibody: ZCE/CHA 111 Indium BLEDTA-iv clinical imaging in colorectal carcinoma, Antibody Immunoconj. Radiopharmaceut., 2:1
Swayne, L.C., Goldenberg, D.M., MaCauley, R., et al, 1991, Radioimmunodetection with Tc 99m-labelled anti-CEA monoclonal antibody Fab, fragments: experience in a large community hospital, Antibody Immunoconj.Radiopharm., 4:553
Thakur, M.L., DeFulvio, J., Richard, M.D., Park, C.H., 1991, Technetium-99m labelled monoclonal antibodies: evaluation of reducing agents, Nucl.Med.Biol., 18:227
VanDongen, G.A., Leverstein, H., Roos, J.C., et al, 1992, Radioimmunoscintigraphy of head and neck cancer using 99mTc-labelled monoclonal antibody E48 F(ab)2, Cancer Res., 52:2569
Vriesendrop, H.M., Shao, Y., Blum, J.E., Quadri, S.M. and Williams, J.R., 1993, Fractionated intravenous administration of ^{90}Y-labelled B72.3 GYK-DTPA immunoconjugate in beagle dogs, Nucl.Med.Biol., 20:571
Ward, B.G., Mather, S.J., Hawkins, L.R., Crowther, M.E., Shepherd, J.H., Granowska, M., Britton, K.E. and Slerin, M.L., 1987, Localization of radioiodine conjugated to the monoclonal antibody HMFG2 in human ovarian carcinoma: Assessment of intravenous and intraperitoneal routes of administration, Cancer Res., 47:4719
Waldman, T.A., Goldman, C.K., Bongiovanni, K.F., et al, 1988, Therapy of patients with human T-cell lymphotropic virus I-induced T-cell Leukemia with anti-Tac, a monoclonal antibody to the receptor for interleukin-2, Blood, 72:1805
Wheldon, T.E., O'Donoghue, J.A., Barett, J. and Michalowski, A.S., 1991, The curability of tumours of differing size by targeted radiotherapy using ^{131}I or ^{90}Y, Radiotherapy and Oncology., 21:91
Wheldon, T.E. and O'Donoghue, 1990, The radiobiology of targeted radiotherapy, Int.J.Radiat.Biol., 58:1
Wheldon, T.E., Amin, A.E., O'Donaghue, J.A. and Barrett, A., 1993, Radiocurability of disseminated malignant disease by external beam irradiation and targeted radionuclide therapy, in: "Monoclonal Antibodies 2: Applications in clinical oncology", A.A. Epenetos, ed., Chapman & Hall, London
Wilson, C.B. and Epenetos, A.A., 1987, Use of monoclonal antibodies for diagnosis and treatment of liver tumours, Bailliere's Clinical Gastroenterology 1:115

TARGETING TRANSCRIPTION FACTORS TO INHIBIT SELECTIVELY GENE EXPRESSION IN PARTICULAR CELL TYPES

A.C. Allison[1] and E.M. Eugui[2]

[1]Dawa Corporation, Belmont, CA and [2]Syntex Research, Palo Alto, CA, USA

INTRODUCTION

This conference is mainly about targeting drugs to desired cell types, although some of the contributions address the difficult problem of facilitating the entry of oligonucleotides into cells. The use of anti-sense oligonucleotides or ribozymes to prevent the expression of particular genes in cultured cells, and of "gene knock-out" experimental animals, provide valuable experimental tools. However, they are not easily applicable to human therapy. We are exploring the use of small molecules to inhibit the activation of transcription factors and thereby suppress the expression of certain genes in particular cell types.

The advantages of this new type of intracellular targeting are shown diagrammatically in Fig. 1. Most conventional drugs are enzyme inhibitors, e.g. of cyclooxygenase or angiotensin-converting enzyme, ion channel blockers, or selective agonists or antagonists of particular receptors (adrenergic, histamine, 5-hydroxytryptamine etc.). It has also been proposed that particular signal transduction systems, e.g. cyclic AMP or GMP, or protein kinase C, can be regulated by inhibiting phosphodiesterases and other means, but little of practical value has emerged so far. As shown in Fig. 1, transcription factors are strategically placed at the point of convergence of signal transduction systems. Moreover each protein transcription factor binds to the promoter-enhancer region of more than one gene, so that by preventing the activation of the obligate set of transcription factors it is possible to suppress the expression of more than one gene. In this paper we illustrate the principle by showing that a subset of antioxidants can prevent the activation of certain transcription factors in cells of the monocyte-macrophage lineage, thereby inhibiting expression of the genes for the pro-inflammatory cytokines TNF-α and IL-1β, as well as IL-6. The same drugs do not inhibit the expression of the IL-6 gene in fibroblasts, and they can actually augment the expression of the IL-1 receptor antagonist in macrophages. Thus the gene selectivity and cell-type selectivity of the effects is demonstrated. The same compounds markedly decrease levels of TNF-α and IL-1β in the circulation of mice following injection of lipopolysaccharide from Gram-negative bacteria (LPS), so they are active in vivo. These observations not only validate the proposed strategy: they open up a new approach to therapy of diseases with inflammatory pathogenesis, and of septic shock.

ANTI-INFLAMMATORY DRUGS

Currently used anti-inflammatory drugs include glucocorticoids and cyclooxygenase

Targeting of Drugs 4, Edited by G. Gregoriadis *et al.*
Plenum Press, New York, 1994

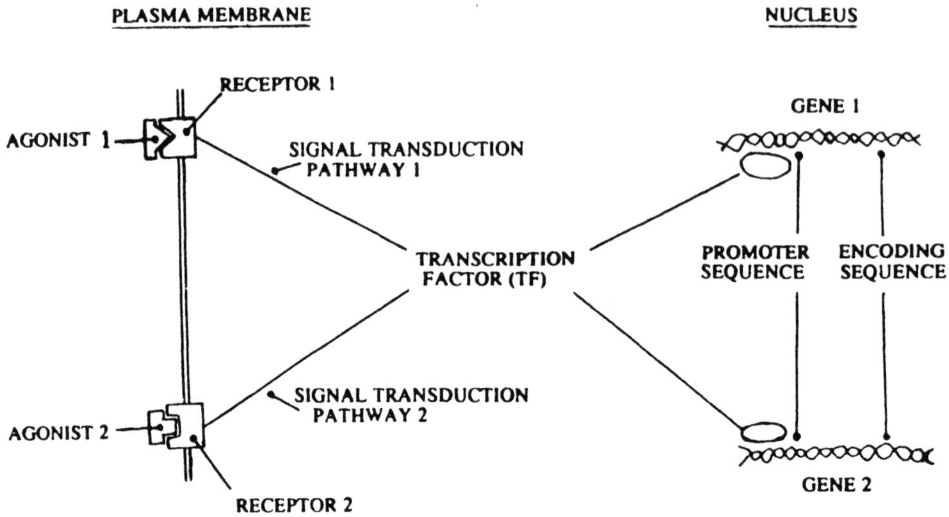

Fig.1. Diagram showing the strategic position of transcription factors. By inhibiting their activation, expression of several genes can be blocked

inhibitors. They have limiting side effects such as the widespread metabolic effects on bone and other target tissues of glucocorticoids and the gastrointestinal erosion produced by nonsteroidal anti-inflammatory drugs. There is need for anti-inflammatory drugs with novel modes of action. Our strategy has been to identify small molecules that can inhibit the expression of genes for pro-inflammatory cytokines and concurrently augment expression of the interleukin-1 receptor antagonist. Cytokines with pro-inflammatory and catabolic effects include TNF-α and IL-1β. These cytokines contribute to the pathogenesis of inflammation, as well as cartilage and bone destruction, by several mechanisms (Arend and Dayer, 1990; Dinarello, 1991; Beutler, 1992). TNF-α and IL-1β induce the expression on endothelial cells of adhesion molecules required for recruitment of leukocytes into inflammatory sites (Bevilacqua et al, 1985). The cytokines induce the production of PGE_2 by synovial fibroblast-type cells (Dayer et al, 1986) and of PGI_2 by endothelial cells (Rossi et al, 1985); these prostaglandins are vasodilators and co-mediators of pain and increased vascular permeability. The same cytokines induce the production by chondrocytes of neutral metalloproteinases that can degrade cartilage matrix (Schnyder et al, 1987), and they induce resorption of bone (Macdonald and Gowan, 1992); these processes contribute to joint erosion in patients with rheumatoid arthritis (RA). The role of IL-6 in the pathogenesis of RA is less well defined: however the cytokine is a co-factor for T-lymphocyte differentiation (Takai et al, 1988) and in the production of immunoglobulins by B-lymphocytes in RA synovial tissue (Nawata et al, 1989). Activated T-lymphocytes (Gaston et al, 1988), and immune complexes (Nardella et al, 1983) are thought to contribute to pathogenesis of RA. Moreover IL-6 can act synergistically with IL-1 in augmenting bone erosion (Ishmi et al, 1990).

Administration of antibodies neutralizing TNF-α (Elliott et al, 1993) or of the interleukin-1 receptor antagonist (IL-ra) (Lebsack et al, 1991) are reported to improve the clinical condition of patients with RA, confirming the importance of TNF-α and IL-1 pathogenesis. However, the antibody and receptor antagonist are expensive and difficult to administer on a continous basis. Antibodies against TNF-α (Fong et al, 1989) and IL-1ra (Fisher et al, 1992) produce only marginal protective effects in a non-human primate model of endotoxic shock, suggesting that inhibiting the action of individual cytokines is insufficient for good therapy. For these reasons our research has been directed towards the identification of a small molecule that can inhibit, in a co-ordinate fashion, the production by monocyte-macrophage lineage cells of TNF-α, IL-1β and IL-6. By "small molecule" we mean a naturally occurring or synthetic organic compound of relative molecular mass 600 or less, which is not a peptide, lipid, oligosaccharide or oligonucleotide.

The potential therapeutic utility of a drug inhibiting cytokine production in a co-ordinate manner is clear: the question arises whether such a strategy is feasible. We chose to investigate antioxidants because of evidence that compounds of this general class can inhibit the activation of transcription factors, NFκB (Schrek et al, 1991; Israel et al, 1992) and AP-1 (Devary et al, 1991; Datta et al, 1992), which are required for induced expression of cytokine genes (Shakhov et al, 1990; Hiscott et al, 1993). Furthermore, it had been reported that the antioxidants butylated hydroxyanisole (Chaudhri and Clark, 1989), N-acetyl cysteine (Peristeris et al, 1992) and probucol (Akeson et al, 1991) suppress the production of TNF-α and IL-1β by mouse peritoneal and splenic macrophages and human promonocytic cell lines. These are examples of small antioxidant molecules inhibiting the production of individual cytokines. We have investigated the problem more systematically, comparing effects of different types of antioxidants on the production of TNF-α, IL-1β and IL-6 in cultured human monocytes activated by LPS and in other ways.

MATERIALS AND METHODS

All the procedures and reagents have been described (Eugui et al, 1994). In brief, peripheral blood mononuclear cells (PBM) form normal human volunteers were separated by Ficoll-paque gradient and enriched in monocytes by adherence to plastic or by rosetting out T-lymphocytes with aminoethylthiouronium bromide-treated sheep erythrocytes. PMB were cultured either in polypropylene tubes ($5x10^5$/1 ml/tube) or in 12 well plates (1 ml/well) in RPMI-1640 supplemented with 5% human AB serum. Cultures were stimulated with lipopolysaccharide (LPS, 20 μg/ml), with or without drugs, and incubated overnight at 37°C in 5% CO_2. The supernatants were harvested and the cell pellets were resuspended in RPMI medium, and freeze-thawed five times to prepare lysates. All samples were stored frozen (-20°C) until ready for cytokine determination by ELISA assays.

To test the effect of drugs on cytokine production in vivo, two assays were used. Mouse peritoneal macrophages were elicited by intraperitoneal injection of thioglycolate medium (40.5g/l, 1 ml), and after 3 days the animals were challenged intraperitoneally with

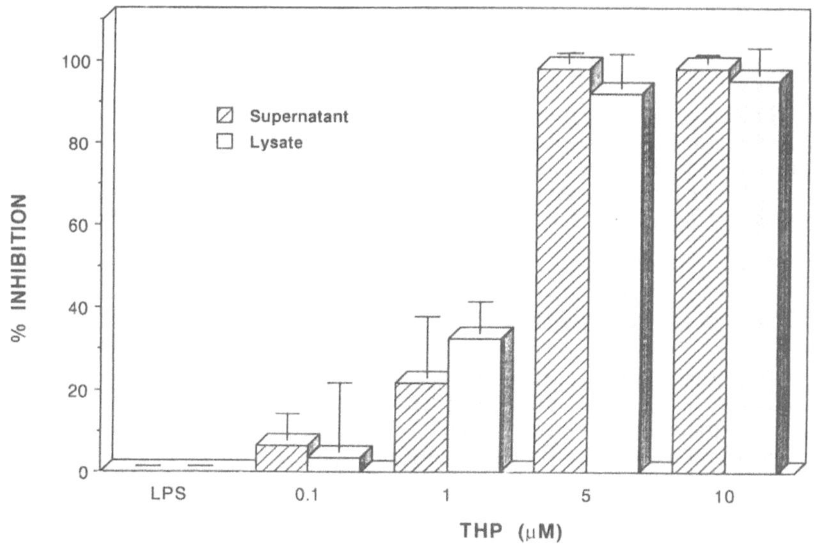

Fig.2. Dose-dependent inhibition by tetrahydropapaveroline (THP) of the production of IL-1β in human peripheral blood mononuclear cells activated by LPS. Means and standard errors of intracellular (lysate) and extracellular cytokines are shown.

Table 1. Antioxidants Vary Widely in Potency as Inhibitors of Cytokine Formation

High activity	(IC_{50} μM)
butylated hydroxyanisole (BHA)	2.9
tetrahydropapaveroline (THP)	1.0
apomorphine	2.6
norapomorphine	1.6
nordihydroguaiauretic acid (NDGA)	1.3
mepacrine	3.0
Low activity (insignificant inhibition in the range 50-200 μM)	
ascorbic acid	
α-tocopherol	
mannitol	
trolox	
butylated hydroxytoluene (BHT)	
quercetin	
N,N'-diphenyl-p-phenylene diamine	
zileuton (5-lipoxygenase inhibitor)	

10-15 μg LPS/mouse. Drugs were administered before and after LPS challenge (-18 hr, -30 min and +2 hr). Four hours after LPS injection the animals were sacrificed and peritoneal exudate cells were collected. Replicate samples ($2x10^6$ cells) were either cultured overnight and supernatants collected for IL-1β determination, or they were frozen and thawed immediately after harvesting to prepare cell lysates. All samples were stored at -20°C until ready to measure IL-1β by ELISA.

In the other assay, circulating levels of TNF-α and IL-1β were measured in plasma collected at intervals following a lethal dose of LPS. Mice were treated with test compounds 30 min before LPS injection (200 μg/mouse, i.p.) and blood samples were collected 1½ hr after LPS administration for TNF-α assay. Mice were given a second dose of drug 2 hr after LPS injection, and blood samples were collected 2 hr later for IL-1β assay.

EFFECTS OF ANTIOXIDANTS ON IL-1β SYNTHESIS

In a first screening, several compounds were tested for capacity to inhibit the production of IL-1β in LPS-stimulated human peripheral blood mononuclear cell (PBM) cultures. Tetrahydropapaveroline (THP), a tetrahydroisoquinoline derivative, was found to be a potent inhibitor of IL-1β production (IC_{50} about 1.5 μM, Fig. 2). The compound decreased intracellular IL-1β as well as cytokine in the culture medium. Other compounds structurally related to THP, 10,11-dihydroxyaporphine (DHA) and norapomorphine, were also tested and found to inhibit IL-1 production efficiently (Table 1). The S(+) and R(-) stereoisomers of DHA were equipotent in this assay, showing separation from dopamine agonist activity; only the R(-) form is a dopamine agonist.

Various antioxidants were tested for capacity to inhibit the production of IL-1β in LPS-activated human PBM. Several moderately lipophilic antioxidants, including butylated

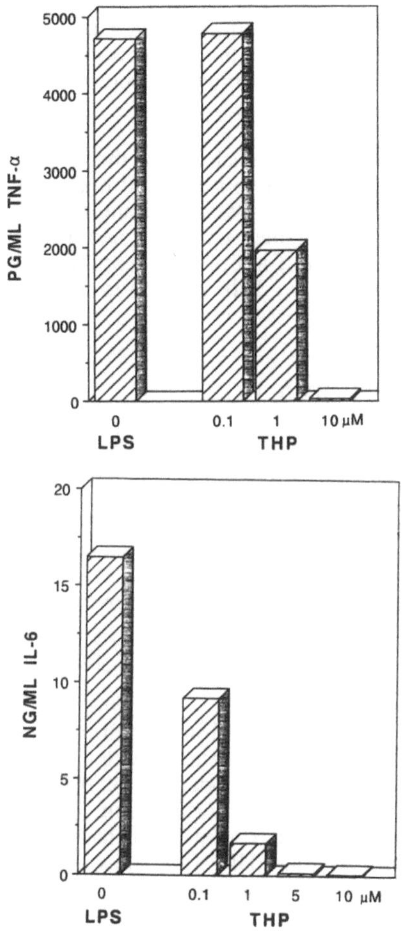

Fig.3. Dose-dependent inhibition by tetrahydropapaveroline (THP) of the production of TNFα and of IL-6 by human peripheral blood mononuclear cells activated by LPS. Cytokines in the supernatant were assayed

hydroxyanisole (BHA) and nordihydroguaiauretic acid (NDGA), were found to be potent inhibitors of IL-1β production (IC_{50} 4 μM or lower, Table 1). NDGA is an inhibitor of 5-lipoxygenase, as well as of lipid peroxidation. However, another redox 5-lipoxygenase inhibitor, Zileuton, did not affect IL-1β formation, suggesting that 5-lipoxygenase products are not involved in the signal transduction system leading to cytokine production in LPS-activated monocytes. The antimalarial drug and phospholipase inhibitor, mepacrine, was also an effective inhibitor of cytokine production.

In contrast the more hydrophilic antioxidants tested, ascorbic and trolox, had no effect on IL-1β production in concentrations up to 200 μM (Table 1). Mannitol, a hydroxyl radical scavenger, was inactive at 100 μM concentrations. The same was true of the physiological lipophilic antioxidant α-tocopherol, as well as some classical antioxidants - butylated hydroxytoluene (BHT), quercetin and N,N'-diphenyl-p-phenylene diamine. N-acetylcysteine had some inhibitory effect (IC_{50} 42 mM) but this was much lower than that of several lipophilic antioxidants (Table 1).

The question arose whether the inhibitory effect of THP is selective for the signalling pathway initiated by LPS or applies also to other highly effective inducers, such as silica and Staphylococcus aureus Cowan I (Pansorbin). THP was found to be equally potent in the inhibition of IL-1β production using a weaker inducer, Zymosan. These observations demonstrate that THP can inhibit the production of IL-1β by human PBM stimulated with several inducers. Similar observations were made with other antioxidants (not shown).

THP was also used to analyse the reversibility of the inhibitory effect. Adherent PBM were cultured with LPS and THP for two hours and repeatedly washed. Fresh medium containing the same amount of LPS was added and the cultures were incubated overnight. Duplicate cultures were maintained with LPS and THP for the same period without washing. Removal of the drug did not reverse its inhibitory effect (not shown). Thus, the inhibition of IL-1β production by THP appears to be irreversible, at least during the 18 hours studied.

EFFECTS ON TNF-α AND IL-6 PRODUCTION

THP, 10,11-dihydroxyaporphine and NDGA were further tested for effects on the production of TNF-α and IL-6 by LPS-stimulated human PBM. All three compounds were found to be approximately equipotent as inhibitors of the production of the three cytokines (Fig. 3). A non-specific effect on protein synthesis was excluded: THP did not decrease the incorporation of ^{3}H leucine into monocytes or U-937 promonocytic cells at concentrations up to 50 μM. The other agents tested required more than 20-fold greater concentrations to inhibit protein synthesis than cytokine production (not shown).

CELL TYPE SELECTIVITY

To ascertain whether this effect is exerted in all cell types, human dermal fibroblasts (HFF) and fibroblast-type cells from synovial tissue of patients with rheumatoid arthritis were studied. In these cells IL-1 induces the production of IL-6. THP, BHA, 10,11-dihydroxyaporphine (apomorphine) and norapomorphine, in concentrations higher than are required to inhibit totally the production of IL-6 in PBM, had no significant effect on IL-6 production in synovial tissue fibroblasts and HFF (not shown). Thus antioxidant-mediated inhibition of cytokine production does not occur in all cell types.

MOLECULAR BIOLOGY

Measurement of messenger RNAs for cytokines by Northern-blot hybridization showed that THP and other antioxidants markedly decreased steady-state levels of TNF-α, IL-1β and IL-6 mRNAs in LPS-activated human monocytes (Eugui et al, 1994). Nuclear transcription assays demonstrated that in unstimulated human monocytes the rate of transcription of the IL-1β gene is very low. It is markedly increased following LPS stimulation of the cells, but THP dramatically suppresses transcription (not shown).

Concentrations of transcription factors in nuclei isolated from monocytes were assayed by gel-retardation assays, using labelled oligonucleotides corresponding to consensus sequences of cognate recognition sites. In unstimulated cells NFκB and AP-1 activity in nuclei was found to be relatively low, and both were increased following LPS stimulation. In the presence of THP and other antioxidants, NFκB and AP-1 activities in the nuclei were very low, even after LPS stimulation. If THP was added to isolated nuclei, it did not inhibit binding of transcription factors to recognition sequences. In cells treated with THP, no effect on several other transcription factors (SP-1, CRE, CTF/NF1, OCT) could be demonstrated (not shown). Hence the antioxidants prevent the activation of some transcription factors, not the binding of these proteins to DNA.

IN VIVO EFFECTS

Some of the compounds in Table 1 were selected for analysis of in vivo effects on cytokine production. Two different assays were used, as described in Methods. In the first assay IL-1β production was measured using elicited peritoneal cells from mice, four hours after LPS challenge. The cytokine was measured in lysates of cells directly recovered from the peritoneal cavity (cell-associated IL-1β), as well as in the supernatants of cells after overnight culture (released IL-1β). Mice were pretreated with two doses of either THP or 10,11-dihydroxyaporphine (apomorphine, 50 mg/kg/dose) before LPS challenge (15 μg/mouse). A third dose of compound was administered two hours later. Both compounds suppressed the production of IL-1β by 53 to 86% (not shown).

In the second assay, TNF-α and IL-1β were measured in serum following a lethal challenge with LPS (200 μg/mouse). Levels of TNF-α in circulating blood peak 1½ hours and those of IL-1β 4 hours after LPS injection. Subcutaneous administration of a single dose of apomorphine, (100 mg/kg), given 30 minutes before challenge, inhibited TNF-α production by 95%. Before IL-1β peak level, mice were given a second dose of apomorphine (50 mg/kg), 2 hours after LPS and 2 hours before bleeding. This treatment reduced the circulating levels of IL-1β by 88% (Fig. 4). Thus in vivo cytokine production is strongly inhibited by THP and by apomorphine.

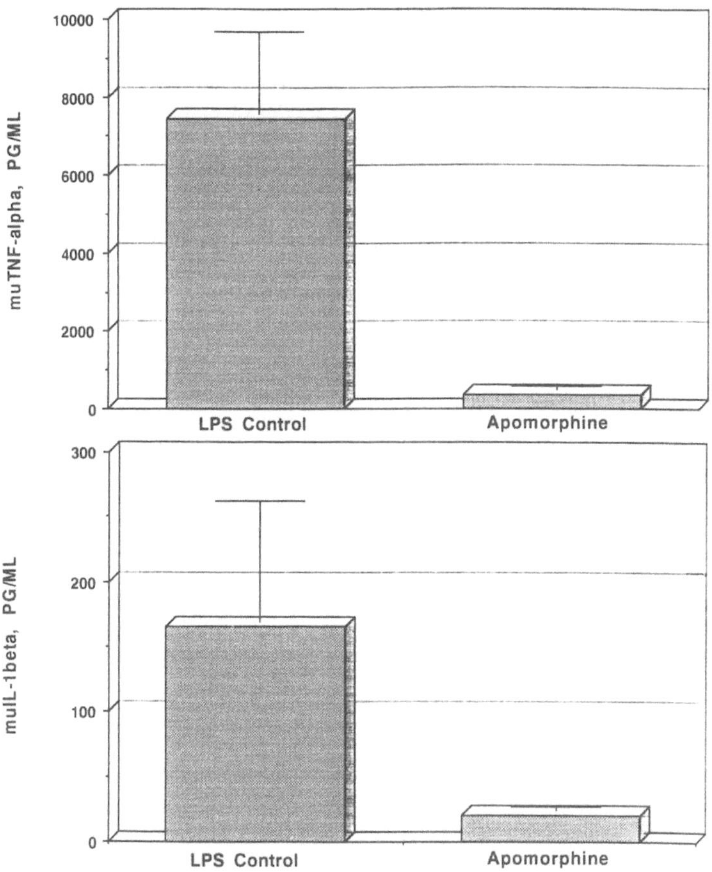

Fig.4 Apomorphine (10,11-dihydroxyaporphine) treatment of mice markedly decreases levels of TNFα and of IL-1β in the circulation of mice challenged with LPS. See methods for technical details

DISCUSSION

We have found that several antioxidants are potent inhibitors of the production of TNF-α, IL-1β and IL-6 in LPS-stimulated human PBM. The same concentrations of antioxidants were equally effective in the inhibition of cytokine production in PBM irrespective of the stimulus (LPS, staphylococci, silica and zymosan). Effects of antioxidants are not due to overall inhibition of protein synthesis and are gene selective. The antioxidants do not affect transcription of the β-actin and c-jun genes (Eugui et al, 1994), and actually increase the production of the interleukin-1 receptor antagonist, as well as of lysozyme and of lysosomal enzymes in PBM (Waters et al, 1993). Inhibition of cytokine production occurs at the level of transcription, and may be related to the observed inhibition of the activation of the transcription factors NFκB and AP-1.

The effects of the antioxidants are also cell-type selective. Antioxidants which are potent inhibitors of IL-6 production in PBM do not inhibit production of the same cytokine in dermal or synovial fibroblasts. The mechanisms stimulating IL-6 production are diverse (Hirano et al, 1990), and presumably the transcription complexes used in fibroblasts are less susceptible to redox regulation than those in PBM. Other cell types should be studied, but our findings raise the possibility that selected antioxidants may be able to suppress the formation of pro-inflammatory cytokines in cells of the monocyte-macrophage lineage without having major effects in other cell types which could lead to unwanted side effects.

Some of the compounds active in cultured cells also show activity in vivo. THP and dihydroxyaporphine were found to inhibit strongly the production of TNF-α and IL-1β in mice challenged with LPS. The antioxidants and mode of administration were not optimized, but the findings are encouraging. They are consistent with reports that 21-octadecylascorbic acid reduces mortality in a mouse model of endotoxemia (Nonaka et al, 1990) and that the 21-aminosteroid tirilazad mesylate, which has antioxidant activity (Braughler et al, 1987), mitigates clinical manifestations of endotoxaemia in neonatal calves (Rose and Semrad, 1992).

Our observations show that the genes for TNF-α, IL-1β and IL-6 function as a casette, being co-ordinately expressed when monocytes are stimulated by LPS or staphylococci. This expression is inhibited by some antioxidants. Remarkably, we have found that the same antioxidants augment the expression of the IL-1ra gene and IL-1ra production in cultures of PBM (Waters et al, 1993). It has been postulated that a balance between the production of pro-inflammatory cytokines, such as TNF-α and IL-1β, and of IL-1ra, plays an important role in determining the course of rheumatoid arthritis (Arend and Dayer, 1990). The same may be true of the arthritis associated with Lyme disease, caused by the bacterium Borrelia burgdorferi. Patients with high concentrations of IL-1ra and low concentrations of IL-1β in synovial fluid had rapid resolution of attacks of arthritis, whereas patients with the reverse pattern of cytokine concentrations had long intervals to recovery (Miller et al, 1993). Using drugs to manipulate the balance of production of cytokines and IL-1ra in such a way as to have anti-inflammatory effects is a novel therapeutic strategy.

The observations now reviewed show that small molecules, which enter cell easily (unlike oligonucleotides), can suppress the activation of transcription factors and inhibit the expression of sets of genes in desired cell types in a therapeutically useful manner. They validate the concept that transcription factors can be used as therapeutic targets.

ACKNOWLEDGMENTS

We are indebted to several colleagues who contributed to the investigations here reported: Barbara DeLustro, Mariola Ilnicka, Simon Lee and Robert Wilhelm. We are also grateful to Linda Miencier for skillful word processing.

REFERENCES

Akeson, A.L., Woods, C.W., Mosher, L.B., Thomas, C.E. and Jackson, R.L., 1991, Inhibition of IL-1β expression in THP-1 cells by probucol and tocopherol, Atherosclerosis, 86:261

Arend, W.P. and Dayer, J.-M., 1990, Cytokines and cytokine inhibitors or antagonists in rheumatoid arthritis, Arthritis & Rheumatism, 33:305

Beutler, B. (Ed.), 1992, "Tumor necrosis factors: the molecules and their emerging role in medicine", Raven Press, New York

Bevilacqua, M.P., Pober, J.S., Wheller, M.E., Cotran, R.S. and Gimbrone, M.A., 1985, Interleukin-1 acts on cultured human vascular endothelium to increase the adhesion of polymorphonuclear leukocytes, monocytes, and related leukocyte cell lines, J.Clin.Invest., 76:2003

Braughler, J.M., Pregenzer, J.F. and Chase, J.L., 1987, Novel 21-aminosteroids as potent inhibitors of iron-dependent lipid peroxidation, J.Biol.Chem., 262:10438

Chaudhri, G. and Clark, J.A., 1989, Reactive oxygen species facilitate the in vitro and in vivo lipopolysaccharide-induced release of tumor necrosis factor, J.Immunol., 143:1290

Datta, R., Hallahan, D.E., Kharbanda, S.M., Rubin, E., Sherman, M.L., Huberman, E., Weichselbaum, R.R. and Kufe, D.W., 1992, Involvement of reactive oxygen intermediates in the induction of c-jun gene transcription by ionizing radiation, Biochemistry, 31:8300

Dayer, J.-M., de Rochemonteix, B., Burners, B., Demczuk, S. and Dinarello, C.A., 1986, Human recombinant interleukin-1 stimulates collagenase and prostaglandin E_2 production by human synovial cells, J.Clin.Invest., 77:645

Devary, Y., Gottlieb, R.A., Lau, L.F. and Karin, M., 1991, Rapid and preferential activation of the c-jun gene during the mammalina UV response, Mol.Cell Biol., 11:2804

Dinarello, C.A., 1991, Interleukin-1 and interleukin-1 antagonism, Blood, 77:1627

Elliott, M.J., Maini, R.N., Feldmann, W., Williams, R.O., Brennan, F.M. and Chu, C.Q., 1993, Treatment of rheumatoid arthritis with chimeric monoclonal antibodies to TNFα: safety, clinical efficacy and control of the acute-phase response, J.Cell Biochem., 17B Suppl.:145

Eugui, E.M., DeLustro, B., Rouhafza, S., Ilnicka, M., Lee, S.W., Wilheim, R. and Allison, A.C., 1994, Some antioxidants inhibit, in a coordinate fashion, the production of tumor necrosis factor-α, IL-1β and Il-6 by human peripheral blood mononuclear cells, Int.Immunol., 6:409

Fischer, E., Marano, M.A., van Zee, K., Rock, C.S., Hawes, A.S., Thompson, W.A., de Forge, L., Kenney, J.S., Remick, D.G., Blaedow, D.C., Thompson, R.C., Lowry, S.F. and Moldawer, L.L., 1992, Interleukin-1 receptor blockade improves survival and hemodynamic performance in Escherichia coli septic shock, but fails to alter host responses to sublethal endotoxemia, J.Clin.Invest., 89:1551

Fong, Y., Tracey, K.J., Moldawer, L.L., Hesse, D.G., Manogue, K.B., Kenney, J.S., Lee, A.t., Kuo, G.C., Allison, A.C., Lowry, S.F. and Cerami, A., 1989, Antibodies to cachetin/tumor necrosis factor reduce interleukin-1β and interleukin-6 appearance during lethal bacteremia, J.Exp.Med., 170:1627

Gaston, J.S.H., Strober, S., Solvera, J.J., Gandour, D., Lane, N., Schurman, D., Hoppe, R.T., Chin, R.C., Eugui, E.M., Vaughan, J.H. and Allison, A.C., 1988, Dissection of the mechanisms of immune injury in rheumatoid arthritis using total lymphoid irradiation, Arthritis and Rheumatism, 31:21

Hirano, T., Akira, S., Tasa, T. and Kishimoto, J., 1990, Biological and clinical aspects of interleukin-6, Immunology Today, 11:443

Hiscott, J., Merois, J., Garoufalis, J., D'Addario, M., Roulston, A., Kwan, I., Pepin, N., Lacoste, J., Nguyen, H., Bensi, G. and Fentin, M., 1993, Characterization of a functional NF-κB site in the human interleukin 1β promoter: evidence for a positive autoregulatory loop, Mol.Cell Biol., 13:6231

Ishimi, Y., Miyaura, C., Jin, C.H., Akatsu, T., Abe, E., Nakamura, Y., Yamaguchi, A., Yoshiki, S., Matsuda, T., Hirano, T., Rishimoto, R. and Suda, T., 1990, IL-6 is produced by osteoblasts and induces bone resorption, J.Immunol., 145:3297

Israel, N., Gougerat-Pocidalo, M.-A., Aillet, F. and Virelizier, J.-L., 1992, Redox status influences constitutive or induced NFκB translocation and HIV long terminal repeat activity in human T and monocytic cell lines, J.Immunol., 149:3386

Lebsack, M.E., Paul, C.C., Bloedow, D.C., Burch, F.X., Sade, M.A., Chase, W. and Catalano, M.A., 1991, Subcutaneous IL-1 receptor antagonist in patients with rheumatoid arthritis, Arthritis Rheum., 34(9, Suppl.):S45

MadDonald, B.R. and Gowan, M., 1992, Cytokines and bone, Br.J.Rheumatol., 31:149

Mannik, M. and Nardella, F.A., 1983, Self-associating IgG rheumatoid factors, in: "New Horizons in Rheumatoid Arthritis", Y. Shiokawa, T. Abe and Y. Yamauchi (Eds.), Excerpta Medica, Amsterdam

Miller, L.C., Lynch, E.A., Isa, S., Logan, J.W., Dinarello, C.A. and Steere, A., 1993, Balance of synovial fluid IL-1β receptor antagonist and recovery from Lyme arthritis, Lancet, 341:146

Nawata, Y., Eugui, E.M., Lee, S.W. and Allison, A.C., 1989, IL-6 is the principal factor produced by synovial of patients with rheumatoid arthritis that induces B-lymphocytes to secrete immunoglobulin, Ann.N.Y.Acad.Sci., 557:230

Nonaka, A., Manabe, T. and Tobe, T., 1990, Effect of a new synthetic free radical scavenger, 2-octadecyl ascorbic acid, on the mortality in mouse endotoxemia, Life Sciences, 47:1933

Peristeris, P., Clark, B.D., Gatti, S., Faggiani, R., Mantovani, A., Mengozzi, M., Orencole, S., Sironi, M. and Chezzi, P., 1992, N-acetylcysteine and glutathione as inhibitors of tumor necrosis factor production, Cell Immunol., 140:390

Rose, M.L. and Semrad, S.D., 1992, Clinical efficacy of tirilazad mesylate for treatment of endotoxemia in neonatal calves, Am.J.Vet.Res., 53:2305

Rossi, V., Brevario, F., Ghezzi, P., Dejana, E. and Mantovani, A., 1985, Prostagladin induced in vascular endothelial cells by interleukin-1, Science, 229:174

Schnyder, J., Payne, T. and Dinarello, C.A., 1987, Human monocytes or recombinant interleukin-1's are specific for the secretion of a metalloproteinase from chondrocytes, J.Immmunol., 138:496

Schrek, R., Rieber, P. and Bauerle, P.A., 1991, Reactive oxygen intermediates as apparently widely used messengers in the activation of NEκB transcription factor and HIV-1, Embo.J., 10:2247

Shakhov, A.N., Collart, M.A., Vassali, P., Nedospasov, S.A. and Jongeneel, C.V., 1990, Kappa B-type enhancers are involved in lipopolysaccharide-mediated transcriptional activation of the tumor necrosis factor alpha gene in primary macrophages, J.Exp.Med., 171:35-47

Takai, Y., Wong, G.G., Clark, S.C., Burkoff, S.J. and Herrmann, S.H., 1988, B cell stimulatory factor-2 is involved in the differentiation of cytotoxic T lymphocytes, J.Immunol., 140:508

Waters, R.V. and Allison, A.C., 1993, Some long-acting antirheumatic drugs and antioxidants induce differentiation of macrophage lineage cells and augment expression of the IL-1 receptor antagonist IL-1ra, Ann.N.Y.Acad.Sci, 696:185

DESIGN, BIOLOGICAL PROPERTIES AND DELIVERY OF ANTISENSE OLIGONUCLEOTIDES

Jack S. Cohen

Pharmacology Department, Georgetown University
Washington DC 20007, USA

INTRODUCTION

Antisense is the term used to describe the application of a nucleic acid of complementary sequence to the sense sequence of a messenger RNA, in order to down-regulate gene expression by intervening in translation (Fig. 1). The antisense molecule can itself be an mRNA, and this has been found to occur as a natural regulatory mechanism (Murray, 1992). Or with more relevance to therapeutic approaches, the antisense molecule can be an oligonucleotide, that is synthesized chemically. Because it is easier to synthesize oligodeoxynucleotides than oligoribonucleotides using an automated procedure, these are the derivates that are used for most antisense applications. Since the area of antisense has been widely reviewed (Ghosh and Cohen, 1992; Murray, 1992), this article will contain only a brief introduction to the topic, and will then focus on the delivery of oligonucleotides as antisense compounds.

Although natural phosphodiester oligodeoxynucleotides have been shown to exhibit antisense effects in down-regulating or inhibiting the expression of a targeted gene, they require high concentrations to be effective. This has been shown to be due to the fact that oligodeoxynucleotides are susceptible to degradation by nucleases that occur ubiquitously in nature (Shaw et al, 1991; Akhtar et al, 1991; Eder et al, 1991). Analogs of phosphodiesters that are resistant to the nucleases have been developed as alternate antisense compounds, particularly those phosphate analogs in which one of the non-bridging oxygen atoms is substituted by another group, thus retaining the major conformational properties of DNA, including the ability to hybridize with the target mRNA sequence. The methylphosphonate (with CH_3-P) and phosphorothiote (with S-P) are the most widely used analogs (Fig. 2). One difference between these analogs is that the former is neutral, while the latter is negatively charged like the phosphate original. These conservatively substituted analogs have been termed first generation analogs.

It is generally considered best to use an oligomer of about 15-20 nucleotides in length because this corresponds both to the approximate length of a single target site in an mRNA, and to the length required for effective hybridization. Shorter sequences (<10) can bind to more complementary sequences, and longer ones (>20) can contain sequences that bind to shorter sequences in the RNA. Clearly it is essential in doing antisense experiments to establish sequence specificity using controls of sense or random sequences. In several cases strong protein binding has been found for oligonucleotides and analogs, and this might be the basis for a different therapeutic modality in which proteins are the targets for

Targeting of Drugs 4, Edited by G. Gregoriadis *et al.*
Plenum Press, New York, 1994

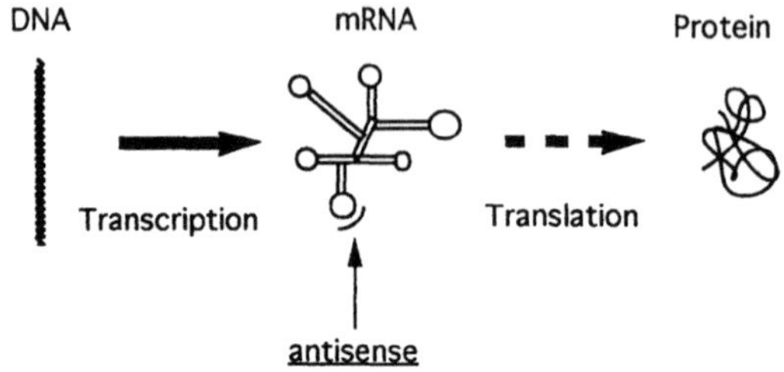

Fig. 1 Schematic representation of the processes of transcription and translation involved in gene expression. Note that an antisense oligonucleotides can bind to the mRNA and cause down-regulation of translation

the oligomer (Tuerk and Gold, 1990). However, as far as antisense is concerned this is an unwelcome side-effect. In the case of phosphorothioate oligomers, binding to polymerases that contain oligomer binding sites, has been shown (Gao et al, 1992), including reverse transcriptase of HIV-1 (Majumdar et al, 1989) and herpes simplex viruses (Gao et al, 1989). Other non-specific effects of oligonucleotides may results from non-specific cleavage of RNA due to RNase H activity (Giles and Tidd, 1992; Ghosh et al, 1992). However, while these non-specific effects reduce the specificity of the oligomers as drugs, they do not prevent antisense inhibitions from being observed, and they only emphasize the need to improve the efficacy of the antisense molecules.

DESIGN OF ANTISENSE COMPOUNDS

The first generation of antisense oligodeoxynucleotide analogs that have been developed in order to possess nuclease resistance have a single non-bridging oxygen atom substitution on the phosphodiester group. These are the methylphosphonate (I) and the phosphorothioate (II) analogs. Other analogs include the phosphoramidate (III) and phosphotriester (IV). The latter is relatively unstable due to the presence of esterases. A major difference between I and II is the absence or presence of negative charge. At present both I and II are being developed for clinical trials, although there are many more

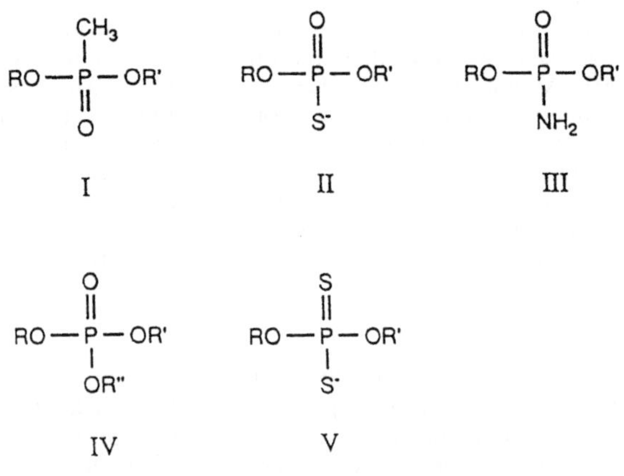

Fig. 2 Structures of modified phosphodiester groups: I, methylphosphonate; II, phosphorothioate; III, phosphoramidate; IV, phospho-triester; V, phosphorodithioate

Table 1. Reported Applications of Phosphorothioate Oligodeoxynucleotides

ANTIVIRAL	
Human Immundeficiency Virus	Matsukura et al, 1987; Agrawal et al, 1989; Matsukura et al, 1989; Kemal et al, 1991; Mag et al, 1991; Kinchington et al, 1992; Lisziewicz et al, 1992
Influenza Virus	Leiter et al, 1990
Hepatitis Virus	Goodarzi et al, 1990; Offensperger et al, 1993*; Wu & Wu, 1992
Human Papilloma Viruses	Storey et al, 1991; Cowsert et al, 1993
Herpes Simplex Virus	Kmetz et al, 1991; Draper et al, 1990
ANTICANCER	
c-*myc*/N-*myc*	Neckers et al, 1992; Watson et al, 1992
ras	Chang et al, 1989
p53	Bayever et al., 1992*
c-*myb*	Ratajczak et al, 1992*
*bcl*2	Reed et al, 1990
MDR	Jaroszewski et al, 1990; Yokozaki et al, 1993
c-*erb* B-2	Schlingensiepen & Brysch, 1992
IL-4	Hikida et al, 1992
ICAM-1	Chiang et al, 1992+
TGF-β	Hatzfield et al, 1991
NF-κB	Kitajima et al, 1992*
p120	Perlaky et al, 1993+
b-FGF	Murphy et al, 1992
OTHER	
Cardiovascular Occlusion (Restenosis/*myc*)	Simons et el, 1992*; Ebbecke et al, 1992
Antimalarial (*Plasmodium Falciparum*)	Rapaport et al, 1992
Develomental (VG1/ *Xenopus* oocytes)	Woolf et al, 1990
Integrin (neural crest attachment)	Lallier et al, 1993

**in vivo* studies, all others *in vitro*

\+ Using DOTMA

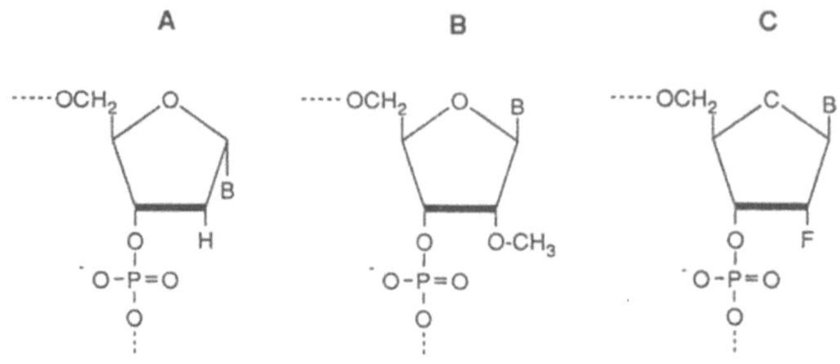

Fig. 3 Structures of modified sugar analogs: A, α-analog; B, 2′-0-methyl; C, 2′- fluoro

biological applications of phosphorothioates at this juncture (Table 1). Phosphodithioates (V) have also been developed to over-come the problem of chirality in that unlike the singly substituted analogs they do not have four different substituents at each phosphorus atom (Marshall and Caruthers, 1993).

Apart from modifications on the phosphate group another area of modifications is the sugar moiety. Deoxyribose analogs include the α-anomer (Fig. 3), in which the glycosidic bond has the opposite configuration to the natural β-anomer, and other 2'-substituted analogs, such as 2'-O-methyl and 2'-fluoro (Fig. 3). These 2'-substituted analogs are also used in relation to ribozymes (Fig. 4), that are antisense ribo-oligomers that contain an active site region that cleaves a complementary RNA (Pieken et al, 1991). Ribozymes can be produced endogenously in cells upon transfection with a suitable vector (plasmid or retrovirus), but can also be added exogenously, as are synthetic antisense oligodeoxynucleotides. Although ribozymes have intrinsic catalytic activity against RNA, while deoxy-oligomers do not, there are several reasons why they might not be as suitable as antisense drugs. First, in order to retain their activity they must retain some ribonucleotides, and these can be degraded by ubiquitous RNases. Second, their catalytic activity is very slow (McSwiggen and Cech, 1989), and so far there are very few reports of in vitro activity of ribozymes (Cooney et al, 1988; Scanlon et al, 1991; Sarver et al, 1990). However, ribozymes offer great promise as understanding of their cellular processing increases (Cech, 1993).

The current phase of the antisense field involves the consolidation of the application of first generation analogs, and the development of second and third generation analogs. Second generation analogs can be defined as those in which the phosphodiester group is completely replaced by a different chemical grouping. Third generation analogs can be defined as those in which both the phosphodiester and deoxyribose are replaced, involving a significant re-designing of the DNA molecule. An example of this is the so-called peptide

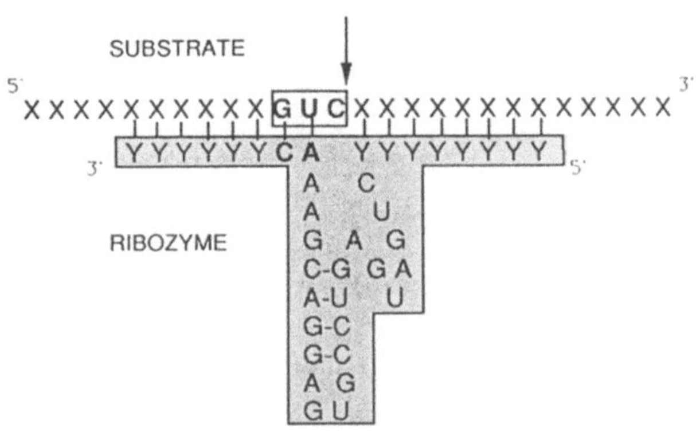

Fig. 4 Representation of a ribozyme. The bases Y are the antisense component, and the cleavage site next to a GUC sequence is shown by an arrow

nucleic acids (PNA)(Egholm et al, 1992), although it is not difficult to envisage that alternative molecules could be designed to bind in their place.

BIOLOGICAL PROPERTIES OF OLIGONUCLEOTIDES

One of the most common targets of an antisense sequence is the 5'-initiation codon region in the mRNA, in order to inhibit formation of the translation complex. Alternative sites include regions of supposed single stranded loops that are predicted by use of a folding computer program (Fig. 5). This can generate the lowest energy structure that has the maximum amount of base pairing, as well as other structures that are of higher energy (Zuker, 1989). Oligomers directed towards these targets can be expected to give increased inhibitory activity (Jaroszewski et al, 1993), but only when RNAse H activity is present (Toulme, 1993). Otherwise the ribosome translation complex has an unwinding activity associated with it that reduces the inhibitory activity of hybrid duplexes towards the 3'-coding region (Wickstrom, 1991). Nevertheless, sites of antisense activity have been observed located in the 3'-region of an mRNA (Chiang et al, 1991). Although RNase H activity is ubiquitous in mammalian cells, there is some question as to whether or not depending on this activity is appropriate for a drug strategy. Alternatively, it could act as a useful targeting method, if it is known that the RNAse H activity is focussed in a certain organ or tissue.

The biological role of antisense oligomers extends beyond their intended targeting of the complementary sense sequence. Protein binding of oligomers has been characterized, that as well as non-specific binding, can include specific binding to DNA binding proteins such as reverse transcriptase (Majumdar et al, 1989) and polymerases (Gao et al, 1992). This protein binding capability is being exploited in the development of active site targeting using a re-cycling approach with PCR (Tuerk and Gold, 1990). However, since this results in an oligodeoxynucleotide as a protein ligand it is beyond the focus of this article. Clearly one advantage of this approach is that an extracellular protein can be targeted, so that the problem of cellular uptake can be avoided. Nevertheless the problem arises of chance antisense activity of a sequence derived to interact with a protein active site.

Another effect can arise when the sequence used results in a sequence-specific effect other than that involving complementary base pairing. Recently we discovered a dramatic

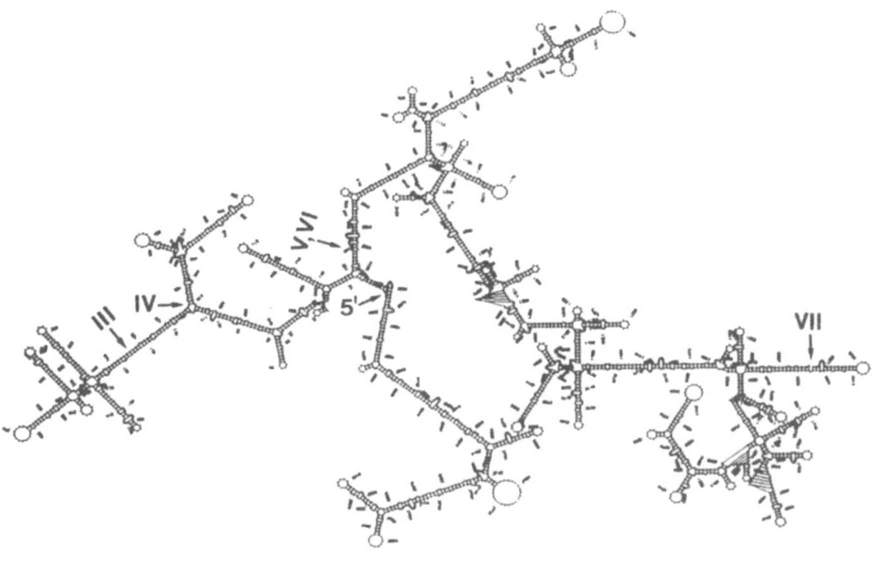

Fig. 5 Folding pattern of a mRNA, in this case for the gene mdr1, showing the first 3,000 bases

effect of oligomers on the morphology and growth of human mammary epithelial cells that apparently resulted from the presence of four contiguous G's (Yaswen et al, 1993). While the origin of this effect is unknown, we presume it results from oligomer binding to a specific receptor or surface protein. This result indicates the cautions that need to be adopted in using multiple controls before declaring an effect seen as a true antisense effect. The best indicator of a true antisense effect is the dose dependent reduction of the target mRNA or protein, although this also depends on the rate of turnover of both species (Ramanathan et al, 1993). Clearly a long-lived protein or mRNA will not make a suitable target for an antisense effect.

DELIVERY OF ANTISENSE OLIGOMERS

Cellular uptake of oligonucleotides is slow (Loke et al, 1988). Some cellular surface binding proteins have been described for oligonucleotides (Loke et al, 1989; Yakubov et al, 1989). Recent research indicates cellular uptake of methylphosphonates occurs via fluid-phase endocytosis and is distinct form the uptake pathway of phosphodiester oligomers (Shoji et al, 1991). The cellular uptake of phosphorothioate oligonucleotides has been examined in some detail (Loke et al, 1989; Marti et al, 1992; Iversen et al, 1992). It should be noted that use of fluorescent conjugated oligomers might cause alterations in the uptake and distribution due to the presence of the highly hydrophobic aromatic group, and radioactive labelling within the oligomer is to be preferred, such as with ^{35}S for phosphorothioates (Stein et al, 1990). Intracellular microinjection of oligonucleotides indicated rapid uptake into the nucleus and complexion with chromatin (Chin et al, 1990). However, in view of other results it is possible that the microinjection technique itself affects the results and is artifactual. It should be noted that while extracellular concentrations of oligomers in vitro in the micro-molar range have been commonly reported, the intracellular concentration may be only 1% of this, but it may be adequate for the inhibition of mRNAs present at nano-molar concentrations.

Delivery of oligonucleotides has been accomplished by many routes (Table 2), many of them traditional methods such as liposomes (Akhtar et al, 1991; Thierry and Dritschilo, 1992). Cationic lipids have been found to greatly enhance cellular uptake of oligonucleotides (Bennett et al, 1992). The problems of targeting of oligonucleotides is in one respect different from that of traditional drugs in that the base sequence is supposed to result in extremely specific site targeting. However, the problems of non-specific side effects are present as for any drug.

The problem of tissue targeting is also essentially the same as for other drugs.

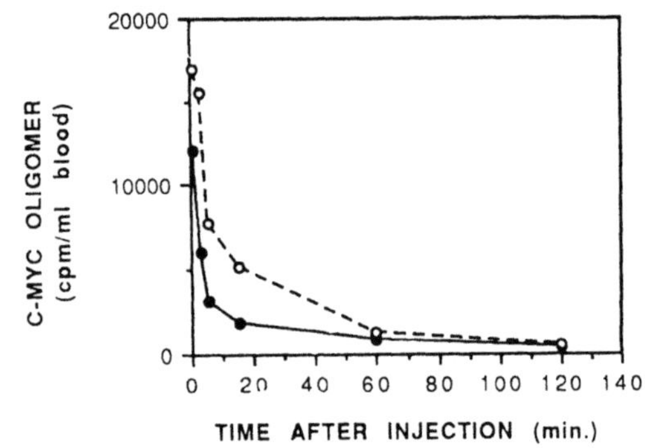

Fig. 6 Time course of ^{32}P-end-labelled c-*myc* oligo after intravenous injection into rat; P0 (solid line) and PS (dotted line)

Table 2. Delivery of Oligonucleotides

1. Direct cellular uptake
2. Liposomes/lipofection
3. Cationic lipids
4. Electroporation
5. Conjugation with lipophilic groups
6. Conjugation with peptide/lysine

Studies so far of in vivo toxicology using ^{35}S-labelled phosphorothioate oligos have shown very little indication of toxicity or organ specific damage (Agrawal et al, 1991; Iversen et al, 1993). Further, an analysis of the pharmacokinetic profiles show the common presence of two plasma excretion rates, one fast and a second slow (Inagaki et al, 1992) that is enhanced for the phosphorothioate analog (Fig. 6). The slow rate indicates that tissue concentrations are adequate for potential antisense dose effects.

Oligodeoxynucleotides have been conjugated with cholesterol in order to improve cellular uptake (Letsinger et al, 1989). Association of antisense oligomers conjugated with cholesterol have also been found to significantly prolong the half-life in plasma in rats, as well as changing the tissue distribution (de Smidt et al, 1991). The mechanism appears to be the formation of complexes with the lipoproteins, that was not found with the control unconjugated oligomer. Another group to which conjugation of oligonucleotides has been shown to improve cellular uptake significantly is poly-lysine (Lemaitre et al, 1987; Clarenc et al, 1993).

Results of infusion of fluorescently tagged phosphorothioate oligonucleotides into rat brains has shown that a concentration gradient is produced and that selective uptake into the glial cells is unexpectedly observed (Neckers et al, personal communication). These results were not found with the phosphodiester oligomers that were degraded. These and other results indicate that we know little of the in vivo selectivity of antisense oligomers and that the results of in vitro cell culture observations may not be indicative of their true in vivo properties. Thus, much more research needs to be done on the tissue distribution of such putative drugs in animals.

CONCLUSION

Several synthetic oligonucleotide analogs are currently in clinical trials; notably phosphorothioates for acute myeloblastic leukemia (Bayever et al, 1992), chronic myelogenous leukemia (Ratajczak et al, 1992), genital warts (human papilloma virus) (Cowsert et al, 1993), and AIDS (HIV) (Agrawal, 1993). These are the harbingers of more such applications to come.

The current cost of producing a gram of oligonucleotide (ca. $6,000) is still very expensive for a drug that would be needed in kilogram quantities. However, the costs of production are being reduced all the time by improvements in the materials and efficiency of the synthetic procedures. However, these compounds are the first generation oligomers. It is very likely that as the first generation analogs go through clinical trials, other analogs with more extensive alterations from the original oligonucleotide structure will be developed. These should be cheaper to produce and have increased efficacy.

Thus, while problems of specificity and delivery common to many new drugs have been experienced with this class of potential drug, there is every hope that these will not prevent the power of the antisense phenomenon from being exploited in order to counteract many human diseases of genetic origin.

REFERENCES

Agrawal, S. and Tang, J.Y., 1993, Guest editorial: GEM91, An antisense oligonucleotide nucleotide phosphorothioate as a therapeutic agent for AIDS, Antisense Res.Devel., 2:261

Agrawal, S., Ikeuchi, T., Sun, D., Sarin, P.S., Konopka, A., Maizel, J. and Zamecnik,P.C., 1989, Inhibition of human immunodeficiency virus in early infected and chronically infected cells by antisense oligodeoxynucleotides and phosphorothioate analogs, Proc.Natl.Acad.Sci.USA, 86:7790
Agrawal, S., Temsamani, J. and Tang, J.Y., 1991, Pharmacokinetics, biodistribution, and stability of oligodeoxynucleotide phosphorothioates in mice, Proc.Natl.Acad.Sci.USA, 88:7595
Akhtar, S., Basu, S., Wickstrom, E. and Juliano, R.L., 1991, Interactions of antisense DNA oligonucleotide analogs with phospholipid membranes (liposomes), Nucl.Acids Res., 19:5551
Akhtar, S., Kole, R. and Juliano, J.L., 1992, Stability of antisense oligodeoxynucleotide analogs in cellular extracts and sera Life Sci., 49:1793
Bayever, E., Iversen, P., Smith, L., Spinolo, J. and Zon, G., 1992, Guest editorial: systematic human antisense therapy begins, Antisense Res.Devel., 2:190
Bennett, C.F., Chiang, M.Y., Chan, H., Shoemaker, J.E.E., Mirabelli, C.K., 1992, Cationic lipids enhance cellular uptake and activity of phosphorothioate antisense oligonucleotides, Molec.Pharmacol., 41:1023
Cech, T., 1993, Ribozymes: Understanding and improving their activity and specificity, Intl.Conf.on Nucleic Acids Med.Applics., p.9
Chang, E.H., Yu, Z., Shinozuka, K., Zon, G., Wilson, W.D. and Strekowska, A., 1989, Comparative inhibition of ras p 21 protein synthesis with phosphorus-modified antisense oligonucleotides, Anti Cancer Drug Des., 4:221
Chiang, M.-Y., Chan, H., Zounes, M.A., Freier, S.M., Lima, W.F. and Bennett, C.F., 1991, Antisense oligonucleotides inhibit intracellular adhesion molecule 1 expression by tow distinct mechanisms, J.Biol.Chem., 266:18162
Chin, D.J., Green, G.A., Zon, G., Szoka, F. and Straubinger, R.M., 1990, Rapid nuclear accumulation of injected oligodeoxyribonucleotides, New Biologist, 2:1091
Clarenc, J.P., Degols, G., Leonetti, J.P., Milhaud, P. and Lebleu, B., 1993, Delivery of antisense oligonucleoties by poly(L-lysine) conjugation and liposome encapsulation, Anti-cancer Drug Design, 8:81
Cooney, M., Czernuszewicz, G., Postel, E.H., Flint, S.J. and Hogan, M.E., 1988, Site specific oligonucleotide binding represses transcription of the human c-myc gene in vitro, Science, 241:456
Cowsert, L.M., Fox, M.C., Zon, G. and Mirabelli, C.K., 1993, In vitro evaluation of phosphorothioate oligonucleotides targeted to the E2 mRNA of papilloma virus: potential treatment for genital warts, Antimicrobial Agents & Chemotherap., 37:171
de Smidt, P.C., le Doan, T., de Falco, S. and van Berkel, T.J.C., 1991, Association of antisense oligonucleotides with lipoproteins prolongs the plasma half-life and modifies the tissue distribution, Nucl.Acids Res., 19:4695
Draper, K.G., Ceruzzi, M., Kmetz, M.E. and Sturzenbecker, L.J., 1990, Complementary oligonucleotide sequence inhibits both Vmw65 gene expression and replication of herpes simplex virus, Antivir.Res., 13:151
Ebbecke, M., Unterberg, C., Buchwald, A., Stohr, S. and Wiegand, V., 1992, Antiproliferative effects of c-myc antisense oligonucleotide on human arterial smooth muscle cells, Basic Res.Cardiol., 87:585
Eder, P.S., DeVine, R.J., Dagle, J.M. and Walder, J.A., 1991, Substrate specificity and kinetics of degradation of antisense oligonucleotides by a 3' exocnuclease in plasma, Antisense Res.Devel., 1:141
Egholm, M., Buchardt, O., Nielsen, P.E. and Berg, R.H., 1992, Peptide nucleic acids (PNA). Oligonucleotide analogues with an achiral peptide backbone, J.Am.Chem.Soc., 114:1895
Gao, W.-Y., Han, F.-S., Storm, C., Egan, W. and Cheng, Y.-C., 1992, Phosphorothioate oligonucleotides are inhibitors of human DNA polymerases and RNase H: Implications for antisense technology, Molec.Pharm., 41:223
Gao, W., Stein, C.A., Cohen, J.S., Dutschman, G.E. and Cheng, Y.-C., 1989, Effect of phosphorothioates homo-oligonucleotides on herpes simplex virus type 2 Induced DNA polyermerase, J.Biol.Chem., 264:11521
Ghosh, M. and Cohen, J.S., 1992, Oligodoexynucleotides as antisense inhibitors of gene expression, Prog.Nucl.Acids Res.Molec.Biol., 42:79
Ghosh, M., Ghosh, K. and Cohen, J.S., 1992, Translation inhibition by phosphorothioate

oligodeoxynucleotides, Antisense Res.Devel., 2:111
Giles, R.V. and Tidd, D.M., 1992, Enhanced RNAses H activity with methylphosphonodiester/phosphodiester chimeric oligodeoxynucleotides, Anti-Cancer Drug Des., 7:37
Goodarzi, G., Gross, S.C., Tewari, A. and Watabe, K., 1990, Antisense oligodeoxyribonucleotides inhibit the expression of the gene for hepatitis B virus surface antigen, J.Gen.Virol., 71:3021
Hatzfeld, J., Li, M.L., Brown, E.L., Sookdeo, H., Levesque, J.P., O'Toole, T., Gurney, C., Clark, S.C. and Hatzfeld, A., 1991, Release of human hematopoeitic progenitors from quiescence by antisense transforming growth factor 1 or Rb oligonucleotides, J.Exp.Med., 174:925
Hikida, M., Haruna, K.-I. and Ohmori, H., 1992, Suppression of interleukin 4 production from type 2 helper T cell clone by antisense oligodeoxynucleotide, Immunol.Letters, 34:297
Inagaki, M., Togawa, K., Carr, B.I., Ghosh, K. and Cohen, J.S., 1992, Antisense oligonucleotides: Inhibition in liver cell proliferation and in vivo disposition, Transplantation Proc., 24:2971
Iversen, P., Mata, J. and Zon, G., 1993, The single-injection pharmacokinetics of an antisense phosphorothioate oligodeoxynucleotide against rev(art/trs) from the human immuno-deficiency virus (HIV) in the adult male rat, J.Pharm.Exp.Therap., in press
Iversen, P.L., Zhu, S., Meyer, A. and Zon, G., 1992, Cellular uptake and subcellular distribution of phosphorothioate oligonucleotides into cultured cells, Antisense Res.Dev., 2:211
Jaroszewski, J.W., Ghosh, M., Ghosh, K. and Cohen, J.S., 1993, Targeting of antisense DNA: comparison of activity of anti-rabbit β-globin phosphorothioate oligodeoxynucleotides with computer predictions of mRNA folding, Antisense Res.Devel., 3:339
Jaroszewski, J.W., Kaplan, O., Syi, J.-L., Sehested, M., Faustino, P.J. and Cohen, J.S., 1990, Concerning antisense inhibition of the multidrug resistance gene, Cancer Commun., 2:287
Kemal, O., Brown, T., Burgess, S., Bishop, J.D. and Leigh-Brown, A.J., 1991,Synthesis of base modified phosphorothioate oligodeoxynucleotides as inhibitors of HIV-1, Nucleosides Nucleotides, 10:555
Kinchington, D., Galpin, S., Jarowszewski, J.W., Ghosh, K., Subasinghe, C. and Cohen, J.S., 1992, A comparison of gag, pol and rev antisense oligodeoxynucleotides as inhibitors of HIV-1, Antivir.Res., 17:53
Kitajima, I., Shinohara, T., Bilakovics, J., Brown, D.A., Xu, X. and Nerenberg, M., 1992, Ablation of transplanted HTLV-1 tax-transformed tumors in mice by antisense inhibition of NF-kB, Science, 258:1792
Kmetz, M.E., Ceruzzi, M. and Schwartz, 1991, Vmw65 phosphorothioate oligonucleotides inhibit HSV KOS replication and Vmw65 protein synthesis, Antivir.Res., 16:173
Lallier, T. and Broneer-Fraser, M., 1993, Inhibition of neural crest cell attachment by integrin antisense oligonuclotides, Science, 259:692
Leiter, J.M.E., Agrawal, S., Palese, P. and Zamecnik, P.C., 1990. Inhibition of influenza virus replication by phosphorothioate oligodeoxynucleotides, Proc.Natl.Acad.Sci.USA, 87:3430
Lemaitre, M., Bayard, B. and Lebleu, B., 1987, Specific antiviral activity of a poly(L lysine)-conjugated oligodeoxyribonucleotide sequence complementary to vesicular stomatitis virus N protein mRNA initiation site, Proc.Natl.Acad.Sci.USA, 84:648
Letsinger, R.L., Zhang, G., Sun, D.K., Ikeuchi, T. and Sarin, P.S., 1989, Cholesteryl-conjugated oligonucleotides: synthesis, properties, and activity as inhibitors of human immunodeficiency virus in cell culture, Proc.Natl.Acad.Sci.USA, 86:6553
Lisziewicz, J., Sun, D., Klotman, M., Agrawal, S., Zamecnik, P. and Gallo, R., 1992, Specific inhibition of human immunodeficiency virus type 1 replication by antisense oligo-nucleotides: an in vitro model for treatment, Proc.Natl.Acad.Sci.USA. 89:11209
Loke, S.L., Stein, C.A., Zhang, Z.H., Mori, K., Nakanishi, M., Subasinghe, C., Cohen, J.S. and Neckers, L.M., 1989, Characterization of oligodeoxynucleotide transport into living cells, Proc.Natl.Acad.Sci.USA, 86:3474
Loke, S.L., Zhang, Z.H., Stein, C.A., Avigan, M., Cohen, J.S. and Neckers, L.M., 1988, Delivery of c-myc antisense phosphorothioate oligodeoxynucleotides to

hematapoietic cells in culture by liposome fusion, Curr.Topics Microbiol.Immunol., 141:282
Mag, M., Muth, J., Lucking, S., Blesert, L. and Engels, J., 1991, HIV Inhibition by antisense oligodeoxynucleotides, Nucleosides Nucleotides, 10:627
Majumdar, C., Stein, C.A, Cohen, J.S., Broder, S.A. and Wilson, S., 1989, Reverse transcriptase stepwise mechanism: Phosphorothioate oligodeoxynuleotides as primer, Biochemistry, 28:1340
Marshall, W.S. and Caruthers, M.H., 1989, Phosphorodithioate DNA as a potential therapeutic drug, Science, 259:1564
Marti, G., Egan, W., Noguchi, P., Zon, G., Matsukura, M., Broder, S., 1992, Oligo-deosyribonucleotide phosphorothioate fluxes and localization in hematopoietic cells, Antisense Res.Devel., 2:27
Matsukura, M., Shinozuka, K., Zon, G., Mitsuya, H., Reitz, M., Cohen, J.S. and Broder, S., 1987, Phosphorothioate analogs of oligodeoxynucleotides: Inhibitors of replication and cytopathic effects of human immunodeficiency virus, Proc.Natl.Acad.Sci.USA, 84:7706
Matsukura, M., Zon, G., Shinozuka, K., Roberts-Guroff, M., Stein, C.A., Mitsuya, H., Wong-Staal, F., Cohen, J.S. and Broder, S.A., 1989, Regulation of viral expression of HIV in vitro by antisense phosphorothioate oligodeoxynucleotide against rev, Proc.Natl.Acad.Sci.USA, 86:4244
McSwiggen, J.A. and Cech, T.R., 1989, Stereochemistry of RNA cleavage by the tetrahymena ribozyme and evidence that the chemical step is not rate-limiting, Science, 244:679
Murphy, P.R., Sato, Y. and Knee, R.S., 1992, Phosphorothioate antisense oligonucleotides against basci fibroblast growth factor inhibit anchorage-dependen and anchorage-independent growth of a malignant cell line, Molec.Endocrinol., 6:877
Murray, J.A.H. (ed.), 1992, Antisense RNA and DNA, Wiley-Liss, Inc., New York
Neckers, L.M., Rosolen, A. and Whitesell, L., 1992, Antisense inhibition of gene expression: a tool for studying the role of N-myc in the growth and differentiation of neuroectoderm-derived cells, J.Immunotherapy, 12:162
Offensperger, W.B., Offensperger, S., Walter, E., Teubner, K., Iglol, G., Blum, H.E. and Gerok, W., 1993, In vivo inhibition of duck hepatitis B virus replication and gene expression by phosphorothioate modified antisense oligodeoxynucleotides, EMBO J., 12:1257
Perlaky, L., Saijo, Y., Busch, R.K., Bennett, C.F., Mirabelli, C.K., Crooke, S.T. and Busch, H., 1993, Growth inhibition of human tumor cell lines by antisense oligonucleotides designed to inhibit p120 expression, Anti-Cancer Drug Des., 7:3
Pieken, W.A., Olsen, D.B., Benselser, F., Aurup, H. and Eckstein, F., 1991, Kinetic characterization of ribonuclease-resistant 2'-modified hammerhead ribozymes, Science, 253:314
Ramanathan, M., MadGregor, R.D. and Hunt, C.A., 1993, Predictions of effects for intracellular antisense oligodeoxyribonucleotides from a kinetic model, Antisense Res.Devel., 3:3
Rapoport, E., Misiura, K., Agrawal, S. and Zamecnik, P., 1992, Antimalarial activities of oligodeoxynucleotide phosphorothioates in chloroquine-resistant Plasmodium falciparium, Proc.Natl.Acad.Sci.USA, 89:8577
Ratajczak, M.Z., Kant, J.A., Luger, S.M., Huiya, N., Zhang, J., Zon, G. and Gerwirtz, A.M., 1992, In vivo treatment of human leukemia in a scid mouse model with c-myb antisense oligodeoxynucleotides, Proc.Natl.Acad.Sci.USA, 89:11823
Reed, J.C., Stein, C.A., Subasinghe, C.A., Haldar, S., Croce, C.M., Yum, S. and Cohen, J.S., 1990, Antisense-mediated inhibition of BCL2 proto-oncogene expression and leukemic cell growth; comparisons of phosophorothioate oligonucleotides, Cancer Res., 50:6565
Sarver, N., Cantin, E.M., Change, P.S., Zaia, J.A., Ladne, P.A. and Stephen, D.A., et al, 1990, Ribozymes as potential anti-HIV therapeutic agents, Science, 247:1222
Scanlon, K.J., Jiao, L., Funato, T., Wang, W., Tone, T., Rossi, J.J. and Kashani-Sabet, M., 1991, Ribozyme-mediated cleavage of c-fos mRNA reduces gene expression of DNA synthesis enzymes and metallothionein, Proc.Natl.Acad.Sci.USA, 88:10591
Schlingensiepen, K.H. and Brysch, W., 1992, Inhibitors of Oncogene expression in tumor cells and tools for gene function analysis, in: Gene Regulation: Biology of Antisense RNA and DNA317-328, R.P. Erickson and J.G. Izant, eds., Raven Press, New York

Shaw, J.-P., Kent, K., Bird, J., Fishback, J. and Froehler, B., 1991, Modified deoxyoligonucleotides stable to exonuclease degradation in serum, Nucl.Acids Res., 19:747

Shoji, Y., Akhtar, S., Periasamy, A., Herman, B. and Juliano, R.L., 1992, Mechanism of cellular uptake of modified oligodeoxynucleotides containing methylphosphonate linkages, Nucl.Acids Res., 19:543

Simons, M., Edelman, E.R., DeKeyser, J.-L. and Rosenberger, R.D., 1992, Antisense c-myb oligonucleotides inhibit arterial smooth muscle accumulation in vivo, Nature, 359:67

Stein, C.A., Iversen, P.L., Subasinghe, C., Cohen, J.S., Stec, W.J. and Zon, G., 1990, Preparation of ^{35}S-labelled phosphorothioate oligodeoxynucleotides by use of phosphonate chemistry, Anal.Biochem., 188:11

Storey, A., Oates, D., Banks, L., Crawford, L. and Crook, T., 1991, Antisense phosphorothioate olignucleotides have both specific and non-specific effects on cells containing human papilloma virus type 16, Nucl.Acids Res., 19:4109

Thierry, A.R., Dritschilo, A., 1992, Intracellular activity of unmodified, phosphorothioated and liposomally encapsulated oligodeoxynucleotides for antisense activity, Nucl.Acids Res., 20:5691

Toulme, J.J., 1993, Antisense oligonucleotides and ribonucleases H, Biochemie, 75:1

Tuerk, C. and Gold, L., 1990, Systematic evolution of ligands by exponential enrichment: RNA ligands to bacteriophage T4 DNA polymerase, Science, 249:505

Watson, P.H., Pon, R.T. and Shiu, R.P.C., 1992, Inhibition of c-myc expression by phosphorothioate antisense oligonucleotide identifies a crical role for c-myc in the growth of human breast cancer, Cancer Res., 51:3996

Wickstrom, E., 1991, Antisense DNA treatment of HL-60 promyelocytic leukemia cells: terminal differentiation and dependence on target sequence, in: Prospects for antisense nucleic acid therapy of cancer and AIDS, Wiley-Liss, New York

Woolf, T.M., Jennings, C.B.G., Rebagliati, M. and Melton, D.A., 1990, The stability, toxicity and effectiveness of unmodified and phosphorothioate antisense oligodeoxynucleotides in Xenopus oocytes and embryos, Nucl.Acids Res., 18:1763

Wu, G.Y. and Wu, C.H., 1992, Specific inhibition of hepatitis B viral gene expression in vitro by targeted antisense oligonucleotides, J.Biol.Chem., 267:12436

Yabukov, L.A., Deeva, E.A., Zarytova, V.F., Ivanova, E.M., Ryte, A.S., Yurchenko, L.V. and Vlassov, V.V., 1989, Mechanism of oligonucleotide uptake by cell: involvement of specific receptors? Proc.Natl.Acad.Sci.USA, 86:6454

Yaswen, P., Stampfer, M.R., Ghosh, K. and Cohen, J.S., 1993, Effects of thioated oligonucleotides on cultured human mammary epithelial cells, Antisense Res.Devel., 3:67

Yokozaki, H., Budillon, A., Totora, G., Meissner, S., Beuacage, S.L., Miki, K. and Cho-Chung, Y.S., 1993, An antisense oligodeoxynucleotide that depletes RIa subunit of cyclic AMP-dependent protein kinase induces growth inhibition in human cancer cells, Cancer Res., 53:868

Zuker, M., 1989, On finding all suboptimal foldings of an RNA molecule, Science, 244:48

LIPOSOMES IN VIVO: CONTROL OF BEHAVIOUR

Gregory Gregoriadis

Centre for Drug Delivery Research, The School of Pharmacy
University of London, Brunswick Square, London WC1N 1AX, UK

INTRODUCTION

The concept of liposomes as a drug carrier put forward nearly 25 years ago (Gregoriadis et al, 1971; Gregoriadis and Ryman, 1972a,b), has recently culminated in the marketing of life-saving pharmaceutical products. The story of liposomes and related impact on drug delivery and targeting has been outlined elsewhere (Gregoriadis, 1993). This chapter describes key developments, mostly in my laboratory, which, have contributed to both the unravelling of vesicle fate in vivo and its control. Such developments refer to the manipulation of liposomal structural characteristics with the purpose of ensuring quantitative retention of entrapped drugs by vesicles during their exposure to blood en route to their destination, and of obtaining vesicle clearance rates that contribute to optimal vesicle distribution to target cells and tissues.

Initial experiments in the early 70's with multilamellar vesicles (MLV) containing a variety of solutes revealed several basic aspects of their behaviour in vivo and in vitro. Thus, the rate of vesicle clearance from the circulation of intravenously injected rats was found to be rapid, dose-dependent and biphasic (Gregoriadis and Ryman, 1972a,b). Subsequently, it was observed (Gregoriadis and Neerinjun, 1974) that the clearance of neutral and positively charged (with stearylamine) MLV was slower than that of negatively charged MLV. (Recent experiments with an alternative positively charged lipid and liposomes of varying bilayer fluidity have shown that a slow clearance for such positively charged liposomes is probably an artefact, masking a normally rapid clearance for vesicles which retain their charge in the circulation; Tan and Gregoriadis, 1989). The effect of vesicle surface charge on clearance was later found (Juliano and Stamp, 1975) to apply to small unilamellar vesicles (SUV) and the same authors were able to show that SUV had a longer circulation time than MLV. An additional observation (Gregoriadis, 1973) made at that time established that water-soluble drugs such as 5-fluorouracil and penicilin G leaked quantitatively from MLV while in circulation. On the basis of current knowledge, this is hardly surprising as MLV used were cholesterol-poor (Gregoriadis, 1973). In contrast, however, liposome-entrapped solutes of large molecular weight (eg. albumin and amylo-glycosidase) did not appear to leak (Gregoriadis and Ryman, 1972 a). Moreover, liposomes with entrapped materials were shown to localise in the macrophages of the reticulo-endothelial system (RES), mostly the liver and spleen (Gregoriadis and Ryman, 1972a, b). Uptake of liposomes by the cells was effected through endocytosis (lysosomotropic pathway) (Gregoriadis and Ryman, 1972b; Segal et al, 1974). Vesicles were seen to lose their organized structure within the lysosomes (probably through the action of phospho-lipases) and release their contents (Segal et al, 1974). In the case of drugs (and depending on molecular size and ability to withstand the hostile milieu of lysosomes), these could then exert their action within the organelles (Gregoriadis and Buckland, 1973) or, following

Targeting of Drugs 4, Edited by G. Gregoriadis *et al.*
Plenum Press, New York, 1994

diffusion from lysosomes, in alternative cell compartments (eg. inhibition of DNA-directed RNA synthesis by liposomal actinomycin D in partially hepatectomized rats; Black and Gregoriadis, 1974). Such early, detailed knowledge of liposomal fate and behaviour enabled us to test, propose or anticipate a number of applications, including the treatment of lysosomal storage diseases (Gregoriadis and Ryman, 1972a; Belchetz et al, 1977) intracellular infections (Gregoriadis, 1973; Bonventre and Gregoriadis, 1978), cancer chemotherapy (Gregoriadis et al, 1974), gene therapy (Gregoriadis and Ryman, 1972b) and, because of the involvement of macrophages in vesicle uptake in vivo, immunopotentiation and vaccine delivery (Allison and Gregoriadis, 1974).

TARGETING OF LIPOSOMES: PREREQUISITES

Work from this laboratory in the early 70's revealed that liposomes with cell-specific molecules (eg. antibodies and asialoglycoproteins) on their surface can interact selectively with cells expressing appropriate antigens and receptors, both in vitro and in vivo (Gregoriadis, 1974; Gregoriadis and Neerunjun, 1975). This led to the concept of vesicle targeting to cells other than those of the RES, for instance accessible (to liposomes) cancer cells and a variety of tissues. Uptake by tumours was, indeed, suggested in work with experimental animals (Gregoriadis et al, 1977). However, it was predicted (Gregoriadis, 1974) that successful targeting would require not only quantitative retention of entrapped drugs by the vesicles (a prerequisite also required for the "passive" uptake of liposomes by the RES) but also sufficiently prolonged presence of vesicles in the circulation so as to enable vesicles to encounter, and eventually associate with, the target cells. Because the behaviour of liposomes in vivo depends on the interaction of their structural features with the biological milieu (Gregoriadis, 1988), knowledge of the effect of the milieu on drug release from vesicles and their rates of clearance was thought essential.

RETENTION OF DRUGS BY LIPOSOMES

Findings from a number of laboratories (Krupp et al, 1976; Sherphof et al, 1978) suggested that liposomes become permeable or destabilized in the presence of blood (thus allowing entrapped solutes to leak out) because plasma high density lipoproteins (HDL) remove phospholipid molecules from the vesicle bilayer. On this basis, it was assumed (Gregoriadis and Davis, 1979) that abrogation of HDL action (which would allow vesicles to remain intact), could be achieved by rendering lipid bilayers packed or "solid". An earlier finding (Papahadjopoulos et al, 1973) of reducing protein-induced vesicle permeability in buffer by the addition of cholesterol into liposomes was encouraging. This assumption was tested experimentally, initially with MLV (Gregoriadis and Davis, 1979) and subsequently with SUV (Kirby et al, 1980). With both types of vesicles, excess cholesterol (eg. phospholipid to cholesterol molar ratio of 1:1) led to bilayer stabilization in the presence of blood or blood plasma with only a minimum of entrapped solute markers or phospholipid being lost in vitro or in vivo after intravenous, intraperitoneal or subcutaneous injection (for a review on related work from this laboratory see Gregoriadis, 1988).

Subsequent experiments (reviewed in Gregoriadis, 1988) demonstrated that reduction of solute leakage from vesicles could also be achieved by substituting the low melting egg phosphatidylcholine (PC) with high-melting phospholipids, especially when equimolar cholesterol was present as well. Thus, liposomes composed of distearoyl phosphatidylcholine (DSPC) alone or sphingomyelin (SM) and equimolar cholesterol were shown to be considerably less leaky to entrapped solutes in the presence of blood plasma (Gregoriadis and Senior, 1980). For instance, liposomes composed of equimolar DSPC and cholesterol were able to completely retain their solute (carboxyfluorescein) content even after 48 hours of exposure to plasma at 37°C (Gregoriadis, 1988).

CLEARANCE OF LIPOSOMES FROM THE BLOOD CIRCULATION

Early work (Gregoriadis and Senior, 1980; Gregoriadis, 1988) demonstrated that liposomal stability in terms of solute retention by cholesterol-rich SUV made of PC and

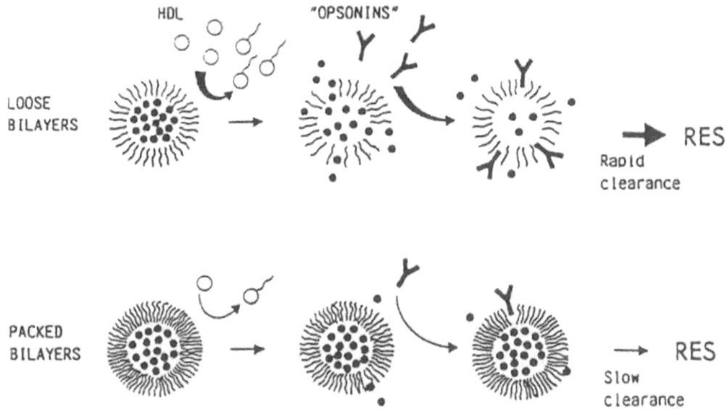

Fig. 1 Correlation between stability of liposomes and clearance from blood circulation. In proposed scheme, the extent of bilayer porosity and leakage of solutes in blood is dependent on the facility with which high-density lipoproteins (HDL) remove phospholipid molecules from the bilayer. Loose bilayers (top) are attacked by HDL more effectively than packed bilayers (bottom). The greater the gaps after HDL attack, the more extensive the opsonin adsorption on vesicles and uptake by reticulo-endothelial system (RES)

equimolar cholesterol was pronounced during relatively short (eg. up to 2 h) periods of vesicle exposure to blood plasma at 37°C but was reduced considerably (eg. to 50% in 24 h) on incubation for longer periods. In contrast, solute retention under the same conditions remained unchanged for vesicles made of equimolar SM and cholesterol even after 48h. This phenomenon was found to correlate with the residence time of the corresponding vesicles in the blood circulation. Thus, the greater the vesicle stability in terms of solute retention, the longer the half-life (Gregoriadis, 1988). However, this relationship between vesicle stability and vesicle clearance was found valid only for neutral SUV (of a size up to about 80 nm diameter): very stable negatively charged SUV or larger neutral vesicles were found to exhibit rapid clearance rates which, in the case of large vesicles (eg. MLV), increased progressively with increasing size, even though bilayer stability remained high in the presence of blood plasma for prolonged periods of time (Gregoriadis, 1988). It therefore follows that the state of bilayer stability in determining vesicle clearance from the circulation is overriden by both vesicle size and surface charge.

A tentative hypothesis (Gregoriadis, 1989) unifying observations on the role of the lipid composition of neutral liposomes in promoting bilayer stability in vitro and their half-lives in vivo was put forward (Fig. 1). The scheme proposes that with liposomes that withstand HDL attack and complete disintegration, their destabilization facilitates the adsorption onto or insertion into the vesicle bilayer of plasma proteins such as opsonins. These are thought to be responsible (Saba, 1970), for the presentation of particles to macrophages. The scheme also predicts that the lesser the degree of bilayer destabilization the less frequent or smaller the "gaps" on it (with a corresponding reduction in the leakage of solutes) and, therefore, the smaller the amount of opsonins associating with the vesicles. As a result, vesicles with decreasing amounts of adsorbed opsonin would exhibit a proportionally reduced uptake by the phagocytic cells of the RES. On the other hand, although probably valid in general terms, the scheme is too simplistic to reflect real events. For instance, a variety of plasma components, including α_2-macroglobulin (Black and Gregoriadis, 1976), components of the complement cascade (Loughrey et al, 1990; Richards et al, 1979), C-reactive protein (Richards et al, 1979) and fibronectins (Rossi and Wallace, 1983) can bind to the surface of vesicles to an extent that depends on their lipid composition and surface properties. Recent work (Moghimi and Patel, 1993) has shown that liposomes interact (or fail to interact) with tissue-specific opsonins that are known to either facilitate or inhibit (dysopsonins) uptake of liposomes with differing lipid compositions by the liver, spleen or bone marrow macrophages. The way by which interaction of the vesicle bilayer surface (exhibiting varying degrees of fluidity or destabilization) with specific opsonins is promoted or inhibited is unknown. However, the inability of a liver-specific opsonin to promote uptake of liposomes that are rich in cholesterol or made of high melting phospholipids, by Kupffer cells is compatible with the sequence of events in Fig. 1.

LONG-LIVED LIPOSOMES

A number of laboratories are currently working on the design of liposomes that are capable of avoiding early interception by the RES macrophages and thus exhibit long half-lives in the blood circulation. Long-lived liposomes, especially when endowed with surface groups that recognize cell targets, are expected to interact effectively with a variety of specific cells both intravascularly and, depending on access, extravascularly. In this respect, long-lived neutral SUV by virtue of their lipid composition, have been shown to deliver drugs to solid tumours or to cure solid-tumour bearing mice (Large and Gregoriadis, 1983). Indeed, the simplicity of this approach, in conjuction with technology (Gregoriadis et al, 1990) which ensures a high drug to lipid mass ratio in vesicles of about 100 nm diameter, renders especially designed SUV the carrier of choice for applications where a small vesicle size is a requirement for effective use (eg. imaging of tumors, delivery of drugs to the bone marrow, etc). On the other hand, in situations where large quantities of drugs (without increasing the lipid dose) or large molecules such as enzymes, haemoglobin, cytokines etc need to be delivered, the design of long-lived larger vesicles is essential.

Work from this laboratory (Senior et al, 1991) and elsewhere (Klibanov et al, 1991; Blume and Cevc, 1990; Papahadjopoulos et al, 1991; Allen, this book) have recently and independently shown that liposomes of about 100 nm in diameter, coated with poly-ethyleneglycol (PEG) and its various derivatives, circulate in the blood for longer periods that similar non-pegylated vesicles. Work with pegylated liposomes is now mushrooming and numerous applications are in various stages of progress. However, improvement are needed in terms of further augmenting vesicle half-lives, applying the concept to even larger vesicles or even replacing PEG with a natural, biodegradable molecule. Evidence from this laboratory (Gregoriadis et al, 1993; McCormack and Gregoriadis, this book) suggests that polysialic acids (harvested from the cultures of certain non-pathogenic bacteria such as Escherishia coli K92) may be an attrative alternative.

REFERENCES

Allison, A.C. and Gregoriadis, G., 1974, Liposomes as immunological adjuvants, Nature, 252:252

Belchetz, P.E., Braidman, I.P., Crawley,J.C.W. and Gregoriadis, G, 1977, Treatment of Gaucher's disease with liposome-entrapped glucocerebroside-β-glucosidase, Lancet, ii: 116

Black, C.D.V. and Gregoriadis, G., 1974, Intracellular fate and effect of liposome-entrapped actinomycin D injected into rats, Biochem.Soc.Trans, 2:869

Black, C.D.V. and Gregoriadis, G., 1976, Interaction of liposomes with blood plasma proteins, Biochem.Soc.Trans., 4:253

Bonventre, P. and Gregoriadis, G., 1978, Killing of intraphagocytic Staph.aureus by dihydrostreptomycin entrapped in liposomes, Antimicrobial Agents and Chemotherapy, 13:1049

Gregoriadis, G., 1973, Drug entrapment in liposomes, FEBS Lett., 36:292

Gregoriadis, G., 1974, Structural requirements for the specific uptake of marcomolecules and liposomes by target tissues, in: "Enzyme Therapy in Lysosomal Storage Disease", J.M. Tager, G.J.M. Hooghwinkel and W.Th. Daems, eds., North-Holland Publising Co., Leiden

Gregoriadis, G., 1988, Fate of injected liposomes: Observations on entrapped solute retention, vesicle clearance and tissue distribution in vivo, in: "Liposomes as Drug Carriers: Recent Trends and Progress", G. Gregoriadis, ed., J. Wiley and Sons Ltd, Chichester

Gregoriadis, G., 1989, The physiology of the liposome, News in Physiological Sciences, 4:146

Gregoriadis, G., 1993, Liposomes: A tale in drug targeting, J.Drug Targeting., 1:3
Gregoriadis, G. and Ryman, B.E., 1972a, Fate of protein-containing liposomes injected into rats. An approach to the treatment of storage diseases, Eur.J.Biochem., 24:485
Gregoriadis, G. and Ryman, B.E., 1972b, Lysosomal localization of β-fructofuranosidase -containing liposomes injected into rats. Some implications in the treatment of genetic disorders, Biochem.J., 129:123
Gregoriadis, G. and Buckland, R.A., 1973, Enzyme-containing liposomes alleviate a model for storage disease, Nature, 244:170
Gregoriadis, G. and Neerunjun, D.E., 1974, Control of the rate of hepatic uptake and catabolism of liposome-entrapped proteins injected into rats. Possible therapeutic applications, Eur.J.Biochem., 47:179
Gregoriadis, G. and Neerunjun, D.E., 1975, Homing of liposomes to target cells, Biochem. Biophys.Res.Comm., 65:537
Gregoriadis, G. and Davis, C., 1979, Stability of liposomes in vivo and in vitro is promoted by their cholesterol content and the presence of blood cells, Biochem.Biophys. Res.Comm., 89:1287
Gregoriadis, G. and Senior, J., 1980, The phospholipid component of small unilamellar liposomes controls the rate of clearance of entrapped solutes from the circulation, FEBS Lett., 119:43
Gregoriadis, G., Leathwood, P.D. and Ryman, B.E., 1971, Enzyme entrapment in liposomes, FEBS Lett., 14:95
Gregoriadis, G., Swain, C.P., Wills, E.J. and Tavill, A.S., 1974, Drug-carrier potential of liposomes in cancer chemotherapy, Lancet, i:1313
Gregoriadis, G., Neerunjun, D.E. and Hunt, R., 1977, Fate of a liposome-associated agent injected into normal and tumour-bearing rodents. Attempts to improve localization in tumour tissues, Life Sciences, 21:357
Gregoriadis, G., da Silva, H. and Florence, A.T., 1990, A procedure for the efficient entrapment of drugs in dehydration-rehydration liposomes (DRV), Int.J. Pharmaceutics, 65:235
Gregoriadis, G., McCormack, B., Wang, Z. and Lifely, R., 1993, Polysialic acids: Potential in drug delivery, FEBS Lett., 315:271
Juliano, R.L. and Stamp, D., 1975, Effects of particle size and charge on the clearance of liposomes and liposome-encapsulated drugs, Biochem.Biophys.Res.Comm., 63:651
Kirby, C., Clarke, J. and Gregoriadis , G., 1980, Effect of the cholesterol content of small unilamellar liposomes on their stability in vivo and in vitro, Biochem.J. 186:591
Klibanov, A.L., Maruyama, K., Torchillin, V.P. and Huang, L., 1990, Amphipathic polyethyleneglycols effectively prolong the circulation time of liposomes, FEBS Lett., 268:235
Krupp, L., Chobanian, A.V. and Brecher, J.P., 1976, The in-vivo transformation of phospholipid vesicles to a particle resembling HDL in the rat, Biochem.Biophys. Res.Comm., 72:1251
Large, P. and Gregoriadis, G., 1983, Phospholipid composition of small unilamellar liposomes containing melphalan influences drug action in mice bearing PC6 tumours Biochem.Pharmacol, 32:1315
Loughrey, H.C., Bally, M.B., Reinish, L.W. and Cullis, P.R., 1990, The binding of phosphatidylglycerol liposomes to rat platelets is mediated by complement, Thromb.Haemost., 64:172
Moghimi, S.M. and Patel, H.M., 1993, Techniques to study the opsonic effect of serum on uptake of liposomes by phagocytic cells from various organs of the RES, in: "Liposome Technology", 2nd Edition, G. Gregoriadis, ed., CRC Press, Boca Raton
Papahadjopoulos, D., Cowden, M. and Kimelberg, H., 1973, Role of cholesterol in membranes. Effect of phospholipid-protein interactions, membrane permeability and enzyme activity, Biochim.Biophys.Acta, 310:8
Papahadjopoulos, D., Allen, T., Gabizon, A., Mayhew, E., Matthay, K., Huang, K., Lee, S.K., Woodle, M.C., Lasic, D.D., Redemann, C. and Martin, F.J., 1991, Sterically stabilized liposomes: improvements in pharmacokinetics, tissue disposition and anti-tumour therapeutic efficacy, Proc.Natl.Acad.Sci.USA, 88:11460
Richards, R.L., Gewurz, H., Seigel, J. and Alving, C.R., 1979, Interaction of C-reactive protein and complement with liposomes, J.Immunol., 112:1185
Rossi, J.D. and Wallace, B.A., 1983, Binding of fibronectin to phospholipid vesicles, J.Biol.Chem., 258:3327

Saba, T.M., 1970, Physiology and physiopathology of the reticulo-endothelial system, Arch.Intern.Med., 126:1031

Segal, A.W., Wills, E.J., Richmond, J.E., Slavin, G., Black C.D.V. and Gregoriadis, G., 1974, Morphological observations on the cellular and subcellular destination of intravenously administered liposomes, Brit.J.Exp.Pathol., 55:320

Senior, J.H., Delgado, C., Fisher, D., Tilcock C. and Gregoriadis, G., 1991, Influence of surface hydrophilicity of liposomes on their interaction with plasma proteins and clearance from the circulation: Studies with polyethylene glycol-coated vesicles Biochim.Biophys.Acta, 1062:77

Scherphof, G., Roerdink, G., Waite, M. and Parks, J., 1978, Disintegration of phosphatidylcholine liposomes in plasma an a result of interaction with high-density lipoproteins, Biochim.Biophys.Acta, 542:296

Tan, L. and Gregoriadis, G., 1989, The effect of positive surface charge of liposomes on their clearance from blood and its relation to vesicle lipid composition Biochem.Soc.Trans., 17:690

PASSIVE TARGETING OF ANTHRACYCLINES ENTRAPPED IN LONG-CIRCULATING(STEALTH) LIPOSOMES IN THE TREATMENT OF CANCER

T.M. Allen

Department of Pharmacology, University of Alberta
Edmonton, AB T6G 2H7

INTRODUCTION

The ability of long-circulating liposomes to act as drug sustained release systems or to passively target to solid tumors has resulted in a number of therapeutic applications which are being tested in animal models and in the clinic. Long-circulating liposomes, also called sterically stabilized liposomes (S-liposomes), have a number of properties which make them suitable for these applications (Table 1). This review will cover applications involving passive targeting to solid tumors of S-liposomes entrapping the anthracyline anticancer drugs doxorubicin (S-DOX) and epirubicin (S-EPI), as these formulations are the closest to being approved for use in the clinic.

PASSIVE TARGETING OF ANTHRACYCLINES TO SOLID TUMORS

Doxorubicin (DOX, adriamycin) and epirubicin (EPI, 4'-epi-doxorubicin) are important antitumor drugs with activity against a wide range of human neoplasms, including many types of solid tumors. DOX is active in the S-phase of the cell cycle and its major side effects are cardiomyopathy, as well as an acute dose-limiting myelosupression. DOX and EPI can be remote loaded efficiently into liposomes using both the pH (Mayer et al, 1986) and ammonium sulfate (Gabizon et al, 1989) gradient techniques and the resulting formulations are stable for over a year. DOX entrapped in conventional liposomes with short circulation half-lives and/or dose-dependent pharmacokinetics (C-DOX) has been previously shown by a number of investigators to result in good therapeutic efficacy against solid tumors in animal models (Gabizon et al, 1982; Herman et al, 1983; Van Hoesel et al, 1984; Mayer et al, 1989), and decreased cardiotoxicity (Gabizon et al, 1982; Herman et al, 1983; Van Hoesel et al, 1984), although liposomal DOX still exhibited myelosuppressive activity (Van Hoesel et al, 1984; Mayer et al, 1989). C-DOX is currently in clinical trials (Treat et al, 1990; Embree et al, 1993). The effects of DOX or EPI entrapped in S-liposomes (S-DOX, S-EPI) have been examined in a number of solid tumors in animal models (Gabizon and Papahadjopoulos, 1988; Ahmad et al, 1993; Huang et al, 1992; Mayhew et al, 1992; Vaage et al, 1992; Papahadjopoulos et al, 1991) and S-DOX (DOX-SL[R], Liposome Technology Inc., Menlo Park, CA) is also in clinical trials (Gabizon et al, 1994).

Targeting of Drugs 4, Edited by G. Gregoriadis *et al.*
Plenum Press, New York, 1994

Table 1. Properties of long-circulating liposomes which make them suitable for application as drug sustained release systems and for passive targeting to solid tumors.

1.	Prolonged survival times in circulation, with plasma half-lives varying from approximately 20 hours in rodents to 45 hours in humans (Allen and Hansen, 1991; Gabizon et al, 1993; Gabizon et al, 1994)
2.	Dose-independent pharmacokinetics (Allen and Hansen, 1991)
3.	Decreased rate and extent of accumulation in the mononuclear phagocyte system (Allen and Hansen, 1991)
4.	Versatility in lipid composition, with a wide variety of drug release rates achievable with maintance of prolonged circulation half-lives (Allen et al, 1991; Woodle et al, 1992)
5.	Ability to migrate across biological barriers, e.g. from subcutaneous or peritoneal cavity into blood (Allen et al, 1991; Allen et al, 1993)
6.	Increased uptake into solid tumors (Gabizon and Papahadjopoulos, 1988; Huang et al, 1992; Gabizon, 1992; Zu et al, 1993) and infected tissues (Bakker-Woudenberg et al, 1993)
7.	Ability to be targeted to specific tissues in vivo (Maruyama et al, 1990; Torchilin et al, 1992; Ahmad et al, 1993).

Pharmacokinetics and Anthracyclines

With the development of techniques to separate liposomal DOX from free DOX in plasma it became possible to look at the pharmacokinetics of the drug separate from its carrier (Gabizon et al, 1989) (Table 2). This is the first liposome-entrapped drug for which this has been possible. Free DOX is rapidly cleared from plasma following i.v. administration with an initial half-life in humans of 0.08 hr (Gabizon et al, 1994) and in mice of 0.013 hr (Gabizon et al, 1989). The elimination of free DOX from plasma is multiphasic with a mean half-life of 20.0 hr and a terminal half-life in humans of 36.7 hrs (Robert et al, 1985), and in mice of 8.6 hrs. Free EPI has a similar plasma pharmacokinetics to DOX, with a mean half-life of 18.3 hr and a terminal half-life of 34.7 hrs, but with a 2-fold higher volume of distribution and a faster clearance (60.2 L $\cdot hr^{-1}$ versus 33.4 L hr^{-1} for DOX) (Robert et al, 1985).

The area under the curve ($AUC_{0\rightarrow\infty}$, μg hr/ml) in mice for DOX, entrapped in long-circulating HSPC:CH:HPI liposomes (LC-DOX) was compared to the AUC for DOX entrapped in liposomes (PC:CH:PG) with short circulation half-lives (C-DOX) and to the AUC for free drug (Gabizon et al, 1989). The AUC was dramatically higher for LC-DOX (1099 μg hr/ml), compared to 58.5 for C-DOX and 3.9 for free DOX. The volume of distribution was 636.5 ml for the free drug versus 4.1 ml for LC-DOX, i.e., only slightly larger than the blood volume of the mice, suggesting that the LC-DOX was confined largely to the central compartment, and that DOX remained in the liposome carrier. In addition the clearance in mice was reduced from 51.3 ml/hr in the case of the free drug to only 0.2 ml/hr in the case of the LC-DOX (Gabizon et al, 1989). These substantial alterations in the pharmacokinetics of LC-DOX, as compared to free DOX, resulted in a 3-fold reduction in the total amount of DOX localizing in heart tissue at 24 hr post-injection, which is useful in reducing cardiotoxicity associated with the free drug (Gabizon et al, 1989). Reductions in the peak levels of DOX localizing in heart were also observed for S-DOX (HSPC:CH:PEG-DSPE, 2:1:0.2) (Gabizon and Papajadjopoulos, 1988).

In another series of experiments, Gabizon et al, have examined the pharmacokinetics of DOX in mice, rats and dogs as a function of liposome composition (Gabizon et al, 1993). Plasma levels of DOX in mice for drug (10 mg/kg) entrapped in liposomes composed of HSPC:CH, 10:8 (C-DOX), HSPC:CH:HPI, 9:8:1 (LC-DOX) and HSPC:CH:PEG-DSPE, 9.25:8:0.75 (S-DOX) were found to be 3.6, 12.5 and 34.8 μg DOX/ml of plasma, respectively at 24 hrs post-injection. For S-DOX, substitution of DPPC for HSPC resulted in a 2-fold reduction in plasma levels of DOX, and substitution of PC

Table 2. A comparison of the half-lives of various doxorubicin formulations in different species

Species	Formulation	T½α (hrs)	T½β (hrs)
Mice	free DOX	0.013	8.6
	LC-DOX	2.0	15.5
Dogs	LC-DOX	13	
	S-DOX	29	
Humans	free DOX	0.08	25
	C-DOX	0.45	13.5
	S-DOX	2	45

for HSPC resulted in a 50-fold reduction in plasma levels (Gabizon et al, 1993). This indicates that the rate of leakage of DOX from liposomes increases as the phase transition of the phospholipid component of the liposomes decreases.

Pharmacokinetic parameters for plasma DOX and liposome-associated DOX have been measured in dogs given doses of 0.5 mg/kg of free DOX, LC-DOX or S-DOX (Gabizon et al, 1993). Free DOX was rapidly cleared from the blood, and because of the low dose of DOX administered, pharmacokinetic parameters could not be determined. For LC-DOX and S-DOX, at least 90% of DOX circulates in the liposome-entrapped form, as no significant amounts of non-liposome-associated drug could be detected in plasma. The S-DOX had a half-life of 29 ± 8 hrs and an $AUC_{0\rightarrow\infty}$ of 276 μg·hr/ml in dogs, compared to 13 ± 1 hrs and 102 ± 29 μg·hr/ml for LC-DOX (Table 2). The half-life of total DOX in plasma, for either S-DOX or LC-DOX was 27 ± 5 hrs and 10 ± 5 hrs, respectively (Gabizon et al, 1993). For the other pharmacokinetic parameters there was also a notable similarity between those obtained for plasma total DOX and those for plasma liposome-associated DOX. As observed with mice, the volume of distribution of both S-DOX and LC-DOX was very similar to the blood volume of the dogs. Therefore both preparations appear to be restricted to the central compartment (Gabizon et al, 1993).

Pharmacokinetics of S-EPI(HSPC:CH:PEG-DSPE) has been examined in rats (Mayhew et al, 1992). As with LC-DOX, S-EPI showed little evidence of the initial rapid distribution characteristic of free DOX or free EPI. S-EPI was eliminated from plasma in a biexponential process with an initial half-life of 2.8 hr and a terminal half-life of 22 hr, which is comparable to the half-lives of LC-DOX in mice of 2.0 hr and 15.5 hrs (Gabizon et al, 1989). Free EPI in rats had a distribution half-life of 0.23 hr and a terminal half-life of 22 hr. The volume of distribution of free EPI was 3700 ml and the volume of distribution of S-EPI was 18 ml, similar to the blood volume of the animal, as also observed for LC-DOX (Gabizon et al, 1989) in mice and S-DOX in dogs (Gabizon et al, 1993). As with LC-DOX and S-DOX, the $AUC_{0\rightarrow\infty}$ was greatly increased for S-EPI (2769 μg·hr/ml) as compared to free EPI (15 μg·hr/ml), and the clearance rate of the S-EPI was substantially reduced compared to free EPI (0.5 ml/hr versus 111.0 ml/hr) (Mayhew et al, 1992).

Therapy of Murine C26 Colon Carcinoma

The therapeutic efficacy of DOX and EPI entrapped in long-circulating DSPC:CH:GM_1 (LC-DOX or LC-EPI) or HSPC:CH:PEG-DSPE (S-DOX or S-EPI) liposomes has been examined in mouse C26 colon carcinoma grown s.c. or in liver (Huang et al, 1992; Mayhew et al, 1992; Papahadjopoulos et al, 1991). Anthracyclines are cytotoxic to this tumor in vitro, but the in vivo tumor is relatively resistant to the action of free DOX up to concentrations of 10 mg/kg (Huang et al, 1992). This model, therefore, mimics the comparative insensitivity of human colon carcinomas to the anthracyclines. Tumor cells were inoculated into the left lobe of the liver (2 x 10^5 cells) or s.c. into the left flank (4 x 10^5 to 1 x 10^6 cells). Mice received up to three weekly injections of free EPI or DOX at doses of 6 mg/kg, or received S-DOX or S-EPI (approximately 0.1 μm diameter liposomes) at doses of 6 or 9 mg/kg, beginning from 1 to 14 days post-injection of tumor. Tumor size was measured, and also survival time was monitored for up to 120 days (Huang et al, 1992).

Treatment of s.c. tumor with S-DOX at 1, 3, 6, 9 or 14 days post-inoculation of tumor showed that maximal tumor growth inhibition was obtained when singly injections of S-DOX (10 mg/kg) were given from 6 to 9 days after tumor, resulting in significant decreases in anti-tumor activity when the liposome injections were given earlier or later than this (Huang et al, 1992). Maximum survival times were obtained when S-DOX was injected from 3 to 9 days after tumor. For all times of injection, S-DOX was superior to free DOX in decreasing mean tumor size and increasing life spans, with 50 to 60% long-term (120-day) survivors in the groups of mice injected with S-DOX from 3 to 9 days post-inoculation of tumor (Huang et al, 1992). The schedule-dependent nature of the therapy with liposome-entrapped drug may be related to the time required for the vascularization of the tumor, at which point the ability of the liposomes to penetrate into and localize in the tumor may be greater (Huang et al, 1992).

In another series of experiments, mice received 3 injections of either S-DOX or S-EPI at 10, 17 and 24 days after s.c. tumor inoculation (Huang et al, 1992). Free DOX or free EPI had little effect on delaying tumor growth in these experiments. Mice receiving either S-DOX or S-EPI had their tumor regress to nonmeasurable sizes, with 90% long-term (120 day) survivors in the groups of mice receiving 6 or 9 mg/kg S-EPI and 100% long-term survivors in the groups receiving 6 or 9 mg/kg S-DOX (Huang et al, 1992). Thus, the antitumor activity of either DOX or EPI entrapped in S-liposomes was very similar at comparable doses.

In order to determine the uptake of liposomes into the tumors, the pharmacokinetics of LC-liposomes (DSPC:CH:GM_1, containing ^{67}Ga) was compared to that for DSPC:CH liposomes (C-liposomes) in tumor-bearing mice (Huang et al, 1992). The plasma clearance was 2.4 fold higher for C-liposomes, and tumor uptake into s.c. tumor was 2.5 fold lower than for LC-liposomes. Uptake of LC-liposomes into liver tumor was 1.3 fold higher than for C-liposomes. The AUCs for ^{67}Ga in tumor tissue were higher than for surrounding normal tissue, possibly reflecting increased vascularization of the tumor tissue (Huang et al, 1992). When PEG-DSPE was substituted for GM_1 in the liposomes (S-liposomes), again s.c. tumor uptake was increased by 2.5 fold relative to C-liposomes (Huang et al, 1992). Treatment of liver tumor at one day post-injection with single injections of LC-DOX (3 or 10 mg/kg) resulted in significant increases in %ILS of 16.7% and 28.7% respectively. Free DOX or C-DOX at the same doses did not result in any significant increases in survival times (Huang et al, 1992).

These results clearly demonstrate the superior therapeutic efficacy of anthracyclines entrapped in long-circulating liposomes in treating a solid tumor which is relatively resistant to treatment with free drug. The superior effect of the drugs, when entrapped in S-liposomes, has been attributed to increased localization of the drug package in the tumor tissue, local release of the drug from the liposomes into the extravascular spaces, and subsequent uptake into tumor (Huang et al, 1992).

The antitumor activity against s.c. murine C26 colon carcinoma has also been compared for EPI entrapped in liposomes (approximately 0.1 μm in diameter) composed of either HSPC:CH:PEG-DSPE (S-EPI) or PC:CH:PG (C-EPI) or to free EPI (Mayhew et al, 1992). The toxicity of single and 3 weekly doses of either free EPI or S-EPI was also determined. The MTD for S-EPI (12.5-15 mg/kg) was slightly higher than for free EPI

(10.0-12.5 mg/kg) for single injections and also for multiple injections (9 mg/kg versus 6 mg/kg).

Mice were implanted s.c. in the flank with 1 x 10^6 C26 cells and treated with S-EPI or C-EPI (6 or 9 mg/kg) on days 10, 17 and 24 post-implantation (Mayhew et al, 1992). In mice treated with C-EPI, tumor growth was slightly retarded as compared to saline controls. In contrast, mice receiving S-EPI had their tumor regress to unmeasurable size prior to the 3rd injection. The superior effect of S-EPI was also apparent in the survival studies, with no animals receiving C-EPI surviving beyond 50 days, while the S-EPI groups of mice had 90% long-term (120 day) survivors, with no evidence of tumor (Mayhew et al, 1992).

In another experiment the antitumor activity of 3 weekly injections of free EPI (6mg/kg) was compared to that of S-EPI (6 or 9 mg/kg) and a mixture of free EPI and empty S-liposomes (same composition as S-EPI, but no drug), beginning at either 3 or 10 days post-implantation (Mayhew et al, 1992). When treatment was begun at 3 days post-implantation, free EPI only slightly delayed tumor growth, while S-EPI at 6 mg/kg resulted in minimal tumor growth, and S-EPI at 9 mg/kg completely suppressed tumor growth. Interestingly, a mixture of free EPI and empty S-liposomes resulted in a slightly decreased rate of tumor growth as compared to saline controls. This may be a consequence of some association of free drug with the empty S-liposomes. When treatment was begun at 10 days post-implantation, free EPI had no effect on tumor growth, but S-EPI either completely suppressed tumor growth, or delayed tumor growth by several weeks (Mayhew et al, 1992). While no mice in the free EPI or free EPI and empty S-liposomes groups survived beyond 65 days, a number of long term (120 day survivors) were observed among the mice receiving S-EPI. When treatment was begun on the third day post-implantation 80-90% long-term survivors were observed, compared to 40-60% when treatment was begun on the tenth day post-implantation (Mayhew et al, 1992). However, when treatment was begun 14 days post-implantation, minimal increases in life span and no long-term survivors were observed in mice receiving S-EPI, although it is possible that the therapeutic effect of S-EPI could be improved were the dosing schedule to be optimized (Mayhew et al, 1992).

Therapy of Murine Mammary Carcinoma

The antitumor activity of S-DOX(HSPC:CH:PEG-DSPE, 85 nm diameter) was compared to the activity of C-DOX(PC:CH:PG, 237 nm diameter) and to free DOX in mice implanted with mammary carcinomas (Vaage et al, 1992). The cell lines used were MC2A, which has a doubling time of 4.5 days; MC2B, with a doubling time of 15 days; and two variants which result in spontaneous metastasis from intra-mammary implants, MC19, with a doubling time of 10 days and a 90% incidence of metastasis, and MC65, with a doubling time of 5.5 days and a 50% incidence of metastasis. Mice were implanted with a 1 mm^3 piece of tumor into each flank in the primary tumor treatment model, and with a 1 mm^3 piece into each 4th mammary gland in the metastasis prevention model. S.C.-implanted mammary carcinoma tumors have been shown previously not to be very susceptible to drug therapy, and fast-growing tumors were less responsive than slow-growing tumors (Vaage and Constanza, 1992). In experiments to treat primary tumor, mice were treated with 3 weekly i.v. injections of 6 or 9 mg/kg of either S-DOX, C-DOX or free DOX, beginning either 3 or 10 days post-implantation of tumor. No observable toxicity due to any of the treatment was observed (Vaage et al, 1992). In treatment of the more rapidly growing MC2A tumor, in all cases S-DOX was significantly better at reducing tumor volume than C-DOX, which was not significantly different than free DOX (Vaage et al, 1992). S-DOX given on days 3, 10 and 17 reduced the tumor incidence, at 69 days post-implantation, to 0% at a dose of 9 mg/kg, and to 25% at a dose of 6 mg/kg. Weekly treatment with S-DOX begun on day 10 post-implantation was less effective than treatment begun on day 3 post-implantation, with a 15% tumor incidence in mice receiving either 6 or 9 mg/kg S-DOX on days 10, 17 and 24 (Vaage et al, 1992). The slower growing MC2B tumor was more responsive to therapy with a tumor incidence, at 69 days, of 0% for treatment with either 6 or 9 mg/kg S-DOX when treatment was begun on day 3 post-implantation. When treatment was begun 10 days post-implantation of MC2B tumors, the results were noticeable inferior, with a 70% or 55% incidence of tumor in mice receiving 6 mg/kg or 9 mg/kg S-DOX, respectively. In these treatment groups S-DOX was not significantly better than either C-DOX or free DOX (Vaage et al, 1992).

The MC2 tumor is immunogenic and this reaction may eventually restrict the growth of tumor, lead to tumor dormancy followed by renewed growth, or cause the rejection of the tumor (Vaage and Harlos, 1991). This makes this tumor a good model for treatment of human breast cancer. Implants, which after initial growth showed at least 3 weeks of slow or arrested growth, were treated with 4 weekly i.v. injections of 9 mg/kg S-DOX or free DOX beginning on day 35-40 post-implantation. Although neither S-DOX or free DOX reduced the incidence of progressive growth of the tumor as compared to controls, the number of surviving mice at 8 weeks after initiation of treatment was significantly greater for S-DOX (71%) than for free DOX (38%) or placebo (44%) (Vaage et al, 1992).

S-DOX was significantly more effective than free DOX in the treatment of metastatic MC26 tumor implanted in the mammary gland when treatment was given on days 7, 14 and 21 post-implantation. The incidence of tumor, on day 69 post-implantation was 3% or 15% for mice receiving 6 or 9 mg/kg S-DOX, respectively (Vaage et al, 1992). In some mice, MC65 or MC19 tumors were surgically removed as each tumor reached a volume of 1.4 to 1.6 cm^3 (days 25.5 or 39, respectively) and the mice received 4 weekly treatments of S-DOX or free DOX beginning on day 22 post-implantation for MC65, or day 58 post-implantation for MC19. Mice with MC19 tumor, treated with S-DOX (6 or 9 mg/kg), had no pulmonary metastasis, compared to 66% small or medium metastasis for mice treated with free DOX (6 mg/kg) and 100% incidence for placebo controls. Mice with MC65 tumor also had good therapeutic responses to treatment with S-DOX with only 1/33 animals showing evidence of metastasis, compared to 4/15 of those treated with free DOX and 11/20 of those treated with saline (Vaage et al, 1992).

The above study is the first to show a significant survival benefit in mice for S-DOX, as compared to free DOX, in the treatment of well-established tumors. In instances where renewed tumor growth occurred after cessation of treatment with S-DOX, the development of drug-resistant cancer cells did not occur (Vaage et al, 1992).

Therapy of Murine J6456 Lymphoma

The tissue distribution, toxicity and antitumor activity of DOX and EPI entrapped in long-circulating liposomes has been assessed in mice bearing an i.m.-implanted (hind-limb muscle) or an ascitic form of J6456 lymphoma (10^6 cells) (Gabizon, 1992). DOX or EPI was associated with HSPC:CH:HPI, 2:1:0.1 liposomes (LC-DOX or LC-EPI) or PC:CH:PG, 2:1:0.1 liposomes (C-DOX or C-EPI) with diameters ranging from 65-100 nm. For tissue distribution of DOX, mice receiving i.m. implants were treated i.v., when the tumor reached 300-700 mg, with 10 mg/kg of free DOX, free EPI, C-DOX, C-EPI, LC-DOX or LC-EPI. For mice receiving LC-DOX or LC-EPI, tumor drug levels peaked at 24 hr post-injection. Uptake of LC-DOX or LC-EPI into the contralateral leg muscle was very low and tumor:muscle ratios of LC-DOX or LC-EPI were 6 or 17-fold higher, respectively, than that seen for the free drugs, indicating selective uptake into tumor (Gabizon, 1992). Tumor levels of C-DOX were actually lower than those seen for free DOX.

Distribution within ascitic fluid of i.v.-injected LC-DOX, as compared to free DOX, was examined 12 days after i.p. inoculation of 10^6 J6456 lymphoma cells when about 2 ml of ascitic fluid had accumulated (Gabizon, 1992). Approximately 10% of the injected dose of 10 mg/kg LC-DOX localized in the peritoneal cavity within 24 hr post-injection, as compared to less than 2% of the free DOX, with more than 90% of the LC-DOX present in the ascitic fluid as liposome-associated, non-degraded DOX. Free DOX and LC-DOX appeared to accumulate in tumor cells to the same extent, but the kinetics of each was very different. The concentration of DOX in tumor cells, after treatment with LC-DOX, reached maximum levels at 24 hrs post-injection, and remained 5-fold lower than LC-DOX levels in the ascitic fluid. Free DOX had reached its maximum concentration in tumor cells by the earliest time point, and drug levels in the ascitic fluid were much lower than in the tumor cells (Gabizon, 1992). In other words, after receiving LC-DOX, there was a delay in uptake of DOX by tumor cells relative to free DOX, but there was a significantly prolonged exposure time of tumor cells to DOX as it was released from LC-DOX. Since no radiolabelled lipid ([^{3}H-cholesterol] from the LC-DOX was associated with tumor cells, it could be concluded that DOX was being taken up by the tumor cells after leaking from the liposomes (Gabizon, 1992). The extent of extravasation of the LC-DOX into the peritoneal cavity increased as the tumor burden (degree of ascites) increased as the tumor burden (degree of ascites) increased and was significantly higher for LC-DOX than for free DOX in mice bearing the ascitic tumor (Gabizon, 1992). This observation again emphasizes the selective localization of long-circulating liposomal formulations of DOX to tumor areas.

The MTD for free DOX, C-DOX and LC-DOX was determined for single i.v. injections in tumor-free mice (Gabizon, 1992). Both C-DOX and LC-DOX resulted in reduced toxicity, as compared to free DOX, with the median survival time for C-DOX and LC-DOX (20 mg/kg) being >90 days compared to 41 days for mice receiving the same dose as free DOX. At a dose of free DOX of 25 mg/kg median survival time was 11 days or less with less than 50% 90-day survivors. Therefore the MTD for single injection of C-DOX and LC-DOX in mice is between 20 and 25 mg/kg, compared to less than 15 mg/kg for free DOX.

The antitumor activity was determined in mice inoculated with 10^6 i.p. J6456 cells and treated i.v. with free DOX, C-DOX and LC-DOX (10 mg/kg) on day 5 post-inoculation (Gabizon, 1992). In these experiments the %ILS of mice treated with C-DOX was not significantly different than for mice receiving free DOX (%ILS of 31% with 3% 90 day survivors), but treatment with LC-DOX resulted in a significant improvement in life-span (%ILS of 103%) over the other two treatment groups with 16% long-term (90-day) survivors. Similar results were obtained when LC-EPI was compared to free EPI with free EPI resulting in an %ILS of 39%, with no long-term survivors, and LC-EPI giving an %ILS of 111% with 15% long-term survivors (Gabizon, 1992). Increasing the dose to 15 mg/kg resulted in toxic deaths in the free DOX group. At a dose of 20 mg/kg large liposomes (210 ± 55 nm in diameter), but not smaller liposomes (80 ± 20 nm in diameter), resulted in toxic deaths. The smaller LC-DOX liposomes also resulted in the best survival times at doses from 5 to 20 mg/kg (Gabizon, 1992).

The mechanism suggested for the improved therapeutic effect of the small, long-circulating liposomes is hypothesized to be related to an increase in microvascular permeability which occurs with the development of ascites associated with the presence of tumor cells in the peritoneal cavity. Long-circulating, but not short-circulating, liposomes will be present in the blood for long enough time periods, and in high enough concentrations, to allow them to extravasate and accumulate in the ascitic fluid in sufficient quantities for therapeutic concentrations of the drug to be released in the presence of the tumor cells. The released drug can then be taken up by tumor cells, or redistributed to other body compartments. Sustained release of the drug from liposomes localized in the ascitic fluid will lead to prolonged exposure of the tumor cells to the drug and improved therapeutic effects (Gabizon, 1992). Experiments supporting the hypothesis that the cytotoxic effect is mediated by release of the drug from liposomes into the extracellular medium, and is not as a result of direct cellular uptake, have been published by Howoritz et al, for LC-DOX and S-DOX using a human ovarian carcinoma cell line (OV-1063) in vitro (Horowitz et al, 1992).

Therapy of Human Lung Cancer in SCID Mice

A recent study examines the effect of S-DOX in the treatment of human lung cancer in severe combined immunodeficient (SCID) mice (Williams et al, 1993). A primary culture of human lung squamous cell carcinoma (TL-1) was established and 2 x 10^6 cells were implanted s.c. in SCID mice. Mice were treated weekly for 10 weeks, beginning at 1 week post-implantation, with i.v. doses of 3 mg/kg of either free DOX, C-DOX (PC:CH:PG) or S-DOX (HSPC:CH:PEG-DSPE). Higher doses of either free DOX or S-DOX caused 100% mortality from toxicity in the SCID mice, and some mortality due to toxicity was observed even at the dose of 3 mg/kg (Williams et al, 1993). Mice treated with free DOX experienced only a slight delay in tumor growth. When treated with C-DOX, mice experienced some delay in tumor growth, but the most effective treatment was S-DOX. In mice receiving S-DOX complete arrest of tumor growth was observed (Williams et al, 1993). In some mice, treatment was delayed until 4 weeks post-implantation, and also in these mice S-DOX treatment resulted in complete growth arrest of tumor. In another experiment the dose of DOX in S-DOX was reduced in order to reduce deaths from toxicity. In these experiments, 2 mg/kg S-DOX resulted in complete arrest of tumor growth with 100% survivors (Williams et al, 1993). Lower doses of S-DOX (1.0 or 0.5 mg/kg) resulted in delayed tumor growth, but not arrest of growth. These experiments provide the first direct evidence that anticancer drugs entrapped in S-liposomes have impressive antineoplastic activity in vivo against human solid tumors.

Human Studies

As noted above, animal experiments indicated that the therapeutic efficacy of passively targeted S-liposomal formulations of anticancer drugs were considerably improved relative to C-liposomal formulations or to free drug. The long-circulation times of the S-liposomal formulations were considered to be central to their efficacy. Human pharmacokinetic experiments, as detailed below, also show an improved pharmacokinetic profile over that observed in animals, and compared to free DOX or C-DOX.

A report on the pharmacokinetics of S-DOX (HSPC:CH:PEG-DSPE,DOX-SL^R) in humans has been published for two dose levels of 25 and 50 mg/m^2. (Gabizon et al, 1994). The pharmacokinetic parameters were independent of the dose given, and nearly all of the DOX measured in plasma was accounted for by liposome-associated DOX. The half-life in humans for S-DOX was even longer than in rodents and dogs, with the majority of the drug being removed with a median half-life of 45 hrs (Gabizon et al, 1994) (Table 2). At 50 mg/m^2 of S-DOX the $AUC_{0\rightarrow\infty}$ in humans was 1096 µg hr/ml (2015 µM·hr), with a volume of distribution of 4.8 L, and a clearance rate of 0.071 L/hr. This can be contrasted with recent data published for 90 mg/m^2 C-DOX (PC:CH.55:45) in humans where, even at a higher dose, the average AUC is significantly lower at 13.6 µg hr/ml (25 µM·hr), and the clearance rate and volume of distribution are significantly higher, averaging 21.8 L/hr and 200 L, respectively (Cowens et al, 1993). C-DOX was cleared in humans with a biexponential half-life, with the α half-life averaging 0.45 hrs and the β half-life averaging 13.5 hrs (Cowens et al, 1993) (Table 2), compared to a primarily mono-exponential half-life of 45 hrs for S-DOX in humans (Gabizon et al, 1994).

S-DOX (DOX-SL^R, Liposome Technology Inc) treatment in humans resulted in a 4- to 16-fold increase in drug levels in malignant pulmonary effusions, relative to those achieved with free DOX, likely by the passive targeting mechanism described above. The dose-limiting toxicity for C-DOX was myelosuppression (Mayer et al, 1989), whereas that for S-DOX the most significant side-effect was stomatitis (Gabizon et al, 1994). Cardiotoxicity, which is dose-limiting toxicity for free DOX, was not seen for liposome formulations.

The ability of DOX-SL^R to localize in significant amounts in skin, following i.v. injection, (which may be contributing to the symptoms of stomatitis observed in some patients) suggested that it may be useful for passive targeting to Kaposi's sarcoma lesions. Experiments involving the Kaposi's sarcoma-like dermal lesions that are expressed in transgenic mice bearing the HIV *tat* gene showed that S-liposomes were highly localized in these lesions (Huang et al, 1993). In multi-center trials of treatment with DOX-SL^R in Kaposi's sarcoma patients, about 70% of the patients had partial responses to treatment, and a few patients had complete responses. Cutaneous lesions responded best, flattening and regressing during the course of the treatment, although responses were also observed in visceral lesions. Dose-limiting toxicities in this patient population were haemotological (neutropenia, leukopenia and anemia) with no cardiotoxicity observed (Northfelt et al, 1993). If these promising results continue to be observed, the treatment of Kaposi's sarcoma may be the first clinically approved applications for S-DOX.

SUMMARY

From a consideration of the above results it would appear that entrapment of DOX or EPI in S-liposomes can result in substantial therapeutic benefit as a result of their ability to passively target to solid tumors. The experiments with S-EPI and S-DOX in a number of animal tumor models demonstrate that these formulations show a considerable improvement in therapeutic efficacy over that seen for the respective free drugs or drugs entrapped in C-liposomes. The experiments also demonstrate the strong schedule-dependent nature of the antitumor effect, with maximum therapeutic efficacy found when therapy was begun between 3 and 10 days post-implantation of the tumor. The schedule dependence is likely related to changes in tumor blood supply and degree of permeability of the tumor vasculature to the liposomes. The mechanism for the increased activity for S-EPI or S-DOX as compared to C-liposome formulations is thought to be related to the increased circulation half-lives for S-EPI or S-DOX which leads to increased localization of drug, associated with S-liposomes, in tumors. Once localized in tumors, free drug is thought to be released in a sustained manner from the liposomes into the interstitial fluid at locally elevated concentrations, from whence it taken up by tumor cells. Therefore, internalization of the S-liposomes together with their associated drug, is not necessary for therapeutic activity, unlike the case for immunotoxins, although internalization of the liposome-drug package may result in increased cytotoxicity against the internalizing cells.

REFERENCES

Ahmad, I., Longenecker, M., Samuel, J. and Allen, T.M., 1993, Antibody-targeted delivery of doxorubicin entrapped in sterically stabilized liposomes can eradicate lung cancer in mice, Cancer Res., 53:1484

Allen, T.M. and Hansen, C., 1991, Pharmacokinetics of stealth versus conventional liposomes: effect of dose, Biochim.Biophys.Acta, 1068:133

Allen, T.M., Hansen, C., Martin, F., Redemann, C. and Yau-Young, A., 1991, Liposomes containing synthetic lipid derivatives of poly(ethylene glycol) show prolonged circulation half-lives in vivo, Biochim.Biophys.Acta, 1066:29

Allen, T.M., Hansen, C.B. and Guo, L.S.S., 1993, Subcutaneous administration of liposomes: a comparison with the intravenous and intraperitoneal routes of injection, Biochim.Biophys.Acta, 1150:9

Bakker-Woudenberg, I.A.J.M., Lokerse, A.F., ten Kate, M.T., Mouton, J.W., Woodle, M.C. and Storm, G., 1993, Liposomes with prolonged blood circulation and selective localization in Klebsiella pneumoniae-infected lung tissue, J.Infect.Dis., 168:164

Cowens, J.W., Creaven, P.J., Greco, W.R., Brenner, D.E., Tung, Y., Ostro, M., Pilkiewicz, F., Ginsberg, R. and Petrelli, N., 1993, Initial clinical (Phase I) trial of TLC D-99 (doxorubicin entrapped in liposomes), Cancer Res., 53:2796

Embree, L., Gelmon, K.A., Lohr, A., Mayer, L.D., Coldman, A.J., Cullis, P.R., Palaitis, W., Pilkiewicz, F., Hudon, N.J., Heggie, J.R. and Goldie, J.H., 1993, Chromatographic analysis and pharmacokinetics of liposome-encapsulated doxorubicin in non-small-cell lung cancer patients, J.Pharmaceut.Sci., 82:627

Gabizon, A.A., Barenholz, Y. and Bialer, M., 1993, Prolongation of the circulation time of doxorubicin encapsulated in liposomes containing a polyethylene glycol-derivatized phospholipid: pharmacokinetic studies in rodents and dogs, Pharm.Res., 10:703

Gabizon, A., Catane, R., Uziely, B., Kaufman, B., Safra, T., Cohen, R., Martin, F., Huang, A. and Barenholz, Y., 1994, Prolonged circulation time and enhanced accumulation in malignant exudates of doxorubicin encapsulated in polyethylene-glycol coated liposomes, Cancer Res., 54:987

Gabizon, A. and Papahadjopoulos, D., 1988, Liposome formulations with prolonged circulation time in blood and enhanced uptake by tumors, Proc.Natl.Acad.Sci.USA, 85:6949

Gabizon, A., 1992, Selective tumor localization and improved therapeutic index of anthracyclines encapsulated in long-circulating liposomes, Cancer Res., 52:891

Gabizon, A., Shiota, R. and Papahadjopoulos, D., 1989, Pharmacokinetics and tissue distribution of doxorubicin encapsulated in stable liposomes with long circulation times, J.Natl.Cancer Inst., 81:1484

Gabizon, A., Dagan, A., Goren, D., Barenholz, Y. and Fuks, Z., 1982, Liposomes as in vivo carriers of adriamycin: reduced cardiac uptake and preserved antitumor activity in mice, Cancer Res., 42:4734

Herman, E.H., Rahman, A., Ferrans, V.J., Vick, J.A. and Schein, P.S., 1983, Prevention of chronic doxorubicin cardiotoxicity in beagles by liposomal encapsulation, Cancer Res., 43:5427

Horowitz, A.T., Barenholz, Y. and Gabizon, A.A., 1992, In vitro cytotoxicity of liposome encapsulated doxorubicin: dependence on liposome composition and drug release, Biochim.Biophys.Acta, 1109:203

Huang, S.K., Lee, K.-D., Hong, K., Friend, D.S. and Papahadjopoulos, D., 1992, Microscopic localization of sterically stabilized liposomes in colon carcinoma-bearing mice, Cancer Res., 52:5135
Huang, S.K., Mayhew, E., Gilani, S., Lasic, D.D., Martin, F.J. and Papahadjopoulos, D., 1992, Pharmacokinetics and therapeutics of sterically stabilized liposomes in mice bearing C-26 colon carcinoma, Cancer Res., 52:6774
Huang, S.K., Martin, F.J., Jay, G., Vogel, J., Papahadjopoulos, D. and Friend, D.S., 1993, Extravasation and transcytosis of liposomes in Kaposi's sarcoma-like dermal lesions of transgenic mice bearing the HIV *tat* gene, Am.J.Pathology, 143:10
Maruyama, K., Kendel, S.J. and Huang, L., 1990, Lipid composition is important for highly efficient target binding and retention of immunoliposomes, Proc.Natl.Acad.Sci.USA, 87:5744
Mayer, L.D., Bally, M.B. and Cullis, P.R., 1986, Uptake of adriamycin into large unilamellar vesicles in response to a pH gradient, Biochim.Biophys.Acta, 857:123
Mayer, L.D., Tai, L.C.L., Dicken, D.S.C., Masin, D., Ginsberg, R.S., Cullis, P.R. and Bally, M.B., 1989, Influence of vesicle size, lipid composition and drug-to-lipid ratio on the biological activity of liposomal doxorubicin in mice, Cancer Res., 49:5922
Mayhew, E.G., Lasic, D., Babbar, S. and Martin, F.J., 1992, Pharmacokinetics and antitumor activity of epirubicin encapsulated in long-circulating liposomes incorporating a polyethylene glycol-derivatized phospholipid, Int.J.Cancer, 51:302
Northfelt, D.W., Martin, F.J., Kaplan, L.D., Russell, J., Anderson, M., Lang, J. and Volberding, P.A., 1993, Pharmacokinetics, tumor localization, and safety of Doxil™ (liposomal doxorubicin) in AIDS patients with Kaposi's sarcoma, Proc.Am.Soc.Clin.Oncol., 12:51
Papahadjopoulos, D., Allen, T.M., Gabizon, A., Mayhew, E., Matthay, K., Huang, S.L., Lee, K.-D., Woodle, M.C., Lasic, D.D., Redemann, C. and Martin, F.J., 1991, Sterically stabilized liposomes: improvements in pharmacokinetics and antitumor therapeutic efficacy, Proc.Natl.Acad.Sci.USA, 88:11460
Robert, J., Vrignaud, P., Nguyen-Ngoc, T., Lliadis, A., Mauriac, L. and Hurteloup, P., 1985, Comparative pharmacokinetics and metabolism of doxorubicin and epirubicin in patients with metastatic breast cancer, Cancer Treat.Rep., 69:633
Torchilin, V.P., Klibanov, A.L., Huang, L., O'Donnell, S., Nossiff, N.D. and Khaw, B.A., 1992, Targeted accumulation of polyethylene glycol-coated immunoliposomes in infarcted rabbit myocardium, FASEB J., 6:2716
Treat, J., Greenspan, A., Forst, D., Sanchez, J.A., Ferrans, V.J., Potkul, L.A., Woolley, P.V. and Rahman, A., 1990, Antitumor activity of liposome-encapsulated doxorubicin in advanced breast cancer: Phase II study, J.Natl.Cancer Inst., 82:1706
Vaage, J., Mayhew, E., Lasic, D. and Martin, F.J., 1992, Therapy of primary and metastatic mouse mammary carcinomas with doxorubicin encapsulated in long circulating liposomes, Int.J.Cancer, 51:942
Vaage, J. and Constanza, M.E., 1992, Effects of combination drug therapy on the subcutaneous and pulmonary growth of a slow- and fast-growing C3H/He mammary carcinoma, Cancer Res., 39:4466
Vaage, J. and Harlos, J.P., 1991, Collagen production by macrophages in tumor encapsulation and dormancy, Brit.J.Cancer, 63:759
Van Hoesel, Q.G.C.M., Steerenberg, P.A., Crommelin, D.J.A., van Dijk, A., van Oort, W., Klein, S., Douze, J.M.C., de Wildt, D.J. and Hillen, F.C., 1984, Reduced cardiotoxicity and nephrotoxicity with preservation of antitumor activity of doxorubicin entrapped in stable liposomes in the LOU/M Wsl rat, Cancer Res., 44:3698
Williams, S.S., Alosco, T.R., Mayhew, E., Lasic, D.D., Martin, F.J. and Bankert, R.B., 1993, Arrest of human lung tumor xenograft grown in severe combined immunodeficient mice using doxorubicin encapsulated in sterically stabilized liposomes, Cancer Res., 53:3964
Woodle, M.C., Matthay, K.K., Newman, M.S., Hidayat, J.E., Collins, L.R., Redemann, C., Martin, F.J. and Papahadjopoulos, D., 1992, Versatility in lipid compositions showing prolonged circulation with sterically stabilized liposomes, Biochim.Biophys.Acta, 1105:193
Zu, N.Z., Da, D., Rudoll, T.L., Needham, D., Whorton, A.R. and Dewhirst, M.W., 1993, Increased microvascular permeability contributes to preferential accumulation of Stealth liposomes in tumor tissue, Cancer Res., 53:3765

LIPOSOMAL DRUG DELIVERY: POSSIBILITIES FOR MANIPULATION

G. Storm[1], I.A.J.M. Bakker-Woudenberg[2], M.C. Woodle[3], G. Blume[4], U.K. Nassander[1], M.H. Vingerhoeds[1], H. Haisma[5], D.J.A. Crommelin[1]

[1]Dept of Pharmaceutics, Utrecht University, PO Box 80 082, 3508 TB Utrecht, The Netherlands; [2]Dept of Clinical Microbiology, Erasmus University, Rotterdam, The Netherlands; [3]Liposome Technology Inc., Menlo Park, CA, USA; [4]Medizinische Biophysik, Urologische Klinik und Poliklinik der Technischen Universität, München, Germany; [5]Dept of Medical Oncology, Free University Hospital, Amsterdam, The Netherlands

INTRODUCTION

Delivery of the drug at a specific site (drug targeting) or controlled and/or prolonged release of the liposome-bound drug are the major considerations for adding liposomes to the existing arsenal of drug delivery systems. In particular the concept of liposomal drug targeting has been evolving rapidly in the past 10 years with the development of "second generation" carriers such as immunoliposomes (liposomes bearing covalently coupled antibodies as homing device) and, more recently, the long-circulating liposomes. This contribution deals with novel approaches in the field of liposomal drug delivery based on the use of long-circulating liposomes and immunoliposomes.

LONG CIRCULATING LIPOSOMES

The recent development of long-circulating liposomes (LCL) showing reduced affinity to the mononuclear phagocyte system (MPS) can be considered as a major breakthrough in liposomal drug delivery. Initially, these LCL formulations contained glycolipids such as monosialoganglioside (G_{M1}) or hydrogenated phosphatidylinositol (HPI) in "rigid" bilayers (Allen et al, 1989; Gabizon and Papahadjopoulos, 1988). Further advances with respect to prolongation of blood circulation times for liposome formulations were achieved by inclusion of the hydrophilic polymer polyethylene glycol (PEG) conjugated to the synthetic phospholipid distearoylphosphatidylethanolamine (PEG-DSPE) (Allen et al, 1991; Blume and Cevc, 1993; Mori et al, 1991). It has been proposed for PEG-containing liposomes (Lasic et al, 1991; Woodle and Lasic, 1992) that steric stabilization of liposomes is responsible for their prolonged circulation times. Stabilization results from local surface concentration of highly hydrated PEG groups that sterically inhibit both electrostatic and hydrophobic interactions of a variety of blood and cellular components at the liposome surface.

Targeting of Drugs 4, Edited by G. Gregoriadis *et al.*
Plenum Press, New York, 1994

For a variety of medical applications, LCL may offer significantly improved treatment options. However, due to the early stage of their development, the therapeutic advantages of LCL have not been widely evaluated yet. The three approaches receiving most attention are:

1. *distribution to non-MPS sites*: Evidence has been presented showing that non-MPS body compartments, such as implanted tumors, sites of infection and inflammation, skin and carcass, are much more accessible to LCL (in particular small ones (<0.1 μm)) than to conventional liposomes;
2. *circulating microreservoir*: Extending the circulation time of relatively large liposomes (0.2 - 0.4 μm) offers the potential to create an encapsulated circulating reservoir. If LCL can be engineered to release entrapped drugs during circulation, the therapeutic availability of drugs which are rapidly degraded or excreted in their free form (e.g. many therapeutic peptides and proteins) may be substantially increased using this approach;
3. *immunospecific targeting*: The combination of the LCL approach for longer circulation with the attachment of antibodies or antibody fragments for drug targeting may provide a means of delivery of drugs to their sites of action with a high degree of specificity (the "magic bullet" revisited). Potential disease candidates include leukemias, lymphomas, graft versus host diseases, and HIV.

In conclusion, LCL offer therapeutic opportunities which were up until now unrealistic due to efficient MPS uptake of conventional liposomes. Below several examples will be given of ongoing research in our laboratories to illustrate the utility of the above mentioned approaches.

Distribution to non-MPS sites: treatment of bacterial infections located outside the MPS

Liposomes, when administered intravenously, rapidly accumulate in MPS cells, particularly those residing in the liver and spleen. This natural behaviour of liposomes can be exploited in the treatment of infectious diseases in two ways. Firstly, liposomes can be used for targeting of immunomodulators giving an enhancement of nonspecific resistance to infections caused by a variety of organisms (Melissen et al, 1992). Secondly, in intracellular infections involving the MPS it has been shown that liposomal encapsulation of antimicrobial agents can result in enhancement of their therapeutic activity, as demonstrated in experimental models of protozoal-, fungal-, and viral infections (Emmen and Storm, 1987; Swenson et al, 1988).

The applicability of liposomes in the treatment of infectious diseases located outside the MPS has been strongly limited by their short residence time in blood due to the fast uptake by the MPS. Targeting of liposome-encapsulated antibiotics to infections outside the MPS would require a prolonged circulation time of liposomes. Several studies aimed at prolonging liposome circulation time revealed that small, neutral, cholesterol-rich liposomes composed of "rigid" lipids can attain prolonged circulation but mostly at relatively high lipid doses (Senior, 1987). Recently, incorporation of specific glycolipids, the most well-known examples being G_{M1} and HPI, has been shown to result in prolonged circulation without the constraint of high lipid doses but still limited to rigid lipids, as described in a recent review on this subject (Woodle and Lasic, 1992). More recently, many reports have shown that PEG-PE incorporated into liposomes also yield enhanced blood circulation times to an extent equal or exceeding that found previously, but without limitations either of lipid dose or of rigid composition (Woodle and Lasic, 1992). These long-circulating liposomes (LCL) are able to oppose rapid uptake by the MPS. Importantly, it was shown that LCL exhibit enhanced localization in a variety of implanted tumors in animals as well as non-MPS body compartments, such as skin and carcass (Gabizon et al, 1990; Woodle and Lasic, 1992).

We have recently demonstrated in a rat model of Klebsiella pneumoniae infection of the left lung that i.v. injected LCL show enhanced localization in infected lung tissue (Bakker-Woudenberg et al, 1992; Bakker-Woudenberg et al, 1993). The liposome types

Table 1. Liposome types studied

"clasical"	LCL
EPG/EPC/Chol (1:10:5) 110 nm	HPI/HSPC/Chol (1:10:5) 100 nm
HSPC/Chol (2:1) 110 nm	PEG-DSPE/PHEPC/Chol (0.15:1.85:1) 85 nm

Mean size is given. EPG, egg-phosphatidylglycerol; EPC, egg-phosphatidyl-choline; Chol, cholesterol; HPI, hydrogenated phosphatidylinositol; HSPC, hydrogenated soy phosphatidyl-choline; PHEPC, partially hydrogenated phosphatidylcholine; PEG-DSPE, distearoylphosphatidylethanolamine derivatized with polyethylene glycol (mol.wt.1900)

(LCL versus "classical" types) studied at a lipid dose of 75 µmol total lipid/kg are shown below.

It was found that the PEG-containing liposomes show the longest circulation time. At 40 h after administration to uninfected and infected rats, 10-20-% of the injected liposomal dose was still circulating.

At 40 h after administration up to 9% of the injected dose of PEG-containing liposomes could be recovered from the infected left lung tissue. Increased localization was highly correlated with the intensity of the infection, and was also dependent on the liposomal dose administered. Interestingly, also the "classical" HSPC/Chol liposome type, being small, neutral and rigid showed prolonged circulation and substantial localization in infected lung tissue at the relatively high lipid dose used. A valuable asset of the LCL containing PEG-PE is that prolonged circulation can be achieved over a wide lipid dose range. Experiments to eveluate the therapeutic activity of LCL loaded with antibiotics are presently being conducted.

Circulating microreservoir: improvements in systemic delivery of peptide drugs

The development of bioactive peptides and proteins as therapeutic agents is strongly limited in many cases by a lack of oral bioavailability and rapid clearance from the blood. LCL might improve their delivery by functioning as a circulating "microreservoir" for sustained release while providing protection and reducing immunogenicity and adverse side effects (Storm et al, 1991). Studies were undertaken to investigate the utility of LCL for sustained i.v. delivery of the antidiuretic hormone vasopressin (VP) (Woodle et al, 1992). VP is representative of low molecular weight peptides that are not orally active and are rapidly cleared. The hormone was encapsulated in large unilamellar liposomes of about

Table 2. Localization in severely infected lungs at times that about 10% of the injected liposome dose is circulating in the blood

"clasical"		LCL	
EPG/EPC/Chol	<<1% (5h)	HPI/HSPC/Chol	1% (16 h)
HSPC/Chol	3% (40 h)	PEG-DSPE/PHEPC/Chol	9% (40 h)

Times after injection between parentheses

250 nm prepared by reverse phase evaporation followed by extrusion. Bioactivity was determined by measurements of diuresis of Brattleboro rats which have a hereditary deficiency in VP production but possess functional VP receptors. The cholesterol content of the LCL was found to have a controlling effect on VP release in serum. Three types of liposomes were selected for the urine production measurements. One contained phosphatidylserine (PS), which was rapidly cleared from the circulation. In the other two types of liposomes, the PS component was replaced by either PG or PEG-DSPE; both types showed prolonged circulation.

I.v. injection of free VP (up to 8 µg/kg) gave reduced urine production for less than 24 h. The PG formulation exhibited a dose-dependent prolonged duration of bioactivity of up to 4 days. When VP was administered in the LCL containing PEG-DSPE, a very different antidiuretic response pattern was observed. Two apparently distinct antidiuretic responses were observed: a modest reduction of diuresis in the first 24 h, followed by a second more pronounced reduced in diuresis beginning on day 3 and persisting for at least four additional days. The reduction in diuresis on the first day after i.v. administration of the PEG-DSPE liposomes could be accounted for entirely by the unencapsulated VP present in the formulation. It was therefore concluded that VP was not released from these liposomes during the first two days post-injection. We reasoned that the lack of VP release from the circulating LCL might be related to the relatively high cholesterol content (33 mol%) of the liposomes. And, indeed, lowering of the liposomal cholesterol content resulted in a prolonged duration of the first antidiuretic response, due to the release of VP from circulating LCL. Interestingly, the duration of the second antidiuretic response appearing on day 3 was not dose dependent and far in excess of the time required to clear these liposomes from the circulation. This was attributed to VP release from liposomes which have distributed intact to another compartment without having been taken up by MPS cells.

In conclusion, the results indicate that by balancing liposome circulation time, release rate, and dose, LCL can be applied to prolong the biological activity of a therapeutic peptide.

Immunospecific targeting: long-circulating proteoliposomes with the homing devices coupled to the ends of the polymeric PEG chains

Perhaps the most significant barrier to be surmounted in the field of liposomal drug targeting is the rapid scavenging by the cells of the MPS. For immunoliposomes and other targeted proteoliposomes, MPS capture suggests that those liposomes which are available initially to interact with their target and fail to do so on their first pass will probably be engulfed before encountering the target for a second time (Wright and Huang, 1989). The recent success in MPS avoidance by utilizing long-circulating liposomes suggests the possibility of overcoming this barrier.

Inhibition of the specific and/or nonspecific interactions between PEG-modified liposomes and the MPS has been invoked to explain the favorable characteristics of long-circulating liposomes. The first tests of this hypothesis have been reported (Blume and Cevc, 1990; Lasic et al, 1991; Woodle et al, 1992). Unfortunately, surface attached PEG-groups can inhibit more than just the uptake of liposomes by the MPS. Long-headed PEG(5000)-PE molecules also prevent the efficient binding of corresponding surface-modified proteoliposomes to their targets (Klibanov and Huang, 1992; Maruyama et al, 1991; Mori et al, 1991; Torchilin et al, 1992). It has, therefore, been suggested to use shorter PEG-segments (Mori et al, 1991). In practice, however, it was observed that even shorter chain molecules (PEG(1900)) on the liposome surface, when used at required high concentrations, hamper the interaction between the target receptors and the homing devices on such pegylated liposomes strongly (Mori et al, 1991). Alternatively, the use of longer spacers with the homing devices at their ends has therefore been discussed with the same goal in mind (Klibanov and Huang, 1992): to develop long-circulating proteoliposomes with

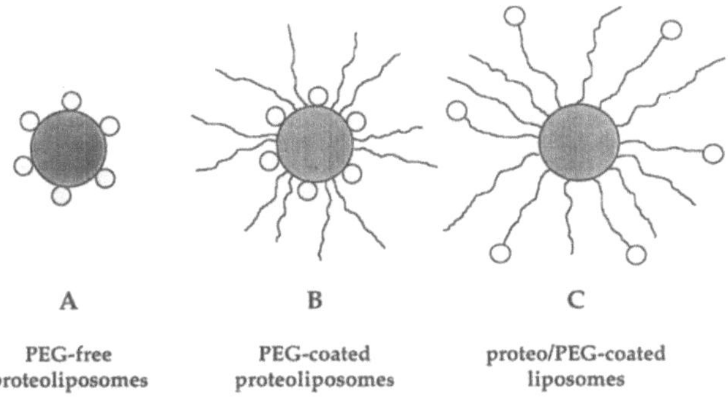

Fig.1 Schematic representation of the proteoliposome types studied

optimal target site affinity. We have followed another approach by coupling the target-recognizing proteins (glu-plasminogen was used) to the ends of the PEG residues which render the liposomes long-lived (Blume et al, 1993).

We have studied the properties of various surface-modified liposomes which are schematically shown in Fig. 1.

This representation makes clear that the PEG-layers on liposomal surfaces may, but need not, mask the interactions between the homing devices and their targets.

The potential problem of PEG-dependent reduction of the liposome-target association was assessed by measuring the binding of the liposome-associated plasminogen to its natural target, fibrin, in vitro (Fig. 2).

The degree of fibrin binding of the proteo/PEG liposomes (type C) appeared to be similar to that of the PEG-free proteoliposomes (type A). This shows that the binding capability of the glu-plasminogen molecules is maintained upon coupling to the ends of the polymeric chains. In contrast to this, the degree of binding of the PEG-coated proteoliposomes (type B) to fibrin is negligible. This loss of binding illustrates the potential

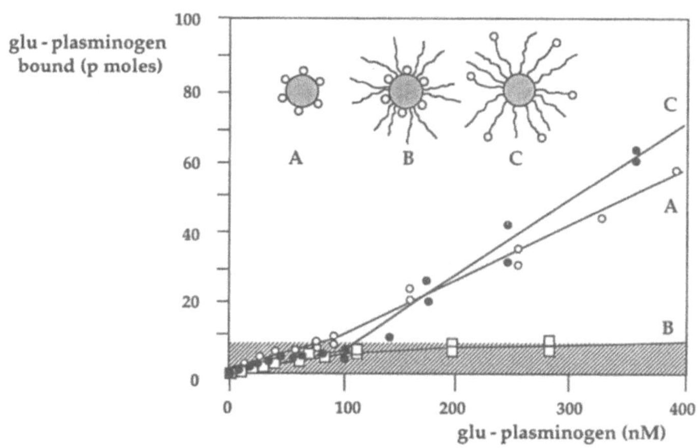

Fig.2 Binding of proteoliposomes to fibrin in vitro

masking problem: the proteins associated with type B proteoliposomes are buried in the PEG-layer such that target recognition is severely hampered.

In addition to target bindingcapability, we also examined whether the glu-plasminogen molecules attached to the outer surface of the pegylated liposomes (types B and C) obviate the prolonged circulation time in vivo (Fig. 3).

In line with expectations, it was found in mice that the PEG-coated proteoliposomes (type B, around 25% of the administered i.v. dose was still circulating at 24 h after i.v. injection) remain in the circulation much longer than the PEG-free proteoliposomes (type A, about 4% of the injected dose at 24 h after injection). This demonstrates the ability of surface pegylation to prolong the circulation time of liposomes even in the presence of the surface-bound proteins. Importantly, the proteo/PEG-coated liposomes (type C) also remained in the blood circulation much longer than the standard, PEG-free proteoliposomes (type A). Typically, around 20% of the injected dose was still circulating at 24 h after adminstration. This demonstrates that the liposome targeting by means of homing devices attached directly to the end of the PEG chains is compatible with the vesicle longevity in the bloodstream. We believe that such pegylated liposomes with the terminally attached targeting moieties will provide valuable assets for the site-specific targeting of the liposomal drug carriers in vivo.

IMMUNOLIPOSOMES IN THE TREATMENT OF OVARIAN CANCER

In certain conditions, the diseased tissue is confined to a particular cavity in the body. This is true, for instance, for ovarian carcinoma, which usually stays throughout its clinical life mainly within the peritoneal cavity. After diagnosis the prognosis for ovarian carcinoma patients is poor. The primary tumors and larger tumor nodules can be removed surgically, but micronodules and floating tumor cells present in the peritoneal cavity cannot be treated adequately by surgery. Intraperitoneal injection of cytostatic solutions has been used with some success, but the residence time in the cavity is short providing insufficient opportunity for deep intratumoral penetration and a major fraction still reaches the systemic circulation (Los et al, 1989). Sustained release of cisplatin, a cytostatic regularly used in

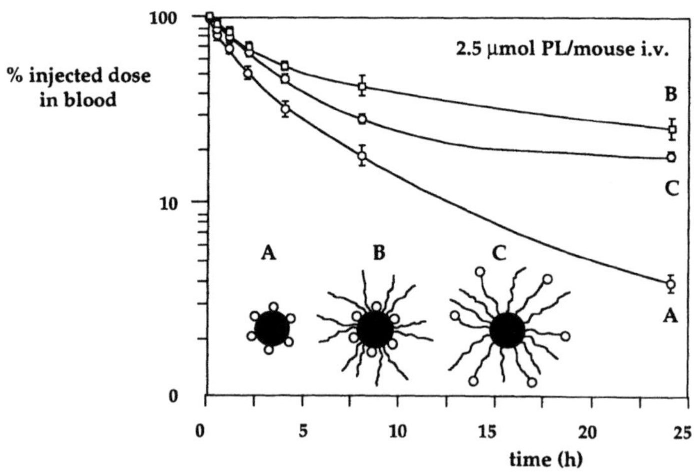

Fig.3 Blood circulation kinetics of [^{3}H]cholesterylhexadecylether-labeled proteoliposomes after i.v. administration to mice (2.5µmol phospholipid per mouse)

the treatment of ovarian carcinomas, from hydrogel rods implanted in the peritoneal cavity reduced the dose-limiting renal toxicity of the drug, but showed also a negative effect on e antitumor response (Los et al, 1990). As sustained release in the peritoneal cavity alone did not improve the therapeutic index of the drug, targeted delivery of cytostatics to the ovarian carcinoma cells with immunoliposomes was a logical next step in the optimization of chemotherapy. Here the aim is to deliver drug-laden liposomes directly at the surface of - or preferably in - the ovarian carcinoma cells (Straubinger et al, 1988). In vitro immunoliposomes bearing methotrexate-gamma-aspartate showed an 8-fold increase in potency against human ovarian carcinoma cells. After intraperitoneal injection, a specific interaction between these human ovarian carcinoma cells and the immunoliposomes was observed qualitatively. We performed similar experiments and quantified in vitro the binding of immunoliposomes with a murine Fab' fragment (OV-TL3) to the human ovarian carcinoma-3 (OVCAR-3) cells (Nässander, 1991). Binding efficacy and specificity could be clearly demonstrated (Fig. 4. RIV1000 served as irrelevant antibody).

Figure 5 shows results on the in vivo interaction between these immunoliposomes and target cells in a xenograft model (athymic nude mice) of i.p. growing OVCAR-3 cells. A rapid, highly efficient and long-lasting adherence to the tumor cells could be demonstrated (Nassander et al, 1992).

However, even though immunoliposomes may be tightly bound to the target cell surface, the generation of sufficient site-specific therapeutic availability of the encapsulated drug is not guaranteed. In principle, two pathways for drug entry into the target cell can be envisaged. Firstly, uptake of the immunoliposomes by the cells or fusion with the cells, followed by intracellular release of the drug may occur (Wright and Huang, 1989). Unfortunately, however, many tumor cells are not capable of internalizing liposomes efficiently. Secondly, release of the encapsulated drug in the close proximity of the target cells may result in a considerable cellular drug uptake. It is necessary that the released

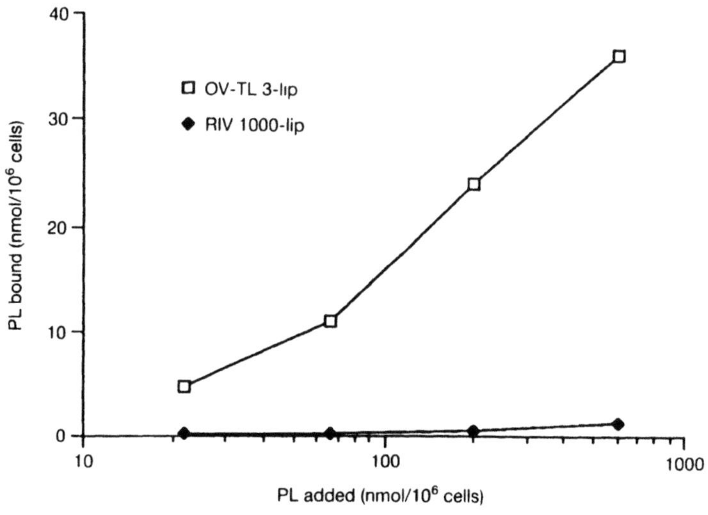

Fig.4 Binding of immunoliposomes to OVCAR-3 cells. Various amounts of OVTL-3 liposomes (11 µg Fab'/µmol phospholipid) and RIV1000 liposomes (7 µg/µmol phospholipid) were incubated with OVCAR-3 cells (2 x 10^6 cells/ml). RIV1000 is a non-relevant antibody. The cells were washed three times by centrifugation to separate bound Fab'liposomes from the unbound vesicles. The amount of cell binding was determined after lysing the liposomes, thereby releasing the entrapped carboxyfluorescein, a marker for the liposome integrity

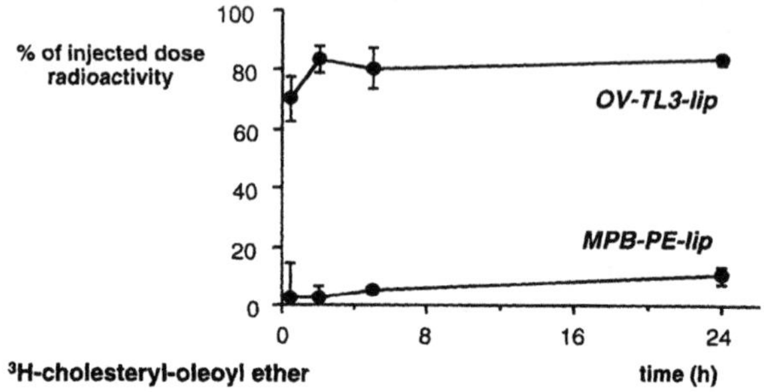

Fig.5 Double radiolabeled liposomes associated with NIH:OVCAR-3-cells in the peritoneal cavity. Degree of binding of OV-TL3 immunoliposomes and unconjugated liposomes (bearing no specific Fab' but only the anchor molecule MPB-PE) to OVCAR-3 cells present in the peritoneal cavity of athymic nude mice. [^{3}H)cholesteryoleylether was incorporated to monitor the target cell binding in vivo

drug is rapidly taken up by the tumor cells, which depends on its physicochemical nature (e.g. lipophilicity). Methods to construct immunoliposomes to release entrapped agents through environmental manipulations (e.g. in response to changes in pH or temperature) are presently being investigated in our laboratory.

IMMUNO-ENZYMOSOMES FOR SITE-SPECIFIC ACTIVATION OF PRODRUGS

In an attempt to avoid potential problems due to insufficient therapeutic availability of the encapsulated drug once an immunoliposome-cell interaction has been established, we started to study an indirect drug targeting approach. This approach concerns immunoliposome-mediated targeting of enzymes, capable of converting relatively non-toxic prodrugs into active cytotoxic agents. The enzyme β-glucuronidase, capable of activating the prodrug epirubicin-glucuronide, was recently coupled to the external surface of immunoliposomes directed towards ovarian cancer cells. These immuno-enzymosomes

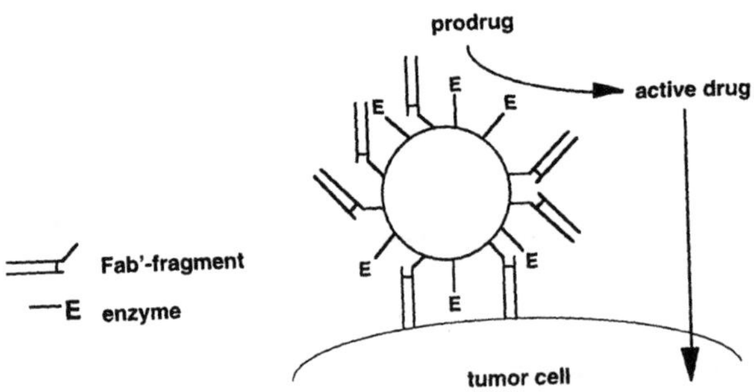

Fig.6 "Drug targeting" with immuno-enzymosomes. Schematic representation of the concept of antibody-directed enzyme prodrug therapy with immuno-enzymosomes. The immuno-enzymosomes are first allowed to bind to the target cells. Then a prodrug is given, which is activated by the cell-bound immuno-enzymosomes in close proximity of the target cell. The active drug can subsequently kill the cell

were tested for their enzymatic activity, stability and capability to bind to the tumor cells. After in vitro binding of these "immuno-enzymosomes" to OVCAR-3 cells, the prodrug was administered. Figure 6 illustrates the strategy of converting a prodrug to the active drug by cell-bound immuno-enzymosomes. Only preincubation of the target cells with immuno-enzymosomes resulted in an enhancement of the cytotoxicity of the added prodrug (Vingerhoeds et al, 1993). Pretreatment with enzymosomes (bearing no specific Fab′ fragments) or immunoliposomes (bearing no enzymes) was clearly ineffective. This demonstrates the ability of the cell-bound immuno-enzymosomes to convert the prodrug to an agent with increased antineoplastic activity.

The use of enzymes for selective prodrug activation at the tumor site has been described earlier as antibody-directed prodrug therapy (ADEPT). In this approach, enzyme is linked to an antibody that binds an antigen preferentially expressed on tumor cells. The few studies carried out with antibody-enzyme conjugates in vitro and in vivo models showed that selective conversion of prodrug into active drug at the tumor site can be obtained (Bagshawe et al, 1988; Haisma et al, 1992). A theoretical advantage of the present immuno-enzymosome approach, as compared to the use of antibody-enzyme conjugates, is that much more than one enzyme molecule can be delivered to the tumor site by a singly targeted liposome. This offers the possibility to increase the enzyme density at the target cell surface, and thereby to induce a higher efficiency of specific prodrug activation.

REFERENCES

Allen, T.M., Hansen, C. and Rutledge, J., 1989, Liposomes with prolonged circulation times: factors affecting uptake by reticuloendothelial and other tissues, Biochim.Biophys.Acta, 981:27

Allen, T.M., Hansen, C., Martin, F., Redemann, C. and Yau-Young, A., 1991, Liposomes containing synthetic lipid derivatives of poly(ethylene glycol) show prolonged circulation half-lives in vivo, Biochim.Biophys.Acta, 1066:29

Bagshawe, K.D., Springer, C.J., Searle, F., Antoniw, P., Sharma, S.K., Melton, R.G. and Sherwood, R.F., 1988, A cytotoxic agent can be generated selectively at cancer sites, Br.J.Cancer, 58:700

Bakker-Woudenberg, I.A.J.M., Lokerse, A.F., Ten Kate, M.T. and Storm, G., 1992, Enhanced localization of liposomes with prolonged blood circulation time in infected lung tissue, Biochim.Biophys.Acta, 1138:318

Bakker-Woudenberg, I.A.J.M., Lokerse, A.F., Ten Kate, M.T., Mouton, J.W., Woodle, M.C. and Storm, G., 1993, Liposomes with prolonged blood circulation and selective localization in Klebsiella pneumoniae infected lung tissue, J.Infec.Dis., 168:164

Blume, G. and Cevc, G., 1990, Liposomes for the sustained drug release in vivo, Biochim.Biophys.Acta, 1029:91-97

Blume, G., Cevc, G., Crommelin, D.J.A., Bakker-Woudenberg, I.A.J.M., Kluft, C. and Storm, G., 1993, Specific targeting with polyethyleneglycol-modified liposomes: coupling of homing devices to the ends of the polymeric chains combines effective target binding with long circulation times, Biochim.Biophys.Acta, 1149:180

Emmen, F. and Storm, G., 1987, Liposomes in treatment of infectious diseases, Pharm.Weekbl., (Sci.Ed.) 9:162

Gabizon, A. and Papahadjopoulos, D., 1988, Liposome formulations with prolonged circulation time in blood and enhanced uptake by tumors, Proc.Natl.Acad.Sci.USA, 85:6949

Gabizon, A., Price, D.C., Huberty, J., Bresalier, R.S. and Papahadjopoulos, D., 1990, Effect of lipid composition and other factors on the targeting of liposomes to experimental tumors: biodistribution and imaging studies, Cancer Res., 50:6371

Haisma, H.J., Boven, E., Van Muijen, M., De Jong, J., Van der Vijgh, W.J.F. and Pinedo, H.M., 1992, A monoclonal antibody-β-glucuronidase conjugate as activator of the prodrug epirubicin-glucuronide for specific treatment of cancer, Br.J.Cancer, 66:474

Klibanov, A.L. and Huang, L., 1992, Long-circulating liposomes: developments and perspectives, J.Liposome Res., 2:321
Lasic, D.D., Martin, F.J., Gabizon, A., Huang, S.K. and Papahadjopoulos, D., 1991, Sterically stabilized liposomes - A hypothesis on the molecular origin of the extended circulation times, Biochim.Biophys.Acta, 1070:187
Los, G., Mutsaerts, P.H.A., Van der Vijgh, W.J.F., Baldew, G.S., Van der Graaf, P.W. and McVie, J.G., 1989, Direct diffusion of cis-diamminedichloroplatinum(II) in intraperitoneal chemotherapy: a comparison with systemic chemotherapy, Cancer Res., 49:3380
Los, G., 1990, Experimental basis of intraperitoneal chemotherapy. Thesis, Free University Amsterdam, The Netherlands,
Maruyama, K., Mori, A., Bhadra, S., Subbiah, M.T.R. and Huang, L., 1991, Proteins and peptides bound to long-circulating liposomes, Biochim.Biophys.Acta, 1070:246
Melissen, P.M.B., Van Vianen, W., Rijsbergen, Y. and Bakker-Woudenberg, I.A.J.M., 1992, Free versus liposome-encapsulated muramyl tripeptide phosphatidyllethanolamine in treatment of experimental Klebsiella pneumoniae infections, Infect.Immun., 60:95
Mori, A., Klibanov, A.L., Torchilin, V.P. and Huang, L., 1991, Influence of the steric barrier activity of amphipathic polyethyleneglycol and ganglioside GM1 on the circulation time of liposomes and on the target binding of immunoliposomes in vivo, FEBS Lett., 284:263
Nassander, U.K., 1991, Liposomes, immunoliposomes and ovarian carcinoma. Thesis, University of Utrecht, The Netherlands
Nassander, U.K. and Crommelin, D.J.A., 1992, In vivo targeting of OV-TL3 immunoliposomes to ascitic ovarian carcinoma cells (OVCAR-3) in athymic nude mice, Cancer Res., 52:646
Senior, J.H., 1987, Fate and behaviour of liposomes in vivo: a review of controlling factors, CRC Crit.Rev.Ther.Drug Carrier Syst., 3:123
Storm, G., Wilms, H.P. and Crommelin, D.J.A., 1991, Liposomes and biotherapeutics, Biotherapy, 3:25
Straubinger, R.M., Lopez, N.G., Debs, R.J., Hong, K. and Papahadjopoulos, D., 1988, Liposome-based therapy of human ovarian cancer: parameters determining potency of negatively charged and antibody-targeted liposomes, Cancer Res., 48:5237
Swenson, C.E., Popescu, M.C. and Rinsberg, R.S., 1988, Preparation and use of liposomes in the treatment of microbial infections, CRC Crit.Rev.Microbiol., 15 (suppl. 1):1
Torchillin, V.P., Klibanov, A.L., Huang, L., O'Donnell, S., Nossif, N.D. and Khaw, B.A., 1992, Targeted accumulation of polyethylene glycol-coated immunoliposomes in infarcted rabbit myocardium, The FASEB J., 6:2716
Vingerhoeds, M.H., Haisma, H.J., Van Muijen, M., Van der Rijt, R.B.J., Crommelin, D.J.A. and Storm, G., 1993, A new application for liposomes in cancer therapy: immunoliposomes bearing enzymes (immuno-enzymosomes) for site-specific activation of prodrugs, FEBS Lett., 336:485
Woodle, M.C., Storm, G., Newman, M.S., Jekot, J., Collins, L.R., Martin, F.J. and Szoka, F.C., 1992, Prolonged systemic delivery of peptide drugs by long circulating liposomes: illustration with vasopressin in the Brattleboro rat, Pharm.Res., 9:260
Woodle, M.C., Mattay, K.K., Newman, M.S., Hidayat, J.E., Collins, L.R., Redemann, C., Martin, F.J. and Papajahdopoulos, D., 1992, Versatility in lipid compositions showing prolonged circulation with sterically stabilized liposomes, Biochim.Biophys.Acta, 1105:193
Woodle, M.C. and Lasic, D.D., 1992, Sterically stabilized liposomes, Biochim. Biophys. Acta, 1113:171
Wright, S. and Huang, L., 1989, Antibody-directed liposomes as drug-delivery vehicles, Adv.Drug Deliv.Rev., 3:343

POLYSIALIC ACIDS: IN VIVO PROPERTIES AND POSSIBLE USES IN DRUG DELIVERY

Brenda McCormack and Gregory Gregoriadis

Centre for Drug Delivery Research, The School of Pharmacy,
University of London, Brunswick Square, London WC1N 1AX, UK

INTRODUCTION

Alongside the development of new drugs and improving the efficacy of established drugs is the need to design ever more effective carrier systems to enhance pharmacological drug action. Conventional use of promising compounds can have many shortcomings in vivo, such as failure to reach target tissues - thus giving rise to undesirable side-effects, inability to penetrate intracellular sites, reduced residence time within the vascular system and extravascular areas (Florence and Gregoriadis, 1991) all pre-empting their therapeutic purpose. During the last two decades a variety of delivery systems have been investigated in an attempt to circumvent these problems and have had variable success in directing drugs to tissues in need of treatment and enabling their release there. Such carriers are antibodies, including monoclonal (Nelson, 1991), ligands designed to bind specifically to cell surface receptors and antigens, polymers with distinct attached ligands, and particulate systems such as liposomes and microspheres. More recent interest has focused on attempts to prolong the presence within vascular and extravascular areas of many established drugs such as antibiotics and cytostatics, as well as a vast range of peptides and proteins which include hormones, cytokines, enzymes and antibodies, all of which are known to be excreted or otherwise removed rapidly, rendering them pharmacologically valueless, unless given in unacceptably toxic amounts. Extending the residence time of liposomes (Gregoriadis, 1988) and other colloidal systems (Roerdink and Kroon, 1989) would also increase their flexibility allowing perhaps contact with tissues other than those of the reticuloendothelial system (RES) by which they are mainly intercepted. (Gregoriadis, 1988; Roerdink and Kroon, 1989). This has been done successfully by coating liposomes and other particles with low molecular weight (750-5000) monomethoxypoly(ethyleneglycol)(mPEG) (Blume and Cevc, 1990; Klibanov et al, 1990; Senior et al, 1991; Papahadjopoulos et al, 1991) or poloxamers (Moghimi et al, 1991; Jamshaid et al, 1988). mPEG seems to act by forming a shell of mPEG molecules (Nucci et al, 1991) around the surface of proteins and particles which sterically interferes with factors in the biological milieu responsible for their removal. However, low molecular weight mPEG, which is known (Nucci et al, 1991) to be excreted rapidly by the kidneys, is of little benefit in extending the half-lives in the circulation of many useful small peptides and conventional drugs.

We have investigated an alternative type of macromolecule which appears to have the same advantages as mPEG. These macromolecules are the naturally occurring polymers of N-acetyl neuraminic acid (NeuNAc), referred to here as polysialic acids, and include the serogroup B capsular polysaccharide from Neisseria meningitidis B and Escherichia Coli

Targeting of Drugs 4, **Edited by G. Gregoriadis *et al.***
Plenum Press, New York, 1994

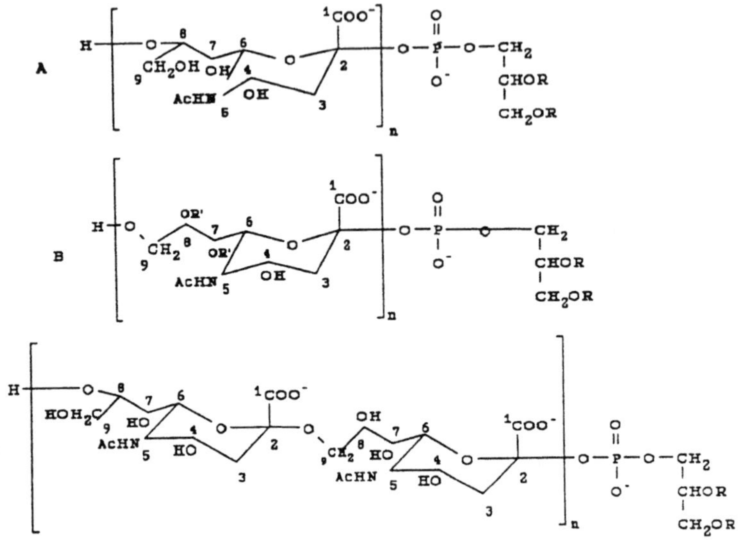

Fig. 1 (A) Serogroup B capsular polysialic acid (PSB) from N.meningitidis B or E.coli K1 is a homopolymer (n=199) of α-(2-8)-linked N-acetyl neuraminic acid. (B) Serogroup C capsular polysialic acid (PSC) from N.meningitidis C is a homopolymer (n=74) of α-(2-9)-linked N-acetyl neuraminic acid; R′:H or Ac. (C) Polysialic acid (PSK92) from E.coli K92 is a heteropolymer (n=78) of alternate units of α-(2-8)-α-(2-9)-linked N-acetyl neuraminic acid. All three polysialic acids contain a phospholipid molecule covalently linked to the reducing end of the polymers

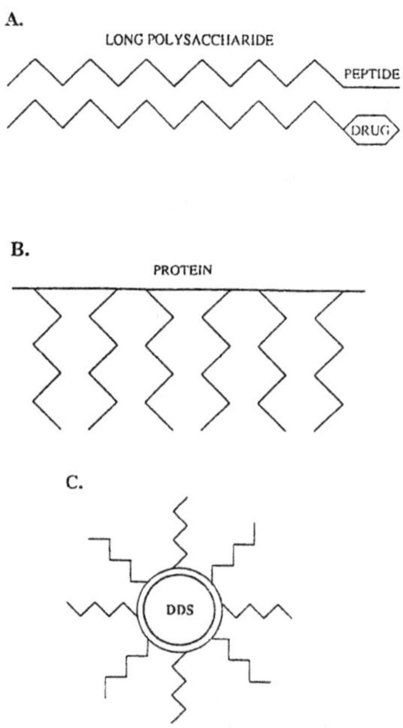

Fig. 2 Schemes of possible uses of polysialic acids in conjuction with small drug and peptides, proteins and microparticles (eg. liposomes)

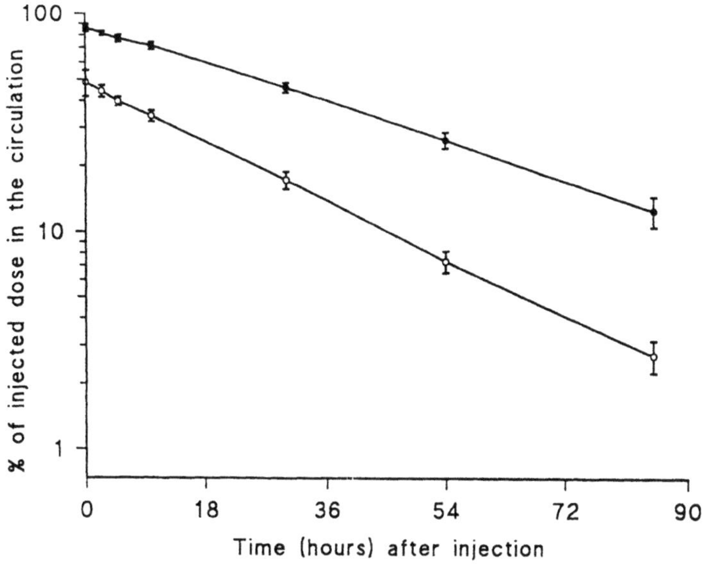

Fig. 3 Clearance of PSB from the blood circulation. In six separate experiments, mice in groups of 3-4 animals were injected intravenously with 1.1-2.0 mg of intact (○) or deacylated (•) PSB and bled at time intervals. NeuNAc in the blood plasma samples are %±S.D. of the dose in total blood. (Values from all groups treated with intact and deacylated PSB, respectively, were pooled).

K1, (PSB), the serogroup C capsular polysaccharide from Neisseria meningitidis C (PSC) and the polysaccharide from Escherichia Coli K92 (PSK 92) as well as their shorter chain derivatives. Structurally they are polymers of repeating 9-carbon NeuNAc units with varying chain length and bearing a phospholipid molecule covalently linked to the reducing end of the chain (Fig. 1). Polysialic acids are highly hydrophilic, have no known receptor (Gregoriadis, 1992) in the body to accelerate their uptake and are not known to be toxic. This suggested that they may exhibit long half-lives in the circulation and, coupled to small peptides and conventional drugs or even linked to large proteins and drug delivery systems, be capable of prolonging their residence times thus potentially augmenting pharmacological action. Fig. 2 shows a schematic representation of three possible separate ways in which these long circulating polysialic acids could be implemented. It is anticipated that in A the peptide or drug would assume the half-life of the carrier, probably a long chain polysialic acid, and in B and C, large proteins and particles coated with shorter chain derivatives would persist in the circulation, sterically hindered from rapid removal.

FATE OF POLYSIALIC ACIDS AFTER INTRAVENOUS INJECTION

A series of experiments were done with each of the three polysialic acids to determine their behaviour in vivo, and with one (PSC), as a conjugate with a low molecular weight model drug (fluorescein) to determine its ability to extend the circulation time of the dye.

Briefly, mice in groups of 3-4 animals were injected into the tail vein with intact or deacylated (see below) polysialic acids and bled at time intervals and the NeuNAc in the plasma assayed by the method of Svennerholm (1975) after precipitation of the native plasma protein-bound NeuNAc (Gregoriadis et al, 1993). The clearance rates resulting from these measurements were found to depend on several factors, e.g. the type of polysialic acid used, its chain length and, in some cases, whether or not it was deacylated. Referring back to Fig. 1 we see that each polysialic acid has a phospholipid molecule covalently bound to

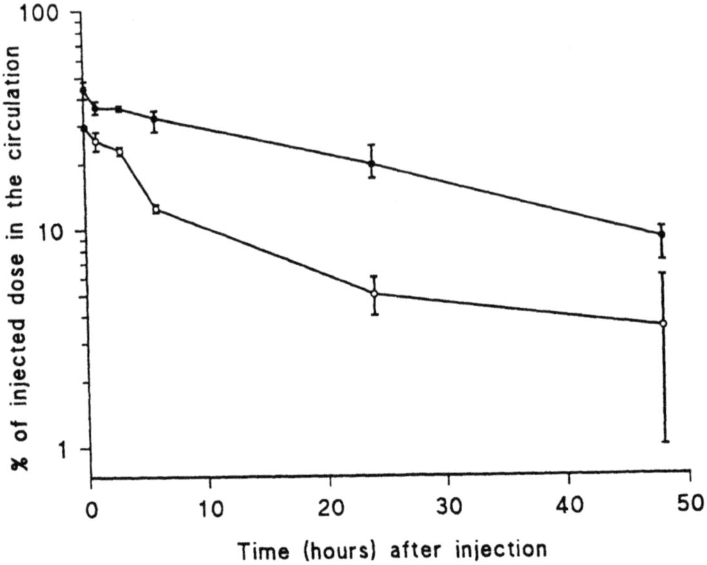

Fig. 4 Clearance of PSC from the blood circulation. Mice in two groups of 5 were injected intravenously with 2 mg of intact (○) or deacylated (•) PSC. For other details see legend to Fig. 3

their reducing end (Gotschlich, 1984). This confers a degree of hydrophobicity and, as a consequence, polysialic acids exhibit varying degrees of micellar behaviour leading to partial formation of aggregates (Gotschlich, 1984). Deaggregation can be easily carried out by removing the acyl groups on incubation at 37°C for 4 hours with 0.1m NaOH and subsequent dialysis (Lifely et al, 1986). Examining the clearance rates of the individual polysialic acids we find that in the case of PSB, clearance was biphasic with approximately 50% of the dose removed after 3 minutes. The remainder was cleared more slowly in a linear pattern giving a half-life of 20 hours. This is presented in Fig. 3 which also shows that, when PSB is previously deacylated, only 5-10% of the injected material was cleared rapidly, the remainder achieving a longer half-life of 30 hours.

The clearance patterns of intact and deacylated PSC were similar to those of PSB except that a greater percentage (70% and 50% respectively) was lost within minutes of injection (Fig. 4). The half-live of the remaining deacylated PSC was 20 hours. On the other hand Fig. 5 shows that the PSK9 had a relatively slow clearance rate during the first 6 hours for both intact and deacylated preparations with a linear clearance thereafter resulting in a half-life of 40 hours.

To interpret these data we note that the molecular weight of the polysialic acids (cf. PSB and PSC), deacylation (see PSB and PSC) and type of linkage of the NeuNAc units [α-(2-8)-α-(2-9)>α-(2-8)>α-(2-9)] can influence clearance rates. However, other factors have yet to be considered. It should be taken into account that deacylation can cause hydrolysis of the 0-acetyl groups of carbons 7 and 8 of PSC (Fig. 1) thus increasing the proportion of shorter chain molecules and contributing to its relatively rapid clearance (Fig. 4). Also, chain lengths of the three polysialic acids shown in Fig 1 are an average so the preparations used are polydisperse with many low molecular weight moieties included. These would tend to be removed rapidly and explain the initial rapid loss.

In fact, experiments with a PSB of short chain length (15 NeuNAc units) have revealed that over 90% of the injected dose is removed from the circulation within 30 minutes (data not shown). The PSK92 pattern is puzzling. The average chain length is only slightly greater (78 NeuNAc units) (Fig 1) than that of PSC yet the clearance rate is halved (40 hours half-life). This could mean that most of the PSR92 polysialic acid molecules were of similar chain length giving an overall slow clearance (and the

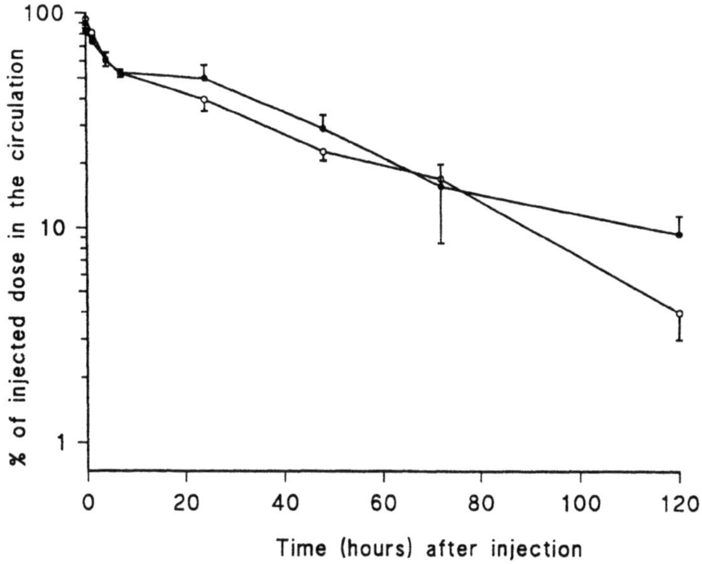

Fig. 5 Clearance of PSK 92 from the blood circulation. Mice in two groups of 4 were injected intravenously with 1.9 mg of intact (∘) or deacylated (•) PSK 92. For other details see legend to Fig. 3

opportunity to achieve the long half-life of 40 hours). We cannot as yet explain why the intact and deacylated PSK92 patterns were so similar unless deacylation had already

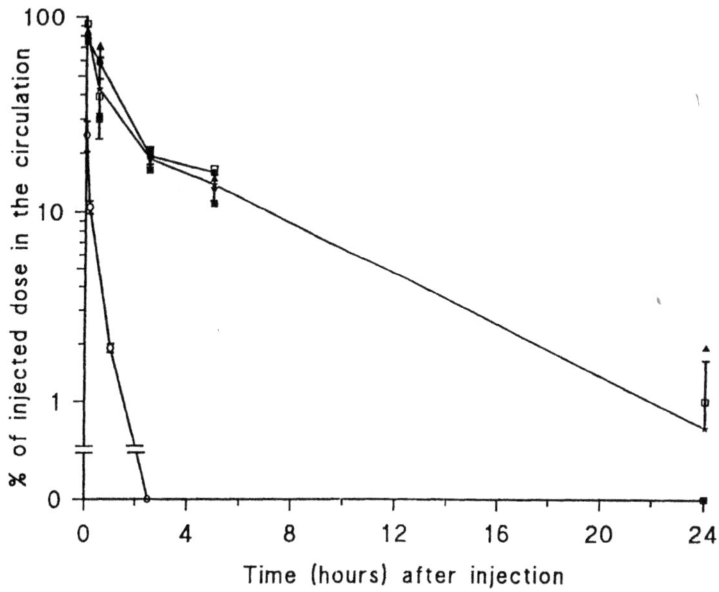

Fig. 6 Clearance of low molecular weight fluorescein-PSB conjugate from the blood circulation. Mice in groups of three were injected intravenously with 28 (▾), 102 (▪), 510 (▴) and 1,528 μg (•) of PSB conjugated to [^{125}I]fluorescein or with 40 μg fluorescein only (∘). Values are means ±S.D. of ^{125}I-radioactivity (closed symbols), NeuNAc (□) or fluorescein. Stars denote the mean of ^{125}I mean values for all doses at each time interval. For other details see legend to Fig. 3

occurred naturally during storage. These results were confirmed in a second experiment with the same preparation. We can only conclude that factors other than those mentioned could also play a role in determining circulatory half-life.

CLEARANCE OF A MODEL DRUG BOUND TO POLYSIALIC ACID

Having determined the clearance patterns of the three polysialic acids we wanted to find out how a model drug coupled to a polysialic acid would behave in vivo. Fluorescein, alone or covalently linked to a low molecular weight deacylated PSB (82 NeuNAc units) was given intravenously in different dose quantities to two groups of mice. Results in Fig. 6 show that over 99% of free fluorescein is removed from the circulation within 1 hour of injection, whereas fluorescein coupled to PSB, following rapid initial loss (80% in 2.5 hours), attained a half-life of 5 hours. As the clearance rate of the PSB itself is nearly identical, we can assume that the fluorescein adopted the half-life of its carrier.

CONCLUSIONS

We have seen that the polysialic acids described here have a long circulatory retention time which they are able to impose on drugs that are normally removed rapidly. Furthermore, the time required for optimum drug effect can be chosen to suit different needs either by the judicial choice of polysialic acid or by the use of hydrolysis products of appropriate chain length. In general, it is expected that polysialic acids with long-lives would be suitable for the delivery of low molecular weight peptides and drugs, while large proteins and liposomes would benefit from coating their surface through conjugation with short chain derivatives (see Fig. 2). To that end, functional groups on polysialic acids are available for conjugation. At the non-reducing end an aldehyde group could be generated by periodate oxidation. Also, carboxyl and hydroxyl groups are possibilities and an amino group could be freed on deacylation (see Fig. 1).

Polysialic acids have many exciting possibilities with few, if any disadvantages. They are biodegradable and their catabolic products are not known to be toxic. Like other polysaccharides, they are T-independent antigens and do not induce immunological memory which has made vaccine development against meningitidis group B or E.Coli K1 difficult (Moreno et al, 1985). On the other hand PSC and PSK92 are known to be slightly immunogenic in humans but only when chain lengths greater that 170 NewNAc units are used (Gotschlich et al, 1981). However, when proteins are coupled to polysialic acids the latter can become T-dependent antigens and induce memory though this is difficult to achieve, especially with PSB. Although antibodies against some of the polysialic acids studied here exist in the blood circulation, these are of low levels and also of low affinity, especially for antibodies against α-(2-8)-linked PSB. (Mandrell and Zollinger, 1982). In fact, it is known that host cell surfaces do have α-(2-8) linked polysialic structures and this is likely to restrict any immunological response (Finne, 1982).

Finally, it is worth noting that as N.Meningitidis is pathogenic whilst E.Coli K1 is only slightly pathogenic and E.Coli K92 non-pathogenic, it may be preferable to use polysialic acids from the latter sources. It is fortunate that PSB and PSK92 have also been shown to exhibit the longest half-lives in the circulation.

REFERENCES

Blume, G. and Cevc, G., 1990, Liposomes for sustained release in vivo, Biochim.Biophys. Acta, 1029:91

Finne, J., 1982, Occurrence of unique polysialosyl carbohydrate units in glycoproteins of developing brain, J.Bio.Chem., 257:11966

Florence, A.T. and Gregoriadis, G., 1991, Recent advances in drug targeting, Trends Biotechnol., 11:295

Gotschlich, E.C., Fraser, B.A., Nischimura, O., Robbins, J.B. and Liu, T.-Y., 1981, Lipid on capsular polysaccharides of gram-negative bacteria, J.Biol.Chem., 256:8915

Gotschlich, E.C., 1984, in: "Bacterial Vaccines", Germanier, R., ed., Academic Press, Orlando

Gregoriadis, G., 1992, International Patent Application No. PCT/GB92/01022

Gregoriadis, G., B. McCormack, Z. Wang and R. Lifely, 1993, Polysialic acids: potential in drug delivery, FEBS Lett., 315:271

Jamshaid, M., Farr, S.J., Kearney, P. and Kellaway, I.W., 1988, Poloxamer adsorption on liposomes: comparison with polystyrene latex and influence on solute efflux, Int.J. Pharm, 68:121

Klibanov, A.L., Murayama, K., Torchilin, U.P. and Huang, L., 1990, Amphipathic polyethyleneglycols effectively prolong the circulation time of liposomes, FEBS Lett., 268:235

Lifely, M.R., Nowicka, U.T. and Moreno, C., 1986, Analysis of the chain length of oligomers and polymers of sialic acid isolated from Neisseria meningitidis group B and C and Eschericcia coli K1 and K92, Carbohydr.Res., 156:123

Mandrell, R.E. and Zollinger, W.D., 1982, Measurements of antibodies to meningoccal group B polysaccharide: low activity binding and equilibrium binding constants, J.Immunol., 129:2172

Moghini, S.M., Porter, C.J.H., Illum, L. and Davis, S.S., 1991, The effect of poloxamer 407 on liposome stability and targeting to bone marrow: comparison with polystyrene microspheres, Int.J.Pharm., 68:121

Moreno, C., Lifely, M.R. and Esdaile, J., 1985, Effect of aluminium ions on chemical and immunological properties of meningoccal group B polysaccharide, Infec.Immun., 47:527

Nelson, H., 1991, Targeted cellular immunotherapy with bifunctional antibodies, Cancer Cells, 3:163

Nucci, M.I., Short, R. and Abuchowski, A., 1991, Advanced Drug Deliv.Rev., 6:133

Papahadjopoulos, D., Allen, T., Gabizon, A., Mayhew, E., Matthay, K., Huang, S.K., Lee, K.D., Woodle, M.C. and Lasic, D.D., 1991, Sterically stabilized liposomes: Pronounced improvement in blood clearance, tissue distribution and therapeutic index of encapsulated drugs against implanted tumours, Proc.Natl.Acad.Sci.USA, 88:11460

Roerdink, F.H. and Kroon, A.M., eds., 1989, "Drug Carrier Systems", John Wiley and Sons, Chichester

Senior, J., Delgado, C., Fisher, D., Tilcock, C. and Gregoriadis, G., 1991, Influence of surface hydrophilicity of liposomes on their interaction with plasma protein and clearance from the circulation: Studies with poly(ethyleneglycol)-coated vesicles, Biochim.Biophys.Acta, 1062:77

RED BLOOD CELLS AS CARRIERS OF DRUGS AGAINST RETROVIRUSES

Mauro Magnani, Luigia Rossi, Laura Chiarantini, Alessandra Fraternale and Anna Casabianca

Institute of Biological Chemistry "G. Fornaini", University of Urbino
61029 Urbino, Italy

INTRODUCTION

Until rcently the only biomedical use of erythrocytes was in transfusion medicine to restore normal oxygen delivery. In recent years a number of procedures have been developed to load these cells with macromolecules including enzymes, drugs, antigens, etc. (Magnani and DeLoach, 1992b). The majority of the methods used are based on the remarkable capacity of the erythrocyte to swell and become leaky when placed in hypo-osmotic solutions. Extracellular substances may enter the erythrocyte at this stage and membrane resealing can then be performed by hyper-osmotic solutions. Upon incubation resealed erythrocytes resume their normal biconcave shape, recover their normal membrane impermeability, and can survive in circulation with a nearly normal life-span (Ihler and Tsang, 1987; Dale, 1987; DeLoach, 1987; Ropars et al, 1987). Thus, after loading with drugs, the erythrocyte can serve as a carrier. Encapsulated drugs, depending on their physico-chemical properties can remain entrapped in the carrier cell, or can permeate the erythrocyte membrane and be released slowly into the circulation or can be metabolized by erythrocyte resident enzymes into diffusible molecules. A number of examples of each of these possibilities have been reported during recent years (Magnani and DeLoach, 1992b; Ropars et al, 1987b). If not metabolized by erythrocyte resident enzymes, non diffusible drugs can be targeted to selective organs or cells by appropriate manipulations of the carrier cell.

TARGETING OF DRUG-LOADED ERYTHROCYTES

Native and carrier erythrocytes end their life span in the reticuloendothelial system (RES). Thus, targeting of drugs to liver and spleen is not a major problem. A number of red cell manipulations shorten the erythrocyte life span and thus, in the case of carrier cells, increase the rate of drug delivery. Unfortunately, these treatments do not usually affect the organ distribution of the carrier cells significantly and thus will not be discussed in this paper. In contrast, other treatments, increase the selective targeting to specific cell types and will be mentioned here as selected examples of erythrocyte targeting. De Flora and co-workers (Zocchi et al, 1989) have shown that erythrocyte encapsulated doxorubicin can be targeted to the liver increasing the therapeutic index of the drug simply by glutaraldehyde treatment. The same treatment also reduces the efflux of encapsulated molecules but should be used carefully since the glutaraldehyde concentration affects the selectivity for spleen or liver targeting. Successful targeting of mycotoxin with carrier erythrocytes treated with glutaraldehyde has also been reported by DeLoach and co-workers (1988) who have extensively investigated the fate of glutaraldehyde treated erythrocytes.

Targeting of Drugs 4, Edited by G. Gregoriadis *et al.*
Plenum Press, New York, 1994

Red blood cell targeting has been also obtained by coupling specific antibodies to the red cell surface through a biotin-avidin-biotin bridge. This method, first introduced by Samokhin and co-workers (Samokhin et al, 1983) consists in the separate biotinylation of intact erythrocytes and the antibody of interest. Coupling of biotinylated antibodies to biotinylated erythrocytes is then obtained through an avidin bridge. By this approach targeting has been obtained to smooth muscle cells (Glukhova et al, 1986) and to denuded areas of vessel wall (Smirnov et al, 1986). Unfortunately, avidin attachment to biotinylated erythrocytes was reported to induce their lysis by homologous complement (Musykantov et al, 1993 and references therein), thus preventing the in vivo use of these modified erythrocytes. Recently we have shown that avidin molecules up to 10^4/cell do not activate the complement pathway (Magnani M., unpublished observation) thus opening new possibilities for the use of these erythrocytes. However, in vivo studies have been shown that coupling of heterologous proteins (i.e. monoclonal antibodies) through a biotin-avidin-biotin bridge to erythrocytes elicits strong immunological responses (Magnani et al, 1992a). Thus, although complement activation can be avoided by coupling a reduced number of biotin (and thus avidin) molecules per cell, these modified erythrocytes are strongly immunogenic preventing their use as carriers except for the delivery of peptides for producing anti-peptide antibodies and possibly to generate peptide vaccines.

Coupling of antibodies to the surface of carrier erythrocytes can be obtained in a number of ways without using the biotin/avidin technology. One of these methods takes advantage of the ability of trivalent chromium to bind protein nonspecifically. This property has been widely used to coat red cell with IgG to be used for passive hemagglutination reactions (Gold and Fudenberg, 1987). Chromic chloride has recently been used to couple anti-Thy 1.2 monoclonal antibody to target erythrocytes to T cells (Chiarantini et al, 1992). The method proved to be effective both in vitro and in vivo. However, scanning and transmission electron microscopy examination revealed erythrocyted binding to T cells but did not show any evidence of membrane fusion. In other words, the ability to target carrier erythrocytes to selected cells does not necessarily mean that transfer of drugs from erythrocytes to the target is feasible. At best a local high drug concentration should be expected loading the carrier erythrocyte with diffusible drugs.

An alternative approach to the coupling of antibodies to the erythrocyte membrane could be the insertion of selected proteins having a membrane spanning sequence (i.e. receptors) into the erythrocyte membrane. Nicolau and co-workers (Mouneimme et al, 1990) have reported the successful electroinsertion of CD4 into human and murine erythrocytes. This approach seems to have a number of potential applications although it suffers the same limitations of the antibody coupled erythrocytes.

RED BLOOD CELLS AS CARRIERS OF DRUGS AGAINST RETROVIRUSES

A broad family of nucleoside analogues, particularly those with a 2',3'-dideoxyribose moiety, are able to inhibit the infectivity and replication of several retroviruses including HIV-1 (Mitsuya et al, 1991a; Mitsuya et al, 1991b). Unfortunately, the use of these drugs is frequently limited by dose-related toxicities and decline of efficacy due to development of resistance to the drug (Larder et al, 1989; Fitzgibbon et al, 1992; Gao et al, 1992). Additional problems include a short half-life in the circulation and a low ability of several cell types to phosphorylate these dideoxynucleosides to 5'-triphosphate. In fact, to be pharmacologically useful as inhibitors of reverse transcriptase the dideoxynucleosides must be phosphorylated by cellular kinases.

Prompted by these considerations we have investigated the possibility of using red blood cells as slow delivery systems for nucleoside analogues or as targeting systems.

Red Blood Cells as Slow Delivery Systems of Nucleosides Analogs

2',3'-Dideoxycytidine (DDC) is the most active of a series of nucleoside analogues against the infectivity and cytophatic effect of HIV-1 (Mitsuya and Broder, 1986; Mitsuya et al, 1987). This drug permeates the human erythrocyte membrane predominantly by nonfacilitated diffusion (Magnani et al, 1989b), thus we have synthesized 2',3'-dideoxycytidine-5' phosphate (DDCMP) as a non-diffusible pro-drug. DDCMP was then encapsulated into human erythrocytes where it is slowly dephosphorylated by an endogenous pyrimidine nucleotidase (Magnani et al, 1989a). Once converted to DDC the drug is released from the carrier erythrocytes. Pharmacokinetic modelling of DDC released from DDCMP carrier reduce the peak of blood DDC that follows the oral or intravenous drug administration and in turn its toxicity. Furthermore, a long-lasting therapeutic effect is expected.

Red Blood Cells as Targeting System of Nucleoside Analogs

The direct administration of nucleoside analogues in phosphorylated form would provide several advantages to the infected host compared to the oral or intravenous administration of the same non-phosphorylated compounds for the reasons mentioned above. Unfortunately, all the phosphorylated derivatives should be administered encapsulated into appropriate carrier systems to avoid their immediate dephosphorylation in body fluids. These carriers should then be efficiently internalized into the infected cells where their content should be released in active (phosphorylated) form.

We have previously shown that autologous red blood cells can be used as carrier system for phosphorylated nucleoside analogues (Magnani et al, 1992c), particularly for the delivery of DDCTP to macrophages. In fact, although the majority of HIV isolates show a preferential tropism for CD4+ T lymphocytes (Yarchoan and Broder, 1989; Gartner et al, 1986; Ho et al, 1986; O'Brien et al, 1990) a growing interest is focused on HIV infection of brain, lung, lymph node and skin macrophages. HIV and other mammalian lentiviruses are not cytophatic in these cells that, consequently, serve as major virus reservoirs with a key role in the propagation of the virus throughout the body as well as in the pathogenesis of AIDS. Thus, targeting active (phosphorylated) nucleoside analogues to macrophages should be considered of great importance in designing new antiviral nucleoside analogues (Richman et al, 1987; Perno et al, 1988).

Targeting of DDCTP-loaded erythrocytes has been obtained in our laboratory by a procedure that does not involve the coupling of any specific macromolecule on the erythrocyte surface, but simply takes advantage of the intrinsic properties of the carrier. In vitro it has been shown that several agents able to induce band 3 clustering (band 3 is the major transmembrane protein of the erythrocyte that function as anion channell) promote autologous IgG binding and complement deposition (Turrini et al, 1991) which in turn promote phagocytosis of treated cells. We have induced band 3 clustering by the addition of Zn. Once the clusters are formed these are made irreversible by the crosslinking agent BS^3. Carrier erythrocytes treated as described above are opsonized by IgG and C3b and actively phagocytosed by macrophages. More interestingly, DDCTP-carrier erythrocytes can deliver DDCTP to macrophages where concentrations higher than 2μM have been found (Magnani et al, 1992c).

Several in vitro and in vivo animal models have been used to evaluate the efficacy of DDCTP-loaded erythrocytes to inhibit retrovirus infection of macrophages and will be summarized, briefly.

In vitro inhibition of HIV infectivity of human macrophages. DDCTP-loaded erythrocytes added overnight to HIV-1 infected human macrophages were found to almost completely inhibit p24 and reverse transcriptase production over three weeks in culture. This treatment was also able to inhibit the infection of T cells co-cultivated with infected macrophages receiving the DDCTP-loaded erythrocytes (Magnani et al, 1990). Finally, PCR analyses showed that this treatment really prevents HIV-1 integration into the macrophage DNA (Magnani et al, unpublished).

In vivo inhibition of LP-BM5 infectivity of C57BL/6 mice. A mixture of murine leukemia viruses known as LP-BM5 cause, in susceptible mice, an immunodeficiency syndrome resembling human AIDS (Jolicoeur, 1991). Administration of DDCTP-loaded erythrocytes treated with Zn and BS^3 at 10 day intervals over a period of 3 months results in reduction of all the typical signs of the disease including splenomegaly, lymphoadenopathy and ipergammaglobulimenia (Magnani et al, 1992).

In vitro and in vivo inhibition of FIV infectivity. We have investigated the efficacy of our dryg delivery system also on a feline immunodeficiency animal model. Three model systems were tested: in vitro infected macrophages from specific pathogen free (SPF) cats, macrophages from naturally infected cats and macrophages from experimentally infected SPF animals. DDCTP-loaded erythrocytes were able to reduce FIV production both by macrophages infected in vitro or obtained from naturally or experimentally infected cats. Furthermore, the administration of DDCTP-loaded erythrocytes protected the majority of peritoneal macrophages during a seven month experimental FIV infection and reduced the percentage of circulating lymphocytes stained by an anti-p24 antibody. It is worth noting that the infected animals were treated with DDCTP-loaded erythrocytes only once every three weeks (Magnani et al, 1993) while usually nucleoside analogues are administered to patients once every 4-8 h.

CONCLUSIONS AND PERSPECTIVES

The use of modified erythrocytes as a delivery system of antiviral nucleoside analogues has two main advantages: the drug is never free in circulation thus its toxicity is reduced, and the drug half-life in circulation becomes a function of the carrier survival that is usually very long compared to other delivery systems. The main disadvantages of such a delivery system are related to the ability of carier erythrocytes to reach all the cell compartments potentially infectable. As documented in this paper targeting to macrophages of DDCTP-loaded erythrocytes is feasible and inhibition of HIV-1, FIV and LP-BM5 infectivity can be obtained. However, this DDCTP-delivery system is not able to directly protect the other cell components infected by HIV-1, namely the CD4+ lymphocytes. It is worth noting that protection of macrophage infectivity reduces the virus spread in the body and the subsequent infection of T lymphocytes as documented in vitro (Magnani et al, 1992c) and in vivo (Magnani et al, 1993). However, this is not a direct protection of these cells by the drug administered.

Attempts to protect as many cellular types as possible should be performed including the oral administration of one or more nucleoside analogues or other antiviral compounds and of phosphorylated nucleoside analogues by delivery systems as described in this paper. Such treatments are expected to protect both the lymphocytes and the macrophages. Preliminary experiments performed in our laboratory (Rossi et al, 1993) suggest that such combined treatment yields an additive response in some but not all the parameters investigated. Thus, the administration of antiviral compounds in different molecular forms (i.e. phosphorylated or not) or with different delivery systems should be carefully evaluated in pre-clinical animal models because of the unpredictability of the effects of these treatments from the conclusion drawn by studies on single treatment.

Finally, due to the increasing diffusion of AIDS and of the appearance of an increasing percentage of drug resistent strains of HIV, the development of new and unconventional therapeutic approaches and appropriate pre-clinical studies on simple and inexpensive animal models of immunodeficiency appear to be of the utmost importance.

Acknowledgements

We are particularly grateful to a number of colleagues that have participated in the studies reported in this paper, in particular we thank Dr. G. Brandi and G.F. Schiavano, Institute Igiene University of Urbino, Dr. L. Silvotti and G. Piedimonte, Institute Pathology, University of Parma and Dr. D. Matteucci and M. Bendinelli, Dept. Biomed., University of Pisa. The work reported has been supported by C.N.R. target project "Biotechnology and Biostrumentation" and Ministerio della Sanita, Istituto Superiore di Sanita - progetto AIDS, 1993 and Progetto Allestimento modelli animali per l'AIDS, Roma, Italy.

REFERENCES

Chiarantini, L., Droleskey, R., Magnani, M. and DeLoach, J.R., 1992, In vitro targeting of erythrocytes to cytotoxic T-cells by coupling of Thy-1.2 monoclonal antibody, Biotechnol.Appl.Biochem., 15:171

Dale, G., 1987, High-efficiency entrapment of enzymes in resealed red cell ghosts by dialysis, in: "Methods in Enzymology", R. Green and K.J. Widder, eds., Acad. Press, San Diego

DeLoach, J.R., 1987, Dialysis method for entrapment of proteins into resealed red blood cells, in: "Methods in Enzymology", R. Green and K.J. Widder, eds., Acad. Press, San Diego

DeLoach, J.R., Andrews, K. and Naqi, A., 1988, Targeting of mycotoxins to reticulo-endothelial system of mice with carrier erythrocytes, Biotechnol.Appl.Biochem., 10:154

Fitzgibbon, J.E., Howell, R.M., Haberzetti, C.A., Sperber, S.J., Gocke, D.J. and Dubin, D.T., 1992, Human immunodeficiency virus type 1 pol gene mutations which cause decreased susceptibility to 2',3'-dideoxycytidine, Antimicrob.Agents Chem., 36:153

Gao, Q., Gu, Z., Parniak, M.A., Li, X. and Wainberg, M.A., 1992, In vitro selection of variants of human immunodeficiency virus type 1 resistant to 3'-azido-3'deoxythymidine and 2',3'-dideoxyinosine, J.Virol., 66:12

Gartner, S., Markovits, P., Markovitz, D.M., Kaplan, M.H., Gallo, R.C. and Popovic, M., 1986, The role of mononuclear phagocytes in HTLV-III/LAV infection, Science, 233:215

Glukhova, M.A., Domagatsky, S.P., Kabakov, A.E., Muzykantov, V.R, Ornatsky, O.I., Sakharov, D.V., Frid, M.G. and Smirnov, V.N., 1986, Red blood cell targeting to smooth muscle cells, FEBS Lett., 198:155

Gold, E.R. and Fudenberg, H.H., 1967, Chromic chloride: a coupling reagent for passive hemagglutination reactions, J.Immunol., 99:859

Ho, D.D., Rota, T.R. and Hirsch, M.S., 1986, Infection of monocyte/macrophages by human T lymphotropic virus type III, 77:1712

Ihler, G.M. and Tsang, H.C.W., 1987, Hypotonic hemolysis methods for entrapment of agents in resealed erythrocytes, in: "Methods in Enzymology", R. Green and K.J. Widder, eds., Acad. Press, San Diego

Jolicoeur, P., 1991, Murine acquired immunodeficiency syndrome (MAIDS): an animal model to study the AIDS pathogenesis, FASEB J., 5:2398

Larder, B.A., Darby, G. and Richman, D.D., 1989, HIV reduced sensitivity to zidovudine (AZT) isolated during prolonged therapy, Science, 243:1731

Magnani, M., Bianchi, M., Rossi, L. and Stocchi, V., 1989a, Human red blood cells as as bioreactors for the release of 2',3'-dideoxycytidine, an inhibitor of HIV infectivity, Biochim.Biophys.Res.Commun., 164:446

Magnani, M., Bianchi, M., Rossi, L. and Stocchi, V., 1989b, 2',3'-dideoxycytidine permeation of the human erythrocyte membrane, Biochem.Int., 19:227

Magnani, M., Chiarantini, L., Vittoria, E., Mancini, U., Rossi, L. and Fazi, A., 1992a, Red blood cells as an antigen-delivery system, Biotechnol.Appl.Biochem., 16:188

Magnani, M. and DeLoach, J.R., 1992b, "The use of resealed erythrocytes as carriers and bioreactors", Plenum Press, New York

Magnani, M., Rossi, L., Brandi, G., Schiavano, G.F., Montroni, M. and Piedimonte, G., 1992c, Targeting antiretroviral nucleoside analogues in phosphorylated form to macrophages: in vitro and in vivo studies, Proc.Natl.Acad.Sci.USA, 89:6477

Magnani, M., Rossi, L., Fraternale, A., Silvotti, L., Quintavalla, F., Piedimonte, G., Matteucci, D., Baldinotti, F. and Bendinelli, M., 1994, Feline immunodeficiency virus infection of macrophages: in vitro and in vivo inhibition by ddCTP-loaded erythrocytes, AIDS Res.Hum.Retrovirus, in press

Mitsuya, H. and Broder, S., 1986, Inhibition of the in vitro infectivity and cytopathic effect of human T-lymphotropic virus type III/Lymphadenopathy associated virus (HTLV-III/LAV) by 2',3'-dideoxynucleosides, Proc.Natl.Acad.Sci. USA, 83:1911

Mitsuya, H., Jarrett, R.F., Matsukura, M., Di Marzo Veronese, F., De Vico, A.L., Sarngadharam, M.G., Johns, D.G., Reitz, M.S. and Broder, S., 1987, Long-term inhibition of human T-lymphotrophic virus type III/lymphadenopathy associated virus (human immunodeficiency virus) DNA synthesis and RNA expression in T cells protected by 2',3'-dideoxynucleosides in vitro, Proc.Natl.Acad.Sci.USA, 84:2033

Mitsuya, H., Yarchoan, R., Kageyam, S. and Broder, S., 1991a, Molecular targets for AIDS therapy, Science, 249:1533

Mitsuya, H., Yarchoan, R., Kageyam, S. and Broder, S., 1991b, Targeted therapy of human immunodeficiency virus-related disease, FASEB J., 5:2369

Mouneimne, Y., Tosi, P.F., Barhoumi, R. and Nicolau, C., 1990, Electroinsertion of full length recombinant CD4 into red blood cell membrane, Biochim.Biophys.Acta, 1027:53

Muzykantov, V.R., Smirnov, M.D. and Klibanov, A.L., 1993, Avidin attachment to biotinylated amino groups of the erythrocyte membrane eliminates homologous restriction of both classical and alternative pathways of the complement, FEBS Lett., 318:108

O'Brien, W.A., Koyanagi, Y., Namazie, A., Zhao, J.Q., Diagne, A., Idler, K., Zack, J.A. and Chen, I.S.Y., 1990, HIV-1 tropism for mononuclear phagocytes can be determined by regions of gp120 outside the CD4-binding domain, Nature, 348:69

Perno, C.F., Yarchoan, R., Cooney, D.A., Hartman, N.R., Gartner, S., Popovic, M.Z., Gerrard, T.L., Wilson, Y.A., Johns, D.G. and Broder, S., 1988, Inhibition of human immunodeficiency virus (HIV-1/HTLV-III Bail) replication in fresh and cultured human peripheral blood monocytes/macrophages by azidothymidine and related 2',2'-dideoxynucleosides, J.Exp.Med., 168:1111

Richman, D.D., Kornbluth, R.S. and Carson, D.A., 1987, Failure of dideoxynucleosides to inhibit human immunodeficiency virus replication in cultured human macrophages, J.Exp.Med., 166:1144

Ropars, C., Avenard, G. and Chassaigne, M., 1987, Large-scale entrapment of drugs into resealed red blood cells using a continous-flow dialysis system, in: "Methods in Enzymology", R. Green and K.J. Widder, eds, Acad. Press, San Diego

Ropars, C., Chassaigne, M. and Nicolau, C., 1987b, "Red blood cells as carriers for drugs. Potential therapeutic applications", Pergamon Press, Oxford

Rossi, L., Brandi, G., Fraternale, A., Schiavano, G.F., Chiarantini, L. and Magnani, M., 1993, Inhibition of murine retrovirus induced immunodeficiency disease by dideoxycytidine and dideocytidine-5'-triphosphate, J.AIDS, 6:1179

Samokhin, G.P., Smirnov, M.D., Muzykantov, V.R., Domogatsky, S.P. and Smirnov, V.N., 1983, Red blood cell targeting to collagen-coated surfaces, FEBS Lett., 154:257

Smirnov, V.M., Domagatsky, S.P., Dolgov, V.V., Hvatov, V.B., Klibanov, A.L., Koteliansky, V.E., Muzykantov, V.R., Repin, V.S., Samokhin, G.P., Shekhonin, B.V., Smirnov, M.D., Sviridov, D.D., Torchilin, V.P. and Chazov, E.I., 1986, Carrier-directed targeting of liposomes and erythrocytes to denuded areas of vessel wall, Proc.Natl.Acad.Sci.USA, 83:6603

Somimano, F., Bischi, G.I., Bianchi, M., Rossi, L. and Magnani, M., 1990, A nonlinear three-compartment model for the administration of 2',3'-dideoxycytidine by using red blood cells as bioreactors, Bull.Math.Biol., 52:785

Turrini, F., Arese, P., Yuan, J. and Low, P.S., 1991, Clustering of integral membrane proteins of the human erythrocyte membrane stimulates autologous IgG binding, complement deposition, and phagocytosis, J.Biol.Chem., 266:23611

Yarchoan, R. and Broder, S., 1989, Immunology of HIV infection. In: "Fundamental Immunology", W.E. Paul, ed., Elsevier, New York

Zocchi, E., Tonetti, M., Polvani, C., Guida, L., Benatti, U. and de Flora, A., 1989, Encapsulation of doxorubicin in liver-targeted erythrocytes increases the therapeutic index of the drug in a murine metastatic model, Proc.Natl.Acad.Sci.USA, 86:2040

NANOPARTICLES FOR THE DELIVERY OF PEPTIDES AND PROTEINS

P. Couvreur and F. Puisieux

Laboratoire de Pharmacie Galénique, URA CNRS 1218,
Université Paris XI - 92296 Chatenay-Malabry Cedex, France

INTRODUCTION

Peptides drugs are gaining more and more interest as their role in physiopathology is better understood, and because of the progress in biotechnology and in conventional chemical synthesis as well. The development of DNA-recombination techniques have made these compounds available on a larger scale than in the past. However, the use of peptides in medicine is partly limited due to their easy degradation by proteolytic enzymes in the gastrointestinal tract, and therefore they have to be administered by injection. In addition, peptides are generally characterized by a short biological half-life, so that repeated injections are generally needed. Even after subcutaneous or intramuscular administration, a low bioavailability is often observed. Finally, most peptides pass poorly through biological barriers, due to their partition coefficient, which is unfavorable to lipidic structures. The foregoing considerations have led to the development of many new drug delivery systems. Among them, micro- and nanoparticles have been proposed for the safe and controlled administration of peptides and proteins. The main advantage of these microdispersed polymeric systems is their easy administration by a single injection whereas dosage forms larger in size may require a surgical incision. On the other hand, the difficulty of their removal from the injection site and their potential for catastrophic release (e.g. dose dumping) have also to be considered. Microparticulate systems may be convenient for other routes of administration (nasal, oral, buccal...). Furthermore, the large diversity of the synthetic and natural polymers available for micro- and nanoparticle preparation allows them to be considered as very flexible. Finally, in spite of their complexity and the possibility of down-regulation of receptors on continued dosing, these systems are of particular interest since they seem able to control the release rate and profile of numerous peptides and proteins and to improve their poor delivery characteristics as well.

The purpose of the present paper is to review and discuss the utility of nanoparticles systems for parenteral and enteral (oral) administration of peptides in different areas of medicine.

NANOPARTICLES FOR THE PARENTERAL DELIVERY OF PEPTIDES

Only a few works have been carried out with nanoparticles for the parenteral administration of peptides and proteins. Polyacrylic starch nanoparticles loaded with dextranase have been prepared by derivatization of polysaccharides (maltodextrin and dextran) with acrylic acid glycidyl ester and emulsion polymerization using N,N,N',N'-

Targeting of Drugs 4, Edited by G. Gregoriadis *et al.*
Plenum Press, New York, 1994

tetramethylenediamine (Artursson et al, 1984). After intravenous administration to mice and rats, the starch nanoparticles prolong the duration of immobilized dextranase and are able to initiate the degradation of the stored dextran in RES (Artursson et al, 1984). In addition, the metabolization of the entrapped enzyme is dependent on the composition of the particle matrix. More cross-linked spheres could better protect the entrapped enzyme. By the same way, L-asparaginase and L-glutaminase are stabilized when immobilized into polyacrylamide nanoparticles (Edman et al, 1987). When placed in the outer compartment of a commercial hollow fiber dialyzer, they give an efficient extracorporeal reactor for decomposing circulating L-asparagine and L-glutamine (Edman et al, 1987).

In another study, Gautier et al, 1992 have shown the efficacy of biodegradable poly-isohexylcyanocrylate (PIHCA) nanoparticles in the release of GRF in vivo. Especially, sustained-release of GRF can be obtained and GRF plasma levels may be maintained constant for approximately 24 hours (Fig. 1). With free GRF, the plasma concentration profile is quite different, ie. the peak corresponding to the maximum plasma concentration appears soon after the administration (2 minutes) and GRF is no longer detected after 100 minutes.

Comparison of the AUC 0-24 h of free GRF and GRF loaded onto nanoparticles shows that, in addition to the "slow release" process, nanoparticles are able to improve the bioavailability of the peptide. This could be attributed to the fact that free administered GRF is very quickly metabolized at the injection site, whereas it is partly protected from a massive enzymatic degradation when it is administered associated with nanoparticles.

This point is confirmed by the fact that 24 hours after a subcutaneous administration, GRF associated with nanoparticles remained intact at the injection site, whereas it had completely disappeard when it is administered in the free form (Gautier et al, 1992).

It has been calculated that, after subcutaneous administration, the bioavailability of GRF is multiplied by 20 when encapsulated in PIHCA nanoparticles.

The fact that nanoparticles are able to deliver GRF in a sustained release manner, is consistent with the in vivo behaviour of radiolabelled nanoparticles, as shown by autoradiography after a subcutaneous administration. It should be noted that autoradiographic experiments have been performed with polyisobutylcyanoacrylate (PIBCA) nanoparticles which, according to previously published results (Grangier et al, 1991), have a quicker degradation time than PIHCA nanoparticles. Hence, PIHCA nanoparticles are expected to remain longer at the site of injection and, effectively transmission electron microscopy studies have confirmed the presence of these nanoparticles at the injection site 24 hours after administration. However, the progressive marking reduction in this tissue suggests that nanoparticles are slowly biodegraded, allowing a progressive and regular release of GRF. These results must be discussed in connection with the in vitro profile of peptide liberation showing that GRF-release from polyalkylcyanoacrylate nanoparticles is a direct consequence of the polymer bioerosion due to the esterase hydrolysis of the ester side chain of the polymer (Lenaerts et al, 1984).

The autoradiography pictures also reveal that the excretion of the biodegradable polymer occurs through urine and faeces after a subcutaneous administration. This observation confirms the results of Grislain et al (1983), concerning the metabolization of nanoparticles after an intravenous administration.

ORAL ADMINISTRATION OF PEPTIDES AND PROTEINS

The non-parenteral administration routes for the delivery of peptides and proteins entrapped into micro- and nanoparticulate systems include the oral and nasal routes. The oral route of administration is usually preferred to the other routes for systemic administration of active compounds since it is the most acceptable and convenient for the patient. However, with peptides and proteins, the bioavailability is very low (generally below 2%). A recent and original strategy to circumvent the enzymatic barrier is to encapsulate peptides into polymeric nanoparticles. This approach is challenging and very

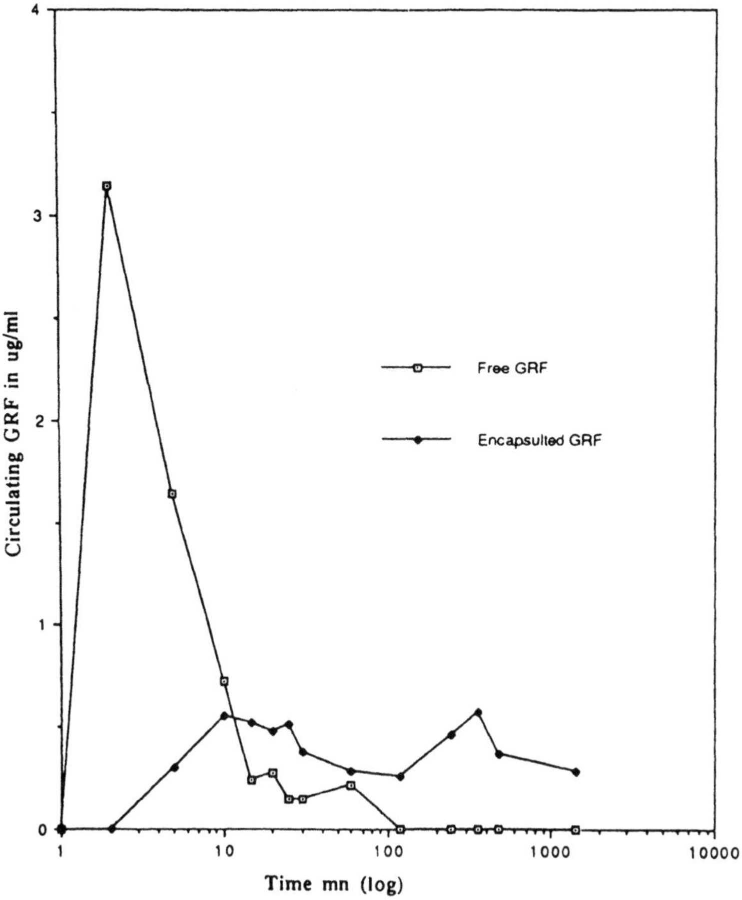

Fig. 1 Concentration of GRF in blood after subcutaneous administration of free or nanoparticle GRF (from Gautier et al, 1992)

attractive but it is still in its infancy because of different limitations which will be discussed below.

1. **Oral delivery of peptides with nanoparticles**

The most unexpected results have been obtained by Damgé et al, 1988, with insulin-loaded polyisobutylcyanoacrylate nanocapsules prepared by interfacial polymerization (Al Khouri et al, 1986). In diabetic rats, insulin-loaded nanocapsules given by a single intragastric administration after an overnight fast, reduce glycemia by 50-60 %. This effect appears 2 days after administration and is maintained for a prolonged period depending on the amount of administered insulin. Thus, with a dose of 50 units of insulin per kg administered under the form of nanocapsules, a reduction of the glycemia is noted for 20 days (Damgé et al, 1988). Free insulin does not affect glycemia when administered orally under the same experimental conditions. In fed diabetic rats, 100 Units/kg of encapsulated insulin were able to reduce glycemia by 25% during 6 days after oral administration. Thus, insulin associated to nanocapsules induces a sustained reduction in glycemia when given orally in diabetic rats, but their efficacy is more obvious in fasted than in fed diabetics animals (Damgé et al, 1988). When insulin nanocapsules are administered intragastrically *in normal rats*, fasted glycemia remains unaffected regardless of the dose. However, the peak of hyperglycemia induced in the normal rat by an oral administration of glucose is reduced by 50% when insulin nanocapsules were administered intragastrically, 2 days before the glucose overloading. It may be that insulin progressively and continously released by the polymeric drug carrier depresses the endogenous secretion of insulin by β-cells, and consequently, the amount of insulin liberated may not be sufficient to affect glycemia. In diabetic rats, this autoregulation phenomenon does not exist, allowing the regulation of fasted glycemia and partly of fed glycemia by the insulin released from nanoparticles. Also, it is difficult to understand the ability of a single oral dose of insulin nanocapsules to maintain hypoglycemia for up to 20 days. It has been hypothesized that the long-term hypoglycemic effect could be due in part to the progressive arrival of nanocapsules from the stomach to the gut, leading to a delayed absorption. Grislain et al, 1983, have found in the rat an intense radioactivity of the gastrointestinal tract 4 h after intubation of radiolabelled polyalkylcyanoacrylate nanoparticles, with the radioactive film still present in contact with the mucosa 24 h later. After arrival in the gut, insulin added to nanocapsules could be protected against proteolytic enzymes and adsorbed by the intestinal mucosa. Then, nanocapsules are probably transported to the liver by the portal way, according to several authors (Couvreur et al, 1986; Verdun et al, 1990), as soon as they are opsonized and taken up by the Kupffer cells of the liver and the macrophages of the spleen Thus, a slow process of redistribution from these cells and/or a slow process of release of insulin from nanocapsules could occur. Finally, because insulin is encapsulated into nanocapsules formed by an interfacial anionic polymerization process it is not out of the question that insulin may be at least partly covalently linked to the polymer which could explain the observed long-term hypoglycemic effect. It is noteworthy that those results first obtained in rats have been confirmed in dogs (Damgé et al, 1991).

In order to determine whether nanocapsules were able to pass through the intestinal epithelium, electron microscopic examinations were performed in rats (Damgé et al, 1990) and dogs (Aprahamian et al, 1987). When lipiodol loaded nanocapsules are administered in an isolated segment of jejunum, they appear as round vesicles associated with intraluminal mucus. Then, they are found in intercellular spaces between enterocytes and finally, they are observed in intravillus capillaries in close contact with red cells or adsorbed to the inner wall of endothelial cells (Damgé et al, 1990). When nanocapsules are administered in the lumen of an isolated ileal segment, they are found preferentially over Peyer's patches in which they pass massively and rapidly (Damgé et al, 1990). Nanocapsules are clearly visible in M cells and in intercellular spaces around lymph cells. This is relevant with another study indicating that the ileum is more effective than the jejunum and especially the duodenum and the colon for insulin nanocapsules absorption (Michel et al, 1991). In summary, although quantitative estimation of nanocapsules absorption is not at all possible using a microscopic investigation, those studies suggest that nanocapsules can cross the jejunal mucosa by a paracellular pathway probably through intercellular spaces formed by the desquamation of well differentiated absorptive cells at the top of the villi. A similar phenomenon could occur in the ileum, but in this segment, a massive and rapid passage of nanocapsules through Peyer's patches seems also to occur.

The uptake and translocation of latex nanospheres and microspheres after oral administration in rats have been confirmed by Jani et al, 1989, using non ionic and carboxylated fluorescent polystyrene nanospheres. In this study, Peyer's patches, villi, lymph nodes and spleen of animals fed with non ionic microspheres show unequivocal evidence of uptake and translocation of the particles (Jani et al, 1989). Carboxylated microspheres are taken up to a lesser degree than non ionised particles but no clear explanation of this difference is given. The extent of absorption of polystyrene nanoparticles by the rat gastrointestinal mucosa has been measured and the size dependency evaluated: the oral absorption of 50 nm particles is 34% and of 100 nm particles is 26% of which total, about 7% (50 nm) and 4% (100 nm) are in the liver, spleen, blood and bone marrow (Jani et al, 1990). Particles larger than 100 nm do not reach the bone marrow, and those larger than 300 nm are absent from the blood. Other authors (Le Fèvre et al, 1985; Eldridge et al, 1990) have shown a similar uptake of particles by the lymphoid aggregates of Peyer's patches.

Polyalkylcyanoacrylte nanospheres (Couvreur et al, 1982) (matricial structure) have been found unable to provide the oral absorption of insulin (Couvreur et al, 1980). However, in this formulation, Dextran 70 was used as a protectant of colloids instead of

Pluronic F68 with polyalkylcyanoacrylate nanocapsules. Thus, very recently, we have tested the hypoglycemic effect in diabetic rats of insulin entrapped into polyalkylcyanoacrylate nanospheres and administered orally with Pluronic and deoxycholate (Michel et al, 1990). In those conditions, a 50% decrease of the glycemia is noted. Therefore, the effect of Pluronic F68 cannot be ignored and may contribute together with deoxycholate to increase the translocation of the particles. A similar approach has been studied recently by Morishita et al, 1992, using Eudragit insulin microspheres containing a protease inhibitor. Strong insulin protective efficiency towards pepsinic degradation has been observed with this preparation (Morishita et al, 1992). When administered orally in diabetic rats, microspheres containing insulin together with aprotinin induce a significant continuous hypoglycemic effect (Morishita et al, 1992). The same insulin microspheres without protease inhibitor produce no marked hypoglycemic response. Thus, the strategy which utilizes a promoter of absorption or a protease inhibitor in association with micro- or nanoparticulates may be useful for enhancing the efficacy of oral insulin administration. Of course, a repeated administration of such cocktail may cause damage to the gastric mucosa and disturb the digestion process of dietary proteins.

The possibility of using nanoparticles (or even microspheres) for oral administration of insulin has been a controversial issue. However, there are abundant reports coming from various independent workers showing evidence of absorption of particulate systems by the gastrointestinal tract (Sanders and Ashworth, 1961; Volkheimer, 1983; Le Fèvre et al, 1985; Maincent et al, 1986). In some cases, a pharmacological effect has even been proved (Damgé et al, 1990; Morishita et al, 1992). Although, a more clear estimation of the quantity of absorbed particles is needed as well as a better understanding of the factors affecting particles uptake, it must be concluded that translocation of small sized particles is possible. The question remains if the extent of particle translocation is compatible with a strategy of peptide drug administration with therapeutic perspectives.

2. Oral delivery of particulate antigens

Oral delivery of antigens may be considered as the most convenient way for producing IgA antibody response. Classically, antigens are captured through pinocytosis by the M Cells localized in the Peyer's patches of the intestine. Then, the antigens are transported intact through those cells and released in the special micro- environment of the lymphoid tissue (Gilligan and Po, 1991; O'Hagan, 1990). Antigen presenting cells are able to internalize foreign antigens and to present them to T-lymphocytes. Thus, IgA antibodies will be secreted not only in the intestinal lymphoid tissue but also in all the other mucosal sites due to the migration of effector β-lymphocytes to those sites where they differentiate to produce IgA antibodies (Kroehenbuhl and Neutra, 1992). However, the oral delivery of antigens is seriously limited by their natural enzymatic break down in the GI tract and by their poor absorption as well (Gilligan and Po, 1991). Thus, it has been postulated that the use of micro- or nanoparticles for the oral delivery of antigens should be efficient since particles promote, as described previously, the uptake by Peyer's patches, achieving simultaneously the protection of the antigenic molecule.

Eldridge et al, 1990, have shown a disseminated mucosal antitoxin antibody response after an oral immunization by toxoid-containing biodegradable PLA microspheres. The size of the polymeric particles is a critical parameter. Tissue penetration occurs specifically through the Peyer's patches and is restricted to microspheres less than 10 um in diameter. Microspheres with a size ranging from 5 um to 10 um remain for a longer time into organized mucosa associated lymphoid tissue. From 5 um to 10 um seems to be the optimal size range to stimulate a purely mucosal response (Eldridge et al, 1990). The majority of PLAGA microspheres <5 um in diameter are transported rapidly through the efferent lymphatics within macrophages (Eldridge et al, 1990). Microspheres of different polymeric composition have also been tested for their ability to concentrate in the Peyer's patches. The results indicate that the polymers with the greatest relative hydrophobicity are absorbed most readily; microspheres made with cellulose and cellulose derivatives are not absorbed (Eldridge et al, 1990). The reason for this "non-recognition" of cellulosic material by the Peyer's patches is not yet clarified. Based on the criteria of biocompatibility, biodegradability, size and adequate absorption by the Peyer's patches, 5 um - 10 um

PLAGA microspheres seem to be the most attractive carrier for the oral administration of antigens (Eldridge et al, 1990).

Biodegradable poly(butyl-2-cyanoacrylate) particles have been shown to enhance the secretory immune response after their oral administration in association with ovalbumin (O'Hagan et al, 1989). This result was not fully repeated in the case of polyacrylamide nanoparticles loaded with the same antigen. It was postulated that much of the antigen was located at the surface of the polymer and could have been degraded during its passage through the gut. The relatively high surface concentration of ovalbumin adsorbed onto poly(butyl-2-cyanoacrylate) particles could have reduced the ability of the proteolytic enzymes in the gut to gain access to and to degrade the antigen, resulting in a greater antigen availability (O'Hagan et al, 1989). Recently, Pappo et al, 1991, have prepared polystyrene nanoparticles conjugated to the anti-M-cell monoclonal antibody 5B11 to target Peyer's patch M cells. Particle conjugates displaying the anti-M cell monoclonal antibody adhere to, and are translocated by M cells more efficiently than particles displaying Ig molecules of unrelated specificity, or than particles displaying the native surface chemistry (Pappo et al, 1991). Although it is not currently known if an intestinal IgA antibody response is generated against 5B11 antibody, these data raise the possibility of specifically target molecules to M cells in order to stimulate the oral immunization against different antigens.

REFERENCES

Al Khouri, N., Roblot-Treupel, L., Fessi, H, Devissaguet, J.P. and Puisieux, F., 1986, Development of a new process for the manufacture of polyisobutylcyanoacrylate nanocapsules, Int.J.Pharm., 28:125

Aprahamian, M., Michel, C., Humbert, W., Devissaguet, J.P. and Damgé, C., 1987, Transmucosal passage of polyalkylcyanoacrylate nanocapsules as a new drug carrier in the small intestine, Biol.of the Cell, 61:69

Artursson, P., Edman, P., Laakso, T. and Sjöholm, I., 1984, Characterization of polyacryl starch microparticles as carriers for proteins and drugs, J.Pharm.Sci., 73:1507

Artursson, P., Edman, P. and Sjöholm, I., 1984, Duration of action of dextranase entrapped in polyacrylstarch microparticles in vivo, J.Pharm.Exp.Ther., 231:705

Couvreur, P., Grislain, L., Lenaerts, V., Brasseur, F., Guiot, P. and Biernacki, A., 1986, Biodegradable polymeric nanoparticles as drug carrier for antitumor agents, in: Polymeric Nanoparticles and Microspheres, P. Guiot and P. Couvreur, eds., CRC Press, Boca Raton

Couvreur, P., Roland, M. and Speiser, P., 1982, Biodegradable particles containing a biologically active substance. US Patent n° 4, 329, 332

Couvreur, P., Lenaerts, V., Kante, B., Roland, M. and Speiser, P., 1980, Oral and parenteral administration of insulin associated to hydrolysable nanoparticles, Acta Pharm. Technol., 26:220

Damgé, C., Michel, C., Aprahamian, M. and Couvreur, P., 1988, New approach for oral administration of insulin with polyalkylcyanoacrylate nanacapsules as drug carrier, Diabetes, 37:246

Damgé, C., Michel, C., Aprahanian, M., and Couvreur, P., 1991, Personal Communication

Damgé, C., Michel, C., Aprahanian, M., Couvreur, P. and Devissaguet, J.P., 1990, Nanocapsules as carriers for oral peptide delivery, J.Control.Rel., 13:233

Edman, P. Artursson, P., Björk, E. and Davidsson, B., 1987, Immobilized L-asparaginase-L-glutaminase from Acinetobacter glutaminasificans in microspheres: some properties in vivo and in an extracorporeal system, Int.J.Pharm., 34:225

Eldridge, J.H., Hammond, C.J., Meulbroek, J.A., Staas, J.K., Gilley, R.M. and Tice, T.R., 1990, Controlled vaccine release in the gut associated lymphoid tissue. I. Orally administered biodegradable microspheres target the Peyer's patches, J.Control.Rel., 11:205

Le Fèvre, M.E., Joël, D.D. and Shidlovsky, G., 1985, Retention of ingested latex particles in Peyer's patches of germfree and conventional mice, Proc.Soc.Exp.Biol.Med., 179:522

Le Fèvre, M.E., Warren, J.B. and Joël, D.D., 1985, Particles and macrophages in murine Peyer's patches, Expl.Cell Biol., 53:121

Gautier, J.C., Grangier, J.L., Barbier, A., Dupont, P., Dussossoy, D., Pastor, G. and Couvreur,P., 1992, Biodegradable nanoparticles for subcutaneous administration of growth hormone releasing factor (hGRF), J.Control.Rel., 20:67
Gilligan, C.A. and Li Wan Po, A., 1991, Oral vaccines: design and delivery, Int.J.Pharm., 75:1
Grangier, J.L., Puygrenier, M., Gautier, J.C. and Couvreur, P., 1991, Nanoparticles as carriers for growth hormone releasing factors (GRF), J.Control.Rel., 15:3
Grislain, L., Couvreur, P., Lenaerts, V., Roland, M., Deprez-Decampeneere, D. and Speiser, P., 1983, Pharmacokinetics and distribution of biodegradable drug-carrier, Int.J.Pharm., 15:335
Jani, P., Halbert, G.W., Langridge, J. and Florence, A.T., 1989, The uptake and translocation of latex nanospheres and microspheres after oral administration to rats, J.Pharm.Pharmacol., 41:809
Jani, P., Halbert, G.W., Langridge, J. and Florence, A.T., 1990, Nanoparticle uptake by the rat gastrointestinal mucosa: quantitation and particle size dependency, J.Pharm.Pharmacol., 42:821
Kroehenbuhl, J.P. and Neutra, M.R., 1992, Molecular and cellular basis of immune protection of mucosal surfaces, Physiol.Rev., in press
Lenaerts, V., Couvreur, P., Christiaens-Leyh, D., Joiris, E., Roland, M., Rollman, B. and Speiser, P., 1984, Degradation of poly(isobutylcyanoacrylate) nanoparticles, Biomaterials, 5:65
Maincent, P., Le Verge, R., Sado, P.A., Couvreur, P. and Devissaguet, J.P., 1986, Disposition kinetics and oral bioavailability of vincamine-loaded polyalkylcyanoacrylate nanoparticles, J.Pharm.Sci., 75:955
Michel, C., Aprahamian, M., Defontaine, L., Couvreur, P. and Damgé, C., 1991, The effect of site of administration in the gastrointestinal tract on the absorption of insulin from nanocapsules in diabetic rats, J.Pharm.Pharmacol., 43:1
Michel, C., Couvreur, P., Vranckx, M., Damgé, C., Puisieux, F. and Balschmidt, P., 1990, Compositions pharmaceutiques contenant de l'insuline pour utilisation dans le traitement du diabète et procédé de préparation de telles compositions pharmaceutiques. Demande de Brevet Belge
Morishita, M., Morishita, I., Takayama, K., Machida, Y. and Nagai, T., 1992, Novel oral microspheres of insulin with protease inhibitor protecting from enzymatic degradation, Int.J.Pharm., 78:1
Morishita, I., Morishita, M., Takayama, K., Machida, Y. and Nagai, T., 1992, Hypoglycemic effect of novel oral microspheres of insulin with protease inhibitor in normal and diabetic rats, Int.J.Pharm., 78:9
O'Hagan, D.T., 1990, Intestinal translocation of particulates-implications of drug and antigen delivery, Adv.Drug Deliv.Rev., 5:265
O'Hagan, D.T., Palin, K. and Davis, S.S., 1989, Poly(butyl-2-cyanoacrylate particles as adjuvant for oral immunization, Vaccine, 7:213
O'Hagan, D.T., Palin, K., Davis, S.S., Artursson, P. and Sjoholm, I., 1989, Microparticles as potentially orally active immunological adjuvants, Vaccine, 7:421
Pappo, J., Ermak, T.H. and Steger, H.J., 1991, Monoclonal antibody-directed targeting of fluorescent polystyrene microspheres to Peyer's patch M cells, Immunology, 73:277
Sanders, E. and Ashworth, C.T., 1961, A study of particulate intestinal absorption and hepatocellular uptake, Exp.Cell.Res., 22:137
Verdun, C., Brasseur, F., Vranckx, H., Couvreur, P. and Roland, M., 1990, Tissue distribution of doxorubicin associated with polyisohexylcyanoacrylate nanoparticles, Cancer Chemother.Pharmacol., 26:13
Volkheimer, G., 1983, Hoematogenous dissemiation of ingested polyvinylchloride particles, Ann.N.Y.Acad.Sci., 409:164

NEW POLY(METHYLIDENE MALONATE 2.1.2) NANOPARTICLES: RECENT DEVELOPMENTS

P. Breton, D. Roy, L. Marchal-Heussler, C. Seguin, P. Couvreur[1] and F. Lescure

Laboratoires UPSA, Laboratoire de Galénique, 128 Rue Danton, 92506 Rueil-Malmaison Cedex, France and [1]Laboratoire de Pharmacie Galénique et Biopharmacie, URA CNRS 1218, Université Paris XI, 92296 Châtenay-Malabry, France

INTRODUCTION

Among the different drug delivery systems which have been developed for the last several years, a few authors have proposed submicronic colloidal polymeric particles (nanoparticles) to improve the pharmacokinetic and tissue distribution of molecules holding a therapeutic interest (Couvreur, 1988). As shown elsewhere, for liposomes (Weinstein et al, 1984; Papahadjopoulos et al, 1984) and other soluble macromolecular drug carriers (Roche et al, 1985; Lemieux et al, 1992) nanoparticles have been shown to be effective in the treatment of certain experimental neoplastic diseases (Chiannilkulchai et al, 1989). Futhermore, it has been demonstrated that this type of vector could be useful in ensuring a better peptide delivery (Damgé et al, 1988).

Biodegradable poly(alkyl-2-cyanoacrylate)(PACA) nanoparticles probably represent one of the most advanced polymeric colloidal system which has extensively been studied (Couvreur and Vauthier, 1991). In addition to their interest as drug vectors, besides experimented in an antitumoral phase I clinical trial (Kattan et al, 1992), PACA nanoparticles offer the very significant advantage of being conveniently and rapidly prepared through anionic polymerization which instantaneously occurs at low pHs as soon as the monomer is introduced into the aqueous medium. In this process, the hydroxide anion acts as the nucleophilic initiator and no other requirement is necessary and any purification step is useless.

However, as already mentioned above, PACA nanoparticles need an acidic pH for their preparation and they revealed themselves relatively toxic in typical cytotoxicity experiments (Lherm et al, 1992). With the aim of reducing these drawbacks, studies were undertaken to propose other kinds of monomer able to undergo a similar aqueous anionic polymerization leading to nanoparticle formation. Recently, De Keyser et al, (1989), based on Ponticello's patent (Ponticello, 1977), developed a symmetric diethyl methylidene malonate (DEMM) monomer, which could form nanoparticles at neutral pH, but which were unfortunately non degradable both in vitro and in vivo (De Keyser et al, 1991).

In the present report, we described preliminary studies about a novel acrylic derivative monomer. The asymmetric methylidene malonate 2.1.2. (MM 2.1.2.) monomer was a good candidate for anionic polymerization in aqueous medium to yield nanoparticles. A few results concerning their perparation, physicochemical characterization, degradability and in vitro cytotoxicity are herein discussed.

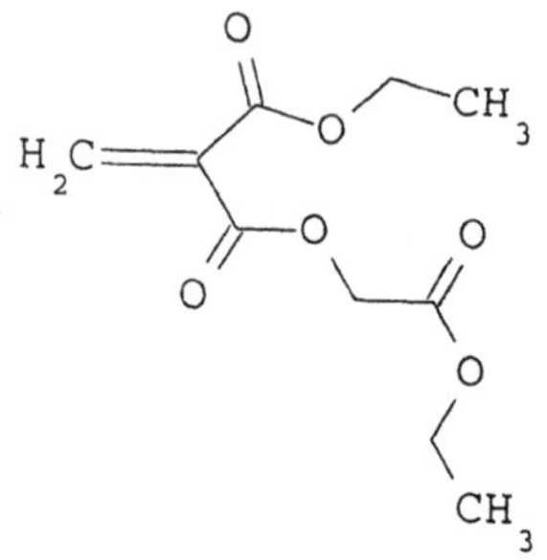

Fig. 1 Molecular structure of Methylidene Malonate 2.1.2. (MM 2.1.2.)

MATERIAL AND METHODS

Chemicals

1-ethoxycarbonyl-1-ethoxycarbonylmethylenoxycarbonyl ethene monomer (Mw = 230) (Fig. 1) also referred to as methylidene malonate 2.1.2. (MM 2.1.2.) was prepared in UPSA Laboratories/Carpibem (Rueil Malmaison, France) according to Bru-Magniez (Bru-Magniez et al, 1988). Sufficient amount of sulfur dioxide (SO_2) was added to prevent polymerization process. Dextran (Mr : 70,000) was purchased from Fluka Chemie AG (Buchs, Switzerland). Most of the buffering salts, reactants and solvents were from Prolabo (Paris, France). All products for tissue culture were provided by Gibco (Cergy-Pontoise, France).

Cells

In vitro cytotoxicity experiments were carried out on murine L929 fibroblast cell line (Flow, Les Ulis, France). They were maintained in MEM medium containing foetal bovine serum (5%), L-Glutamine (2mM), penicillin (50 µg/mL) and streptomycin (50 µg/mL) and were split every 3-4 days after a 10 min-treatment with 0.25% trypsin/0.02%EDTA at 37°C.

Nanoparticle Preparation And Characterization

Each step of the protocol was completed under sterile condition. Poly(methylidene malonate 2.1.2.) nanoparticles were prepared, at different pH, through anionic polymerization in phosphate buffer ($KH_2PO_4/Na_2HPO_4$0,066M) supplemented with 1% dextran (Lescure et al, 1991). One hundred microliters of monomer maintained under vacuum for 3 h were rendered SO_2-free prior to being added dropwise under magnetic stirring to 10 mL of polymerization medium kept at 25°C. Particle formation, followed by measuring the suspension turbidity at 400 nm, was completed within about 23 hours. Afterwards, possible aggregates were eliminated by filtration on 8 µm filter paper (Whatman, Maidstone, England). Polymerization yield was calculated by pelleting (280,000 g for 5 min) (Hitachi CS100 ultracentrifuge, Tokyo, Japan), drying and finally weighing the solid fraction of an initial suspension aliquot. Nanoparticles were subsequently freeze-dried and, so, could be stored for a long time at room temperature. They were usually resuspended just before use by adding an appropriate volume of sterile distilled water to give concentration of 6 to 8 mg polymer per mL. Nanoparticle characterization was done according to their size, measured with a N4 submicrometer particle analyzer (Coulter Electronics, Hialeah, Florida) and their constitutive polymer molecular weight distribution estimated by gel permeation chromatography (GPC) (Polymer Laboratories, Church Stretton, England) as previously described by Lescure (Lescure et al, 1992) for poly(isohexylcyanoacrylate) (PIHCA) and poly(isobutylcyanoacrylate) (PIBCA). PIHCA and PIBCA nanoparticles were prepared according to Couvreur (Couvreur et al, 1979).

HPLC Analysis And NMR Spectra

Kinetic of polymerization was determined using an isocratic Gold HPLC system (Beckman, Fullerton, CA) equipped with a C-18 Hypersil ODS reversed-phase column (25x0.46 cm) (Touzart et Matignon, Vitry-sur-Seine, France) maintained at 30°C. The mobile phase was water/acetonitrile (70/30 ; v/v) plus 0.1% trifluoroacetic acid (Merck, Darmstadt, Germany). These HPLC conditions were chosen upon the calculation of MM 2.1.2. total solubility parameter (Barton, 1991). Ultra-violet detection was achieved at 210 nm. One milliter sample of a nanoparticle suspension was ultracentrifuged (280,000 g for 5 min) and, while the supernatant was discarded, the pellet was dissolved in 2.5 mL eluant prior to injection (20 µl per injection). Proton nuclear magnetic resonance (NMR) spectra were recorded with a Bruker AC200 spectrometer (Wissembourg, France). Samples were dissolved in deutered chloroform or in DMSO-d_6 from CEA (Saclay, France). Chemical shifts were expressed as ppm from tetramethylsilane (Me_4Si=0.00).

Scanning Electron Microscopy

Environmental scanning microscopy (E-SEM) photomicrographs were performed on an Electroscan equipment (Wilmington, MA). A drop of nanoparticle suspension was applied on to a 220 nm filter (Millipore, Saint-Quentin-en-Yvelines, France) and directly observed without any drying and metal-coating step. Primary electron acceleration voltage, working distance and water vapor pressure were respectively 15 keV, 6mm and 2.9 Torr.

PMM 2.1.2. Nanoparticle Degradation Product Assay

Degradation of PMM 2.1.2. nanoparticles mainly occured through hydrolysis of the ester functions to generate ethanol and, to a lesser extent, glycolic acid. Ethanol assay was performed at pH 9.0 for 15 min with a kit (Sigma, La Verpilliere, France) based on the conversion of nicotinamide adenine dinucleotide (NAD) into its reduced form (NADH) in the presence of ethanol and alcohol dehydrogenase (ADH). NADH was quantified at 340 nm and was directly correlated to the ethanol concentration which was easily calculated by referring to a standard curve established in the same experimental conditions.

Ethanol release was expressed in a percentage of the theoretical ethanol amount, which could be obtained if all monomeric units were degraded, each yielding two ethanol molecules:

$$\text{Ethanol release (\%)} = \frac{\text{Measured concentration}}{\text{Theoretical concentration}} \times 100$$

Kinetic Of PMM 2.1.2. Nanoparticle Degradation

Ethanol production was measured at different pH values by means of the assay described above. Two milliliters of PMM 2.1.2 nanoparticles were suspended in 2 mL 1M glycine-NaCI/HCI buffer at about 3-4 mg polymer per mL, at various pHs, and incubated at 37°C for different periods of time, in sealed ampoules which avoided any ethanol loss by evaporation. The suspension was then pelleted (280,000 g for 5 min) and a 10 µl supernatant aliquot was mixed at room temperature with 3mL of ethanol assay kit (NAD + ADH). Optical density was read on a Model 25 spectrophotometer (Beckman).

In Vitro Cytotoxicity

L929 fibroblasts were distributed into 96 well plates at 1.25 x 10^4 cells per mL of culture medium (200 µl per well), and incubated for 24 hours at 37°C in a humidified and

CO_2-enriched (5%) atmosphere, to allow adherence and proliferation. At that time, wells were drained off and different amount of nanoparticles (PMM 2.1.2., PIHCA or PIBCA) or degraded PMM 2.1.2. nanoparticles in complete MEM, were added and incubated for a 24, 48 or 72 h period at 37°C in the same conditions as above. Cytotoxicity was determined according to the methodology described by Borenfreund and Puerner (Borenfreund and Puerner, 1985). Briefly, culture supernatant was eliminated and cells were washed twice with a washing medium (culture medium supplemented with 20 mM HEPES) prior a 3 h treatment with 200 µl/well of neutral red solution (50 µg per mL of washing medium) to ensure dye uptake exclusively into viable cells. Neutral red was removed and cells were further fixed in a solution of formaldehyde (3.7%) and $CaCl_2$ (1%) (100 µl/well) for 1 min, and homogenized in a water/ethanol/acetic acid mixture (49.5/49.5/1.0 - v/v/v) (200 µl/well). Optical density was read at 540 nm on a EL 310 Bio-Tek Instruments microplate reader (Winooski, VT).

The percentage of cytotoxicity was calculated in accordance with the formula:

$$\text{Cell viability (\%)} = \frac{\text{OD control - OD exp.}}{\text{OD control}} \times 100$$

RESULTS

Each PMM 2.1.2. nanoparticle batch was prepared in the same way to yield a milky suspension. Nanoparticle physicochemical characteristics were studied especially according to the proton concentration of the polymerization medium. Both particle size and polymer molecular weights were shown to be influenced by the pH variations (Fig. 2A). As previously mentioned by De Keyser (De Keyser et al, 1991) for PDEMM nanoparticles, size measurements roughly gave a U-like curve. Around the neutrality, in the pH range 5.5-7.0, the particle diameter was about 250-275 nm, with a good batch-to-batch reproductibility, while, for lower (acidic) and higher (basic) pH, this value respectively increased up to 450 nm at pH 5.0 and approximately 350 nm at pH 9.0. These diameters, first measured by light scattering analysis, were subsequently confirmed by environmental scanning electron microscopy (E-SEM) (Fig. 3). As expected, since hydroxide anion was described as the polymerization initiator (De Keyser et al, 1991), molecular weights, Mw, strongly increased with the pH (Fig. 2A). In the pH range studied, no evident correlation was noticed between the PMM 2.1.2. molecular weight and the nanoparticle size (Fig. 2B).

Polymerization and particle formation kinetics were followed by measuring turbidity at 400 nm. This parameter increased for 5 h after initiating the process and was then roughly stabilized until the end of the reaction by 23 h (data not shown) (Lescure et al, 1994). Adequate HPLC conditions were also very convenient for monitoring the polymerization kinetics: the three successive chromatograms, recorded at 2, 4 and 22 h (Fig. 4), clearly evidenced the progressive disappearance of the monomer (retention time: 6 min) and the simultaneous emergence of a new major species (probably oligomeric) (approximate retention time: 9-10 min) as the result of the polymerization reaction. It is worth to mention that the increase in turbidity was strictly correlated with appearance of this new compound. Even though the exact molecular structure(s) of this (these) molecule(s) has (have) not been elucidated yet, the proton NMR spectrum, displayed on Figure 5A, was typical of oligomeric compounds and all the signals were partially attributed by comparison with the monomer NMR spectrum (Fig. 5B).

In vitro degradation of PMM 2.1.2. nanoparticles versus pH is displayed on Fig. 6. At pH 9.0 and above, ester hydrolysis rapidly occured and reached a plateau after 24 h incubation; about 35-40% hydrolysis was noted after 144 h. Similarly to the data published by Lenaerts (Lenaerts et al, 1984) for PIBCA nanoparticles, PMM 2.1.2. underwent a much lower degradation rate with decreasing pH close to neutrality (only 10-15% hydrolysis at pH 7.0) and it appeared perfectly stable when maintained at pH 5.5. The PMM 2.1.2. degradation profile in complete culture medium was very similar to what obtained in protein-free phosphate buffered saline and further experiments conducted in presence of porcine liver esterases or in rat plasma (Lescure et al, 1994) showed, in both cases, a degradation level comparable to that obtained at basic pH (data not shown).

Finally, PMM 2.1.2. nanoparticle in vitro cytotoxicity was assayed on adhesive L929 cell line. Results are reported on Figure 7. Poly(methylidene malonate 2.1.2.) nanoparticles were significantly less toxic than their PIHCA- or PIBCA-made counterparts: the PMM 2.1.2. dose required to observe 50% cell death (ID50) at day 2 was regularly shown to be about 60 µg/mL, while it was 10 µg/mL for PIHCA and PIBCA respectively. As far as they were concerned, PMM 2.1.2. nanoparticle degradation products appeared much less toxic than intact nanoparticles with an ID50=350 µg/mL at day 2 and 250 µg/mL at day 3.

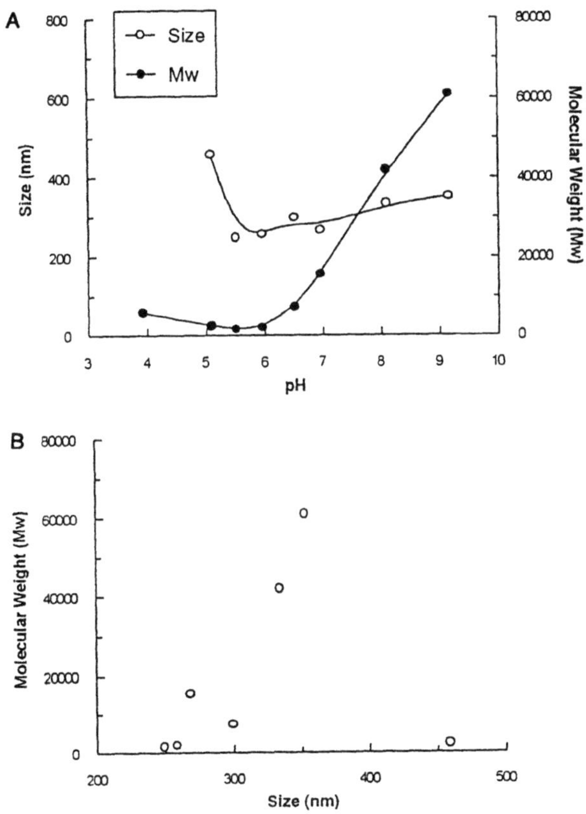

Fig. 2 Influence of polymerization medium pH on PMM 2.1.2. nanoparticle size and on molecular weight (Mw) of intrinsic polymers (A). Correlation plot between molecular weight and size (B)

DISCUSSION

Methylidene malonate 2.1.2. (MM 2.1.2.), added in dextran-containing phosphate buffer promptly underwent an anionic polimerization to finally produce nanoparticles. Concerning polyalkylcyanoacrylate preparation, De Keyser et al, (1991) underlined that hydroxide anion acts as a polymerization initiator. In this regard, it was interesting to observe the influence of polymerization medium pH on PMM 2.1.2. polymer molecular

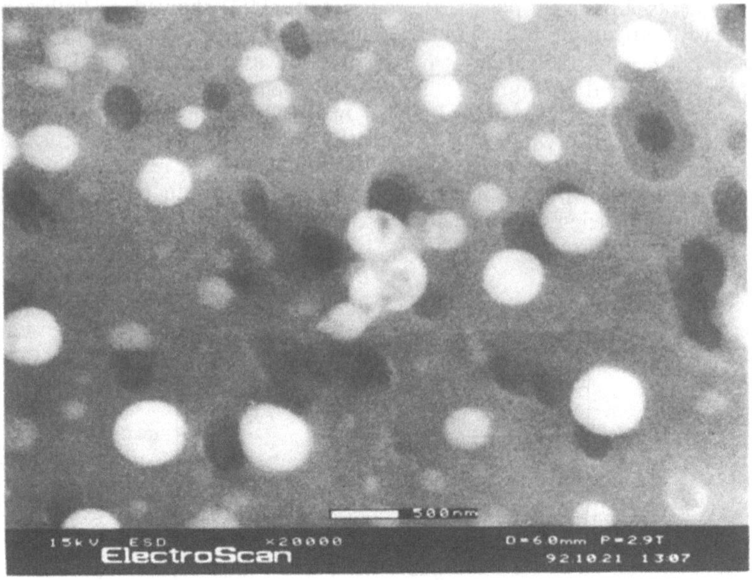

Fig. 3 Environmental scanning electron microscopy (E-SEM) picture of PMM 2.1.2. nanoparticles

weight as well as on nanoparticle size. As published elsewhere for other kind of polymers (De Keyser et al, 1991; Leyh et al, 1984; Vansnick et al, 1985), proton concentration was proved to be a crucial parameter having a very significant effect on these two physical characteristics which greatly modulate the body distribution of vectors as well as the degradation rate of the polymer (Vezin et al, 1980) and, thus, the bioavailability of any loaded drug. Our experimental data (Fig. 2) confirmed the results of De Keyser (De Keyser et al, 1991) obtained with PDEMM and highlighted the important influence of proton concentration on particle volume, especially at acidic pH range between 5.0 and 6.0. Curve profiles displayed on Figure 2, showing the average molecular weight of the polymer increasing with the pH (as described before for PACA) (Leyh et al, 1984; Vansnick et al, 1985), is in agreement with the concept describing the proton as an antagonist of hydroxide anion, responsible for the polymerization termination. However, unlike results obtained by Leyh and Vansnick (Leyh et al, 1984; Vansnick et al, 1985) from pH 2.8 to 4.3, in our present pH range working conditions, there was no linear inversely proportional correlation between the particle diameter and the constitutive polymer length; the smallest polymers (Mw<1,000), produced at pH 5.1, generated the largest nanoparticles, but PMM 2.1.2. with Mw=20,000-30,000 did lead to nanoparticles that were slightly smaller than those obtained from much longer polymers, with Mw equal to 60,000.

In addition to turbidity measurements (Lescure et al, 1994), we also took advantage of our high performance liquid chromatography facilities to follow the kinetics of polymerization. When polymerization was achieved at pH 5.5 (our standard conditions), turbidity and HPLC methodologies both gave a very similar assessment of the nanoparticle formation rate. Successive HPLC profiles showing the emergence of a single major peak suggest the monodispersity of the polymerization-resulting structures. Futhermore, gel permeation chromatography analysis always determined a main peak molecular weight (Mp) included between 650 and 700 (trimer) (Lescure et al, 1991) giving rise to the possibility of a cyclization process occuring, according to a Michael's reaction, as soon as a trimer was formed, to yield a stable 1,3,5-trisubstituted cyclohexane ring. Moreover, cyclization kinetics would be much higher than polymerization kinetics and, consequently, would forbid any further chain elongation. It appeared quite interesting to take up the challenge and try to identify the molecular structure(s) of this HPLC-purified fraction: however, despite the first proton nuclear magnetic resonance investigations, we were unable to confirm this hypothesis and determine the precise structure(s) involved in PMM 2.1.2. nanoparticles. Obviously, additional experiments, presently under progress, are required to answer this difficult question. Indeed, this objective will not be easy to reach as it can be assumed that asymmetric MM 2.1.2. polymerizes in yielding several optical isomer couples (enantiomers and/or diastereoisomers) of a unique molecular entity, and this racemization process indisputably renders NMR analysis more difficult (Breton et al, 1990).

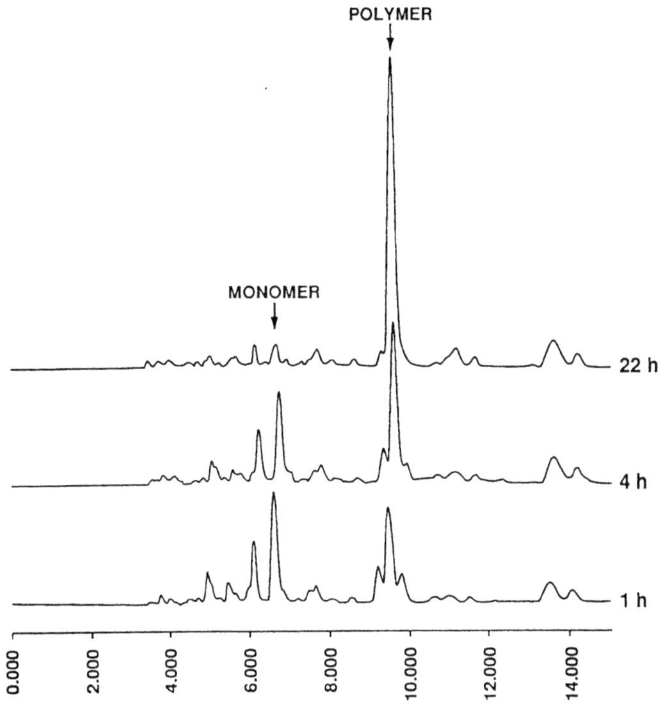

Fig. 4 Kinetics of PMM 2.1.2. nanoparticle formation followed by HPLC

Scanning electron microscopy (SEM) was a precious tool in nanoparticle physical characteristic analysis. In this regard, environmental-SEM (E-SEM) provided a very convenient way to observe nanoparticles, in situ, suspended in phosphate buffer and, even if the resolution was not as good as in classical SEM, observations were done in soft conditions such as objects did not undergo any impairing treatment. Especially, E-SEM could be considered to evaluate nanoparticle size dispersity (Fig. 3) and, thus, was a valuable instrument to confirm the mean particle diameter, first obtained by light scattering.

Methylidene malonate 2.1.2. has been designed as an improvement for the symmetric diethyl methylidene malonate (DEMM) initially developped by De Keyser (De Keyser et al, 1989; De Keyser et al, 1991). Consequently, it was essential to demonstrate that, unlike PDEMM, PMM 2.1.2. was degradable. Poly(methylidene malonate 2.1.2.)-made

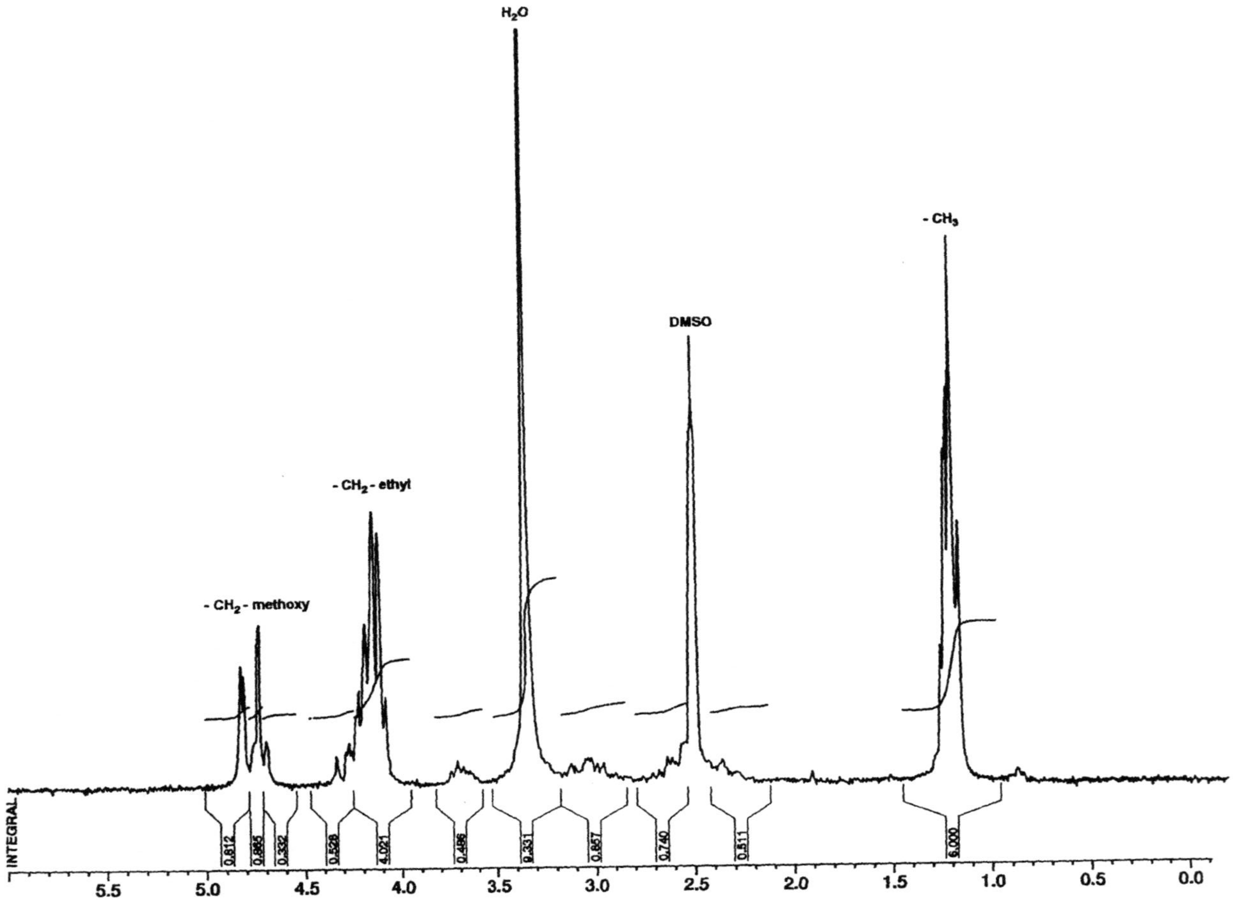

Fig. 5A 200 MHz proton NMR spectrum of a putative PMM 2.1.2. oligomer (about 1 mg/mL) recorded at 300K in DMSO-d_6

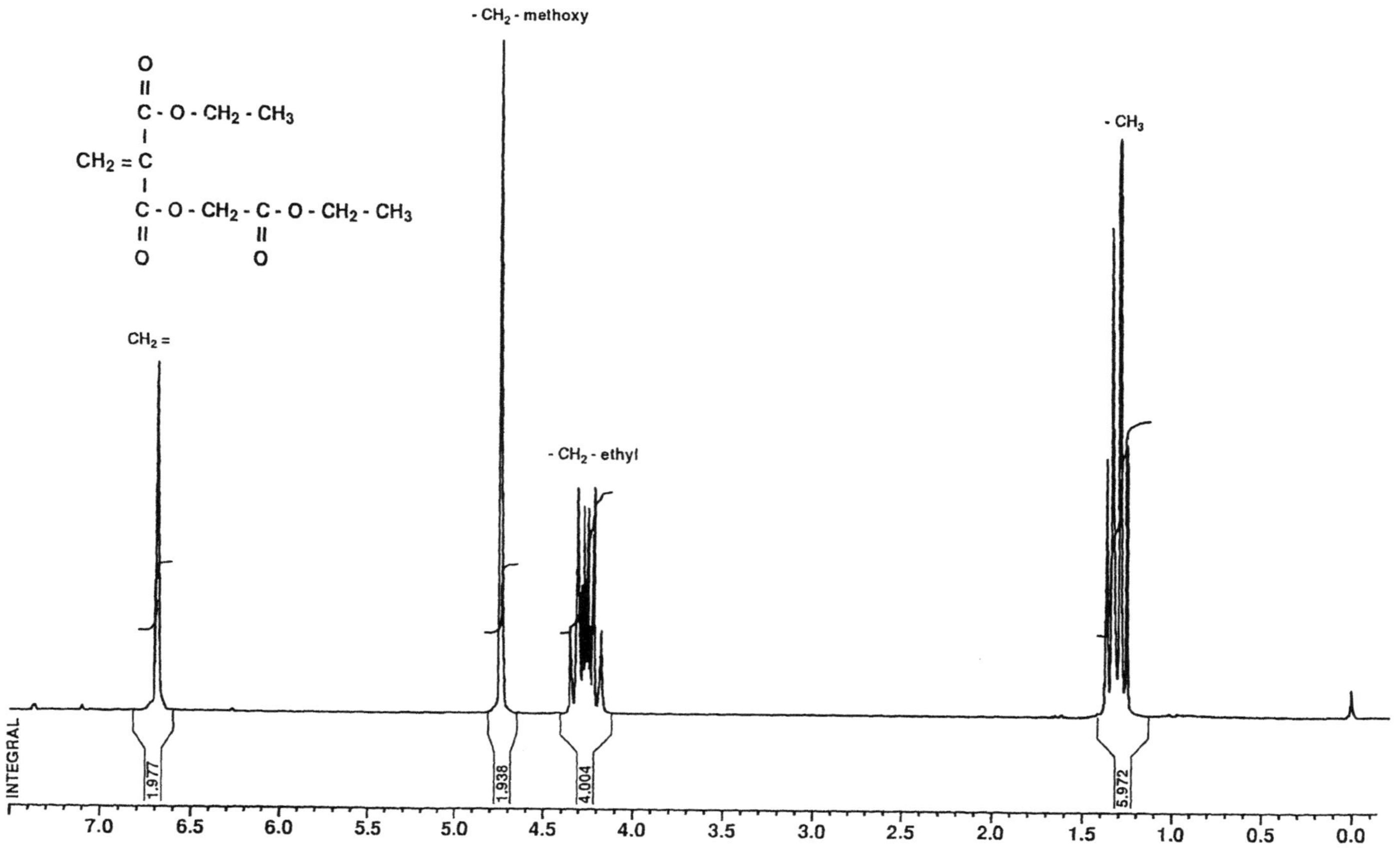

Fig. 5B 200 MHz proton NMR spectrum of MM 2.1.2. (3 mg/mL) recorded at 300K in $CDCl_3$

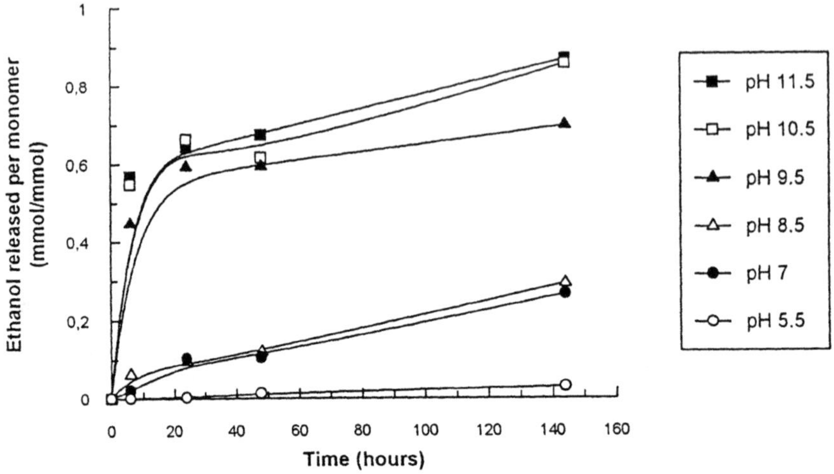

Fig. 6 Degradation rate of PMM 2.1.2. nanoparticles incubated at 37°C at various pH values

nanoparticles were shown to be very basic pH-sensitive and their degradation occured through ethanol release. Like PACA (Lenaerts et al, 1984; Leyh et al, 1984), the conventional formaldehyde-producing degradation route had a very low efficiency (data not shown). In fact, the main PMM 2.1.2. degradation pathway consisted in an alkaline erosion resulting from ester-bearing side chain hydrolysis. Based on the greater susceptibility of esters of ester to hydrolysis such results were expected and, at least in this respect, MM 2.1.2. could be considered as a real advancement in the use of methylidene malonate as a polymeric colloidal biomaterial. As a matter of fact, the first pharmacokinetics data of unloaded PMM 2.1.2., obtained in rat, were rather encouraging (Lescure et al, 1994).

Preliminary investigations concerning in vitro cytotoxicity of PMM 2.1.2. nanoparticles clearly demonstrated that L929 fibroblasts had a better tolerance towards this new polymeric carrier than the one observed in the case of PIHCA and PIBCA nanoparticles. Present results also showed that pre-degraded PMM 2.1.2. nanoparticles were only very weakly toxic (Fig. 7). (Lherm et al, 1992) have recently proposed that PIHCA and PIBCA nanoparticle cytotoxicity, closely related to their degradation velocity, could be following on a close contact between cells and nanoparticles. Consequently, degradation-resulting compounds, highly concentrated at close proximity of the cell surface, would be responsible for either L929 cell growth inhibition or death. According to these data and, regarding the rapid erosion kinetics of PMM 2.1.2. compared to the PIBCA slower one (Couvreur et al, 1989), it can reasonably be assumed that PMM 2.1.2. nanoparticles could be solubilized before extensive interaction with cell plasma membrane. Thus, higher PMM 2.1.2. nanoparticle amounts would be necessary to ensure a critical microscopic concentration of cytotoxic degradation products close to the cell surface.

In short, the new advanced MM 2.1.2. monomer could be described as the first example in the use of methylidene malonate to design potential nanoparticular carriers, candidates for drug targeting. Indeed, poly(methylidene malonate 2.1.2.) nanoparticles would represent a good compromise between biodegradability, which is an important requirement for drug-controlled release systems and, toxicity properties.

Acknowledgement

The authors gratefully thank Drs. T. Gungor and J.M. Teulon, from UPSA Laboratories - CARPIBEM, for performing MM 2.1.2. synthesis and NMR spectra. Also, a special thank should be addressed to Prof. C. Roques-Carmes and Ms C. Millot, from InterUnec (Besançon, France), for their helpful technical support in scanning electron microscopy.

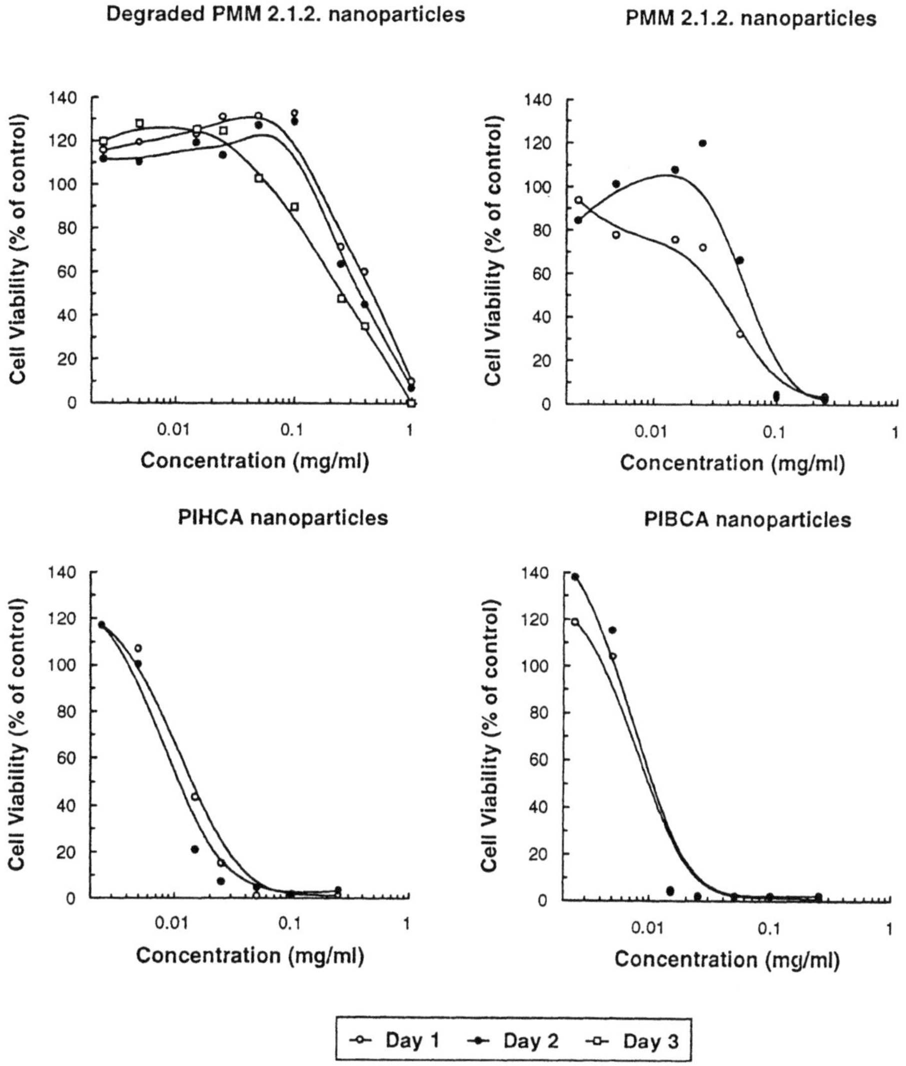

Fig. 7 In vitro cytotoxicity of degraded or intact PMM 2.1.2. nanoparticles and P1HCA- or PIBCA-made nanoparticles. Experiments were carried out on L929 cell line as described in "Material and Methods"

REFERENCES

Barton, A.F.M., 1991, CRC handbook of solubility parameters and other cohesion parameters - Second Edition, CRC Press Inc., Boca Raton, FL

Borenfreund, E. and Puerner, J.A., 1985, Toxicity determined in vitro by morphological alterations and neutral red absorption, Toxicol.Lett., 24:119

Breton, P., Monsigny, M. and Mayer, R., 1990, Synthesis of new acylpseudopeptides analogous to N-acetylmuramyl dipeptide (MDP), Tetrahedron, 46:4265

Bru-Magniez, N., De Cock, C.J., Poupaert, J., De Keyser, J.L. and Dumont, P., 1988, Procédés de préparation de monoesters ou diesters de l'acide endoéthano-9, 10 dihydro-9, 10 anthracène dicarboxylique-11, 11, nouveaux monoesters ou diesters ainsi préparés et utilisation de ceux-ci pour la préparation de méthylidène-malonates

symétriques ou asymétriques, European Patent, EP 0 283 364 A2
Chiannilkulchai, N., Driouich, Z., Benoit, J.P., Parodi, A.L. and Couvreur, P., 1989, Doxorubicin-loaded nanoparticles: Increased efficiency in murine hepatic metastases, Select Cancer Ther., 5:1
Couvreur, P., 1988, Polyalkylcyanoacrylates as colloidal drug carriers, CRC Crit.Rev.Ther.Drug Carrier Syst., 5:1
Couvreur, P. and Vauthier, C., 1991, Polyalkylcyanoacrylate nanoparticles as drug carrier: present state and perspectives, J.Control Release, 17:187
Couvreur, P., Kante, B., Roland, M., Guiot, P., Bauduin, P. and Speiser, P., 1979, Polycyanoacrylate nanocapsules as potential lysosomotropic carriers: preparation, morphological and sorptive properties, J.Pharm.Pharmacol., 31:331
Couvreur, P., Vranckx, H., Brasseur, F. and Roland, M., 1989, Toxicité des nanosphères à base de polycyanocrylates d'alkyle, S.T.P. Pharma, 5:31
Damgé, C., Michel, C., Aprahamian, M., Couvreur, P., 1988, New approach for oral administration of insulin with polyalkylcyanoacrylate nanocapsules as drug carrier, Diabetes, 37:246
De Keyser, J.L., De Cock, C.J., Poupaert, J.H. and Dumont, P., 1989, Synthesis of ^{14}C labelled acrylic derivatives: diethyl [3-^{14}C] methylidene malonate and isobutyl[3-^{14}C] cyanoacrylate, J.Label.Comp.Radiopharm., 27:909
De Keyser, J.L., Poupaert, J.H. and Dumont, P., 1991, Poly(diethyl methylidenemalonate) nanoparticles as a potential drug carrier: Preparation, distribution and elimination after intravenous and peroral administration to mice, J.Pharm.Sci., 80:67
Kattan, J., Droz, J.P., Couvreur, P., Marino, J.P., Boutan-Laroze, A., Rougier, P., Brault, P., Vranckx, H., Grognet, J.M., Morge, X. and Sancho-Garnier, H., 1992, Phase I clinical trial and pharmacokinetic evaluation of doxorubicin carried by polyisohexylcyanoacrylate nanoparticles, Invest.New Drugs, 10:191
Lemieux, P., Page, M. and Noël, C., 1992, In vivo cytotoxicity and antineoplastic activity of a transferrin-daunorubicin conjugate, In vivo, 6:621
Lenaerts, V., Couvreur, P., Christiaens-Leyh, D., Joiris, E., Roland, M., Rollman, and B. Speiser, P., 1984, Degradation of poly(isobutylcyanoacrylate) nanoparticles, Biomaterials, 5:65
Lescure, F., Zimmer, C., Roy, D. and Couvreur, P., 1992, Optimization of polyalkylcyanoacrylate nanoparticle preparation: influence of sulfur dioxide and pH on nanoparticle characteristics, J.Colloid Interface Sci., 154:77
Lescure, F., Seguin, C., Breton, P., Roy, D. and Couvreur, P., 1994, Preparation and characterization of novel poly(methylidene malonate 2.1.2.)-made nanoparticles, Pharm.Res., in press
Lescure, F., Zimmer, C., Roy, D., Teulon, J.M. and Couvreur, P., 1991, Synthesis and evaluation of a new biodegradable monomer, Proceed.Intern.Symp.Control.Rel.Bioact. Mater., 18:325
Leyh, D., Couvreur, P., Lenaerts, V., Roland, M. and Speiser, P., 1984, Etude du mécanisme de dégradation des nanoparticles de polycyanoacrylate d'alkyle, Labo-Pharma - Probl.Tech., 32:100
Lherm, C., Muller, R.H., Puisieux, F. and Couvreur, P., 1992, Alkylcyanoacrylate drug carriers: II. Cytotoxicity of cyanoacrylate nanoparticles with different alkyl chain length, Int.J.Pharm., 84:13
Papahadjopoulos, D., Allen, T., Gabizon, A., Mayhew, E., Matthay, K., Huang, S.K., Lee, K.D., Woodle, M.C., Lasic, D.D., Redemann, C. and Martin, F.J., 1991, Sterically stabilized liposomes: Improvements in pharmacokinetics, and anti-tumor therapeutic efficacy, Proc.Natl.Acad.Sci.USA, 88:11460
Ponticello, I.S., 1977, Substituted acrylates, US Patent 4 056 543
Roche, A.C., Bailly, P. and Monsigny, M., 1985, Macrophage activation by MDP bound to neoglycoproteins: Metastasis eradication in mice, Invas.Metastasis, 5:218
Vansnick, L., Couvreur, P., Christiaens-Leyh, D. and Roland, M., 1985, Molecular Weights of free and drug-loaded nanoparticles, Pharm.Res., 36
Vezin, W.R. and Florence, A.T., 1980, In vitro heterogeneous degradation of poly(n-alkyl α-cyanoacrylates), J.Biomed.Mat.Res., 14:93
Weinstein, J.N. and Leserman, L.D., 1984, Liposomes as drug carriers in cancer chemotherapy, Pharmacol.Ther., 24:207

ORAL UPTAKE AND TRANSLOCATION OF NANOPARTICLES: A REAL BUT USEFUL PHENOMENON ?

A.T. Florence, A.M. Hillery, N. Hussain and
P.U. Jani

Centre for Drug Delivery Research, The School of Pharmacy
University of London, Brunswick Square, London WC1N 1AX, UK

INTRODUCTION

Peptide and proteins represent an increasingly important group of drugs which frequently require specialized modes of delivery. A great challenge is now presented to deliver these substances by the oral route. While the wall of the mammalian gastrointestinal tract has been commonly assumed to be a largely impenetrable barrier to the passage of particulate matter, work carried out in several laboratories (Florence and Jani 1993., Jani et al., 1989, 1992a, Le Fevre et al. 1978a,b, 1980, 1989., Eldridge et al., 1990) has shown that the uptake of microspheres through specialised regions of the GI tract (eg Peyer's patches) is possible. Experimental evidence in animal models has confirmed that sub-micron particulates can be taken up into the intestinal Peyer's patches following oral administration, and transported, if below a critical size, via the lymphatics to organs such as the liver and spleen. This phenomenon could be exploited such that these particles could serve as drug carriers to deliver therapeutic peptides and proteins to the systemic circulation, following the oral uptake of the intact drug-carrier complex. But, in spite of the evidence there is controversy over the nature and extent of the phenomenon. With peptide and protein antigens it is clearer that the oral route holds promise for oral immunisation as immune responses are not subject to conventional dose response relationships. Since the Peyer's patches are the sites of induction of the secretory immune response (Bjerke et al. 1993., McStecky and McGhee 1989), antigens have been encapsulated in microparticles and delivered orally to induce enhanced secretory immunity in attempts to develop effective oral vaccines. Additional targeting of the antigen loaded particles to mesentery lymph nodes following translocation from Peyer's patches, should boost the induction of systemic immunity.

We have been investigating the extent and rates of uptake of non-biodegradable fluorescent microparticles in the range of 50nm to 3μm in diameter to characterise the factors affecting particulate carrier uptake, described below and have also conducted initial studies in the absorption of a particle forming peptide polymer conjugate (Hillery et al. to be published).

POSSIBLE ROUTES OF ENTRY OF PARTICULATE

Although the subject of particulate uptake across the gut is gaining more widespread acceptance, it is not without controversy. Part of this stems from the conflicting mechanisms

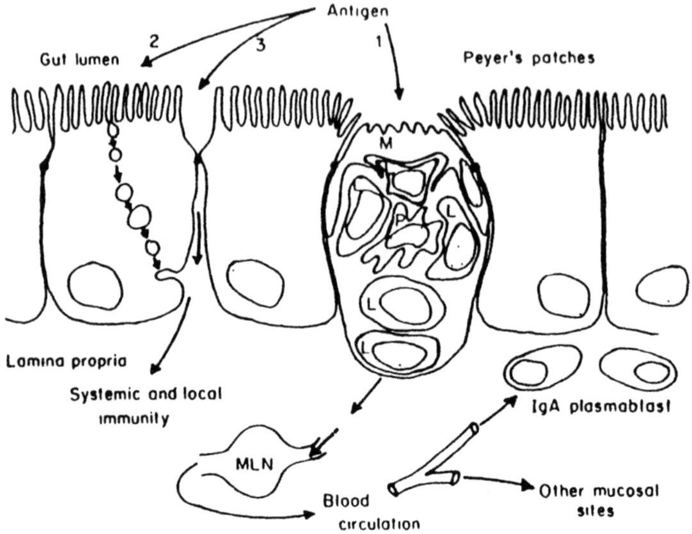

Fig. 1 Schematic representation of the mode of entry of food antigens across the intestinal mucosa. The antigens in the gut lumen enter via (1) the specialised microfold epithelial (M) cells of the Peyer's patches in the normal intestine, (2) the enterocytes in the immature intestine, or (3) the interepithelial tight junction in osmotically stressed intestine.

that have been put forward to explain the phenomenon, which include (a) persorption (Volkheimer 1968, 1975) (b) endocytosis by ordinary enterocytes (Bockman and Winborn, 1966) (c) paracellular transport (Aprahamian et al., 1987) (d) uptake by intestinal macrophages (Wells et al., 1988) and (e) uptake through the GALT. (Le Fevre et al., 1978a, b, 1980, 1989; Eldridge et al., 1990; Jani et al., 1989, 1990, 1992a, b; Urbanski et al., 1989). Further controversy concerning the oral uptake of particulate is due to the different approaches to quantifying the extent of uptake, with much variability reported in determinations of the amount taken across the gut (Alpar et al., 1989; Jani et al., 1990; Jenkins et al. 1994). In a recent review O'Hagan et al., (1992) referred to the "extraordinarily high" level of uptake obtained in work published by our group (Jani et al., 1990). Proof of the true extent of uptake is obviously crucial in determining the viability of the route, as uptake of the carrier must be commensurate with the required dose for activity of the peptide or protein.

There are three possible modes of uptake of antigens and colloidal particles, some less efficient than others: transport across intact epithelial membranes (transcellular route) possibly by endocytosis, transport through endothelial tight junctions (the paracellular route) and transport via the modified epithelial cells (M-cells, see Wolf and Bye 1984) of Peyer's patches (Fig.1). Certainly Peyer's patches (Fig.2) are the site of adherence and uptake of polymer nanoparticles, as well as bacteria and reovirus particles. Absorption of macromolecules and particulate matter involving the process of endocytosis is complex and involves formation of pinocytotic vacuoles, a process dependent on both the hydrophobicity of the membrane and the molecule or particle, in addition to other factors such as the charge of the particle.

The present study provides further evidence both on the site of maximum uptake in the gut and the extent of uptake of polystyrene model particles in the rat. It also deals with a comparative and quantitative study of lymphoid and non-lymphoid uptake of 60nm fluorescent polystyrene particles. While histological studies could be criticised for the lack of quantitation, direct analysis of tissues for polystyrene, we posit, cannot be gain said.

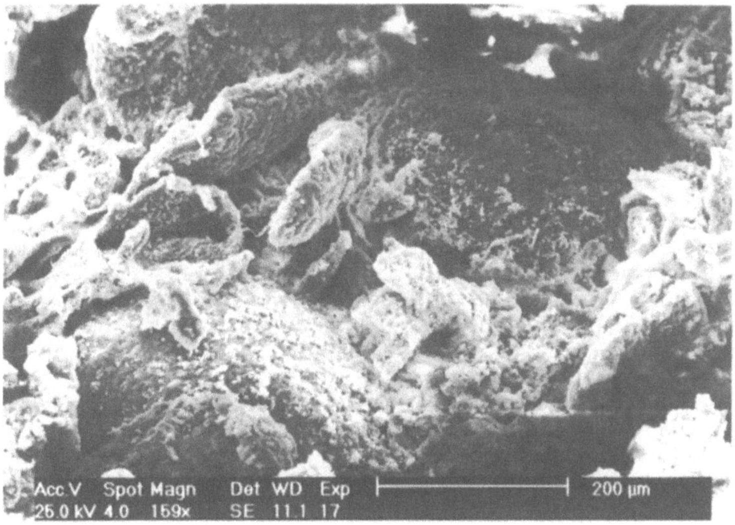

Fig. 2 An electron micrograph of a portion of one Peyer's patch as seen by a scanning electron microscope, note the surrounding villi, amongst the bulging lymphoid follicles, some of which overhang the follicular surfaces

MATERIALS AND METHODS

Microspheres

Monodisperse non-ionized polystyrene microspheres with covalently linked fluorescein (mean diameter: 59nm ±3nm; 2.5% solids-latex) were obtained from Polysciences Ltd (Northampton, UK). The size of the microspheres was confirmed using photon correlation spectrometry: (Multi 8 Computing Correlator Type 7032CE with Liconix He/Cd Laser. Malvern, UK)

Rats

Female Sprague-Dawley rats (average weight 180g, 9 weeks old) were housed in metabolic cages to prevent coprophagia and to allow monitoring of growth curves. The rats were fed Maintenance Expanded Rat and Mouse Pellet Diet (Bantin and Kingman, Hull U.K.) and maintained in an air conditioned environment at 20^0C under a 12 h light: 12 h dark schedule.

Oral dosing and sample collection

The rats were dosed by gavage every morning for 5 days (total dose 12.5mg/kg; daily volume 0.1 mL volume), using a blunt tipped feeding needle inserted into the stomach. The animals were fasted for 12 h prior to dosing and food was returned 3 h later, except after the final dose. Free access to water was given at all time. After the final dose, the animals were fasted for 12 h to clear the gut of food and unabsorbed microspheres. The animals were sacrificed using the excess ether method and the stomach, intestine, liver, spleen and kidney removed. The stomach and intestinal tissue was cut longitudinally and the contents removed; the gut tissues were then washed gently with distilled water in an attempt to remove any particles that were not absorbed, but were adhering to the surface of the tissue. All dissected tissues were weighed and stored separately at -70^0C.

Histology

For histological analysis, frozen sections of the various organs were prepared using a cryostat and the particles viewed with a fluorescent microscope as described by Jani et al. (1992a).

Gel Permeation Chromatography

In some experiments the gut was divided into the following sections for analysis: 1. Lymphoid Small Intestine (LSI): excised Peyer's patches and mesenteric connective tissue; 2. Non Lymphoid Small Intestine (NLSI): the remaining tissue of the small intestine; 3. Lymphoid Large Intestine (LLI): appendix and visible follicular aggregates from the colonic region; and 4. Non Lymphoid Large Intestine (NLLI): the remaining tissue of the large intestine.

The organs selected for analysis were finely cut and lyophilized. The lyophilisate was ground to a fine powder at -70°C and extracted with chloroform (3 x 20mL) over a period of 3 days. The solvent was removed in vacuo and the extracted polystyrene made up to 5 mL with tetrahydrofuran (THF).

The GPC analysis was carried out on 2 x 30cm PLGel Mixed B 10Å columns with a PLGel Mixed B 10Å guard column (Polymer Laboratories Ltd, Shropshire UK); using a Waters 510 HPLC pump, a Waters 484 Tunable Absorbance Detector (λ = 254nm) and a Hewlett Packard HP3396A Integrator. THF (BDH AnalaR grade) was used as the solvent and mobile phase at a flow rate of 1mL/min. Column calibration was carried out using polystyrene standards (EasiCal PS-1; Polymer Laboratories Ltd) of molecular weight 580-8,500,000.

Polystyrene concentrations were determined from a plot of Area Under the Curve against known concentrations of polystyrene (Range: 0.025mg - 2.5µg). The amount of polystyrene in each test sample was then calculated from the standard curve. The results were expressed as percentage uptake of the administered dose in the tissues.

Extraction efficiency

The efficiency of the extraction procedure for polystyrene was estimated by injecting a known concentration of microspheres into a perfused rat liver and subjecting the liver to the above extraction procedure (using the method of Jani et al., 1990). 70-80% of the injected dose was recovered from the liver and the results from the gut tissue were adjusted to account for a 70-80% recovery and detection sensitivity.

Conjugation of tomato lectin to microsphere surfaces

The methodology used and tests to ensure the maintenance of lectin activity are described by Hussain et al. (1994).

RESULTS

UPTAKE OF POLYSTYRENE PARTICLES (60NM)

The percentage of the administered dose of 60nm polystyrene particles taken up by lymphoid and non-lymphoid regions of the small and large intestine is presented in Table 1. Approximately 10% of the administered dose was recovered from the entire GI tract, after 5 days dosing. GPC analysis for polystyrene was only carried out on gut tissue and not on sites beyond the GI tract. This highly sensitive method used to quantitate uptake cannot, in our view, be subject to substantial error and we argue that the hydrophobic nature of polystyrene and the small diameter of particles used, 60nm, assists the uptake process (Eldridge et al., 1990).

Although prior to GPC analysis, the gut tissues were washed in an attempt to

Table 1 The percentage of the administered dose of 50nm polystyrene particles taken up by lymphoid and non-lymphoid regions of the small and large intestine

	LSI	NLSI	LLI	NLLI
n	3	3	4	4
WEIGHT (g)	0.53 ± 0.02	2.3 ± 0.6	0.86 ± 0.1	1.0 ± 0.2
AREA(cm^2)	1.4 ± 0.03	271.18[a]	NM	NM
(%)UPTAKE	3.38 ± 0.41	2.32 ± 0.48	3.0 ± 0.21	2.15 ± 0.12

LSI = Lymphoid Small Intestine
NLSI = Non-Lymphoid Small Intestine
LLI = Lymphoid Large Intestine
NLLI = Non-Lymphoid Large Intestine
NM = Not Measured.
a = from Fisher and Parsons (1950)

remove particles which were merely adhering to the suface, it was possible that some of the detected uptake represented particles which were adsorbed to the surface rather than translocated to the serosal side.

In a previous study, Jani et al. (1990) used GPC to analyze polystyrene uptake in the gut as an entire unit. Here we have attempted to subdivide the various gut regions and separate the lymphoid tissue, including the Peyer's patches, from the non-lymphoid tissue. On average, 15 Peyer's patches were observed in the small intestine of each rat in this study. The patches were oval in shape, situated generally in the antimesometrial wall and uniformly distributed over the small intestine. The diameter of the Peyer's patches was measured by callipers and the corresponding gross surface area available for absorption calculated. Approximately 60% of the polystyrene detected in the small intestinal tissue was recovered from the Peyer's patches, although the patches constitute less than 1% of the area of the entire small intestine (Table 1). The measured uptake of polystyrene particles through the **total** lymphoid tissue of the gut (ie small plus large intestinal lymphoid tissue) was statistically much greater than through total non-lymphoid tissue (the mean of paired differences using a paried t test was 0.9789 ± 0.2, two-tailed p value = 0.0031)Particles taken up by absorptive cells may be subject to lysosomal degradation, which is not important for non-biodegradable polystyrene, but is obviously an important consideration for biodegradable drug carriers and the labile molecules they containe and (ii) discharge back into the lumen, on desquamation of the cell. There is a marked absence of lysosomal degradation by M cells and material taken up may subsequently be transported through the lymphatics, as discussed above.

With this work, we have again provided quantitative proof of the importance of the GALT for the uptake of particles and confirmed the role of the Peyer's patches in the uptake of particles through the small intestine. The importance of the lymphoid tissue in the large intestine was also shown: a significant amount of the total uptake occurring here, particularly in the lymphoid sections of this tissue (Table 1). Targeting of colloidal carriers to the lymphoid regions of the large intestine could therefore represent a significant method of enhancing particle/carrier uptake *via* the oral route.

The preferential uptake of polystyrene particles through the GALT was confirmed through the use of fluorescence microscopy. A high density of microspheres were concentrated in the serosal layer of the Peyer's patches, i.e. uptake and translocation from the gut lumen and mucosal layer had taken place. Particles traversed the mesentery *via* the mesentery lymph vessels and were observed within the mesenteric connective tissue and mesenteric lymph nodes. Absorbed microspheres were subsequently transported from the lymphatic circulation into the venous circulation and were taken up by other tissues. A high density of particles were observed in the liver, with lesser amounts visible in the spleen; the

Table 2 Oral uptake of Tomato lectin-congugated and plain microspheres as a function of total dose administered.

Tissue	TL conjugates %	Plain[1] Polystyrene %
Intestine	12.8 ±3.00	
PP	0.89 ±0.04	4.28 ±0.52
Colon	0.90 ±0.02	6.53 ±0.75
Liver	2.64 ±1.70	1.38 ±0.35
Spleen	1.17 ±0.80	0.51 ±0.03
Blood[2]	3.04 ±0.64	n/d
Heart	0.29 ±0.03	n/d
Kidney	0.67 ±0.16	n/d
Systemic[3]	23.0 ±6.53	1.89 ±0.38

n=4 PP = Peyer's patches n/d = not detectable

[1] Data from Jani et al 1990 using plain microspheres (500nm) and chronic administration for 10 days. Whole intestinal used for quantitation, histological examination indicating the Peyer's patches to be highly enriched with microspheres compared to villous tissue.
[2] 2ml blood sample taken via cardiac puncture.
[3] Includes extrapolation to whole blood volume (12mL for 200g rat).

particles were able to penetrate the fenestrations of the kidney and were seen in the outer cortex.

LECTIN INDUCED UPTAKE AND TRANCYTOSIS IN GUT

Intestinal absorption is significantly enhanced by the adhesion of particulates to the intestinal epithelium as illustrated by certain enterobacteriaceae whose attachment is mediated through lectin-like adhesins on their outer surface (Trier 1990). An increasing number of plant lectins, glycoproteins of non-immune origin capable of recognising glycosidic residue with high specificity, have been shown to be resistant to GIT degradation and to bind to the intestinal epithelium (Pustzai et al., 1990; Lehr et al, 1990) Tomato lectin (TL; from Lycopersicon esculentum) is non toxic upon epithelial binding (Kilpatrick et al., 1985) and highly resistant to proteolytic degradation. Hitherto, only in vitro work with tomato lectin for oral drug delivery purpose has been conducted.

To investigate the in vivo potential of TL, carboxylated fluorescent polystyrene microspheres of 500nm diameter size with tomato lectin coupled to their surface were administered to female Wistar rats of 10 weeks daily for 5 days by oral gavage. After the final dose rats were fasted overnight in a microsphere free environment but allowed water to clear the gut of any unabsorbed microspheres. The removal of organs and quantification of tissue extracted latex by gel permeation chromatography was accomplished by the method described above. Comparison of the results (Table 2) with those for plain non-ionised 500nm microspheres as determined previously in our laboratory (Jani et al., 1990) revealed 3 times greater intestinal absorption and internalisation of TL-conjugates and 9-12 fold greater systemic absorption of TL-conjugates compared to their nonionized counterparts. Non-ionized particles were not detected in blood in our previous work, possibly due to their lower absorption and capture by the reticuloendothelial system. TL conjugates however appeared to avoid capture due to their modified surfaces, thus yielding elevated blood levels. In addition small amounts of TL-conjugates were seen in the heart and kidneys indicating extensive systemic distribution. Fascinatingly, the uptake of TL-conjugates was 13 times greater through villous tissue than via Peyer's patches demonstrating the re-direction and the relative unavailability of complementary glycosidic binding structures over rat lymphoid tissue.

Although the M-cells are considered to be one of the main port of entry of colloidal particulates, the most abundant intestinal cell is the enterocyte which is still capable of receptor mediated endocytosis via class II receptors, similar to the hepatocyte asialoglycoprotein receptor in which endocytosis is an obligatory step following ligand binding. Lectins are thought to interact with these receptors, which may explain the uptake of our 500nm microspheres. The work (to be reported in full elsewhere) demonstrates the potential use of tomato lectin as a bioadhesive component of oral drug delivery systems and in enhancing intestinal translocation of colloidal carriers incorporating either antigens for oral vaccination or drug molecules that are either susceptible to degradation or those that are intrinsically poorly absorbed.

DISCUSSION AND CONCLUSIONS

It was concluded that particulate uptake across the gut is possible, reproducible and occurs to a reasonable extent. The uptake of polystyrene latex particles (60nm) across the murine GI tract was demonstrated via chromatographic (GPC) and histological (fluorescent microscopy of frozen sections) techniques. The extent of uptake, quantified using GPC, was determined as 10% of the administered dose, across the entire GI tract. Quantitative proof of the importance of the GALT for the uptake of particles was provided and the important role of the Peyer's patches in the uptake of particles was confirmed.

The use of tomato lectin has shown not only the significant intestinal absorption of large colloidal particles than hitherto thought possible but the re-routing through villous tissue and the avoidance of reticuloendothelial capture. The present study demonstrates that the use of non-toxic epithelial binding lectins as agents which may increase epithelial trancytosis of orally administered antigens and therapeutic colloidal systems are ready for further exploration.

The reproducibility of uptake of other colloidal particles, particularly biodegradable systems, needs to be established. We have also clearly demonstrated that the process in the small intestine at least can be inhibited by poloxamer adsorption and can be increased, probably by greater involment of non-GALT tissue, by particulate covalent attachment of tomato lectin's to the polystyrene particle surface. Several independent studies in our laboratories have confirmed the extent of the uptake; the discrepancies between other data and ours needs to be addressed.

Acknowledgements

Several agencies have supported the work described in this paper. Syntex Research Centre (AMH, PUJ), C.W. Maplethorpe bequest of the University of London for the provision of a postdoctoral fellowship to P.U.J., The Royal Pharmaceutical Society of Great Britain /CRISP award shared with Syntex (NH) and Mr. David Mc Carthy for his assistance with the histology, Mr Adrian Rogers, Department of Toxicology, for expert photographic assistance.

REFERENCES

Alpar, H.O., Field, W.N., Hyde, R. and Lewis, D.A., 1989, The transport of microspheres from gastro-intestinal tract to inflammatory air pouches in the rat. J.Pharm. Pharmacol., 41:194

Aprahamian, M., Michel, C., Humbert, W., Devissaguet, J.P. and Damge, C., 1987, Transmucosal passage of polyalkylcyanoacrylate nanocapsules as a new drug carrier in the small intestine. Biol.Cell., 61:69

Bjerke, K., Halstensen, T.S., Jahnsen, F. and Brandtzaeg, P., 1993 Distribution of macrophages and granulocytes expressing L1 protein (calprotectin) in human Peyer's patches compared with normal ileal lamina propria and mesenteric lymph nodes. Gut, 34:1357

Bland, P.W. and Britton, D.C., 1984, Morphological study of antigen sampling structures in the rat large intestine. Infect.Immunol., 43:693

Bockman, D.E. and Winborn, W.B., 1966, Light and intestinal microscopy of intestinal ferritin. Observations in sensitised and non-sensitised hamsters. Anat.Rec., 155:603

Eldridge, J.H., Hammond, C.J., Meulbroek, J.A., Staas, J.K., Gilley, R.M., Tice, T.R., 1990, Controlled vaccine release in the gut-associated lymphoid tissues: orally administered biodegradable microspheres target the Peyer's patches. J.Controlled Release., 11:205

Evans, J. M., 1973, Gel permeation chromatography; a guide to data interpretation. Polymer Engineering Science., 13:401

Fisher, R.B. and Parsons, D.S., 1950, The gradient of mucosal surface area in the small intestine of the rat. J.Anat., 84:272

Florence, A.T. and Jani, P.U., 1993, Particulate delivery: the challenge of the oral route. In: "Pharmaceutical Particulate Carriers, Therapeutic Applications", A. Rolland, ed., Marcel Dekker Inc., New York

Hussain, N., Jani, P.U. and Florence, A.T., 1994, Oral uptake of Lectin-Conjugated microspheres through the murine intestine, in press

Jani, P.U., Florence, A.T. and Mc Carthy, D.E., 1992a, Further histological evidence of the gastrointestinal absorption of polystyrene nanospheres in the rat. Int.J.Pharm., 84:245

Jani, P.U., Mc Carthy, D.E. and Florence, A.T., 1992b, Nanosphere and microsphere uptake via Peyer's patches: observation of the rate of uptake in the rat after a single dose. Int.J.Pharm., 86:239

Jani, P.U., Halbert, G.W., Langridge, J. and Florence, A.T., 1990, Nanoparticle uptake by the rat gastrointestinal mucosa: Quantitation and Particle size dependency, J.Pharm.Pharmacol., 42:821

Jani, P.U., Halbert, G.W., Langridge, J. and Florence, A.T., 1989, The uptake and translocation of latex Nonospheres and Microspheres after oral administration to rats, J.Pharm.Pharmacol., 41:809

Jani, P.U., Florence, A.T., McCarthy, M.E., 1992, Further histological evidence of the gastrointestinal absorption of polystyrene nanospheres in the rat, Int.J.Pharm., 84:245

Jenkins, P.G., Howard, K.A., Blackhall, N.W., Thomas, N.W., Davis S.S. and O'Hagan, D.T., 1994, The quantitation of absorption of micropaticles into the intestinal lymph of wistar rats. Int.J.Pharm., 102:261

Kilpatrick, D.C., Pustzai, A. and Grant, A., 1985, Tomato lectin resists digestion in the mammalian alimentary canal and binds to intestinal villi without deleterious effects. FEBS Letts., 185:299

Lee, V., 1991, Changing needs in drug delivery in the era of peptide and protein drugs. In: "Peptide and Protein Drug Delivery", V. Lee, ed., Marcel Dekker Inc., New York

LeFevre, M.E., Boccio, A.M., and Joel, D.D., 1989, Intestinal barrier to large particulate in mice, Proc.Soc.Exp.Biol.Med., 190:23

LeFevre, M.E., Vanderhoff, J.W., Laissue, J.A. and Joel, D.D., 1978a, Accumulation of 2μm latex particles in mouse Peyer's Patches during chronic latex feeding. Experientia., 34:120

LeFevre, M.E., Olivo, R., Vanderhoff, J.W. and Joel, D.D., 1978b, Accumulation of latex in Peyer's patches and its subsequent appearance in villi and mesenteric lymph nodes. Proc.Soc.Exp.Biol.Med., 159:298

LeFevre, M.E., Hancock, D.C. and Joel, D.D., 1980, Intestinal barrier to large particulate in mice. J.Tox.Environ.Health., 6:691

Lehr, C.M., Bouwstra, J.A., Kok, W., Noach, A.B.J., de boer, A.G. and Junginger, H.E., 1990, Bioadhesion by means of specific binding of tomato lectin. Pharm.Res., 9:547

McStecky, J. and McGhee J.R., eds., 1989, New Strategies for oral immunization. SpringerVerlag: Berlin.

O'Hagan, D.T., Christy, N.M. and Davis, S.S., 1992, Particulate and Lymphatic Drug Delivery. In: "Lymphatic Transport of Drugs", W.N. Charman and V.J. Stella, eds., CRC Press

Pustzai, A., Ewen, S.W.B., Grant, A., Peumans, W.J., Van Damme, E.J.M., Rubio, L. and Bardocz. S., 1990, Relationship between survival and binding of plant lectins during small intestinal passage and their effectiveness as growth factors. Digestion, 46 (suppl2): 309.

Trier, J.S., 1990, Structure and function of intestinal M-cells. Gastroenter.Clinics.North Am., 20:531

Urbanski, S.J., Arsenault, L., Green, F.H. and Haber, G., 1989, Pigment resembling atmospheric dust in Peyer's patches. Modern Pathol., 2:222
Volkheimer, G. and Schulz, F.H., 1968, The phenomenon of persorption. Digestion, 1: 213.
Volkheimer, G., 1975, Hematogenous dissemination of ingested polyvinylchloride particles. Ann.N.Y.Acad.Sci., 246:164
Wells, C.L., Maddaus, M.A., Erlandsen, S.L. and Simmons, R.L., 1988, Evidence for the phagocytic transport of intestinal particles in dogs and rats. Infect.Immunol., 56:278
Wolf, J.L. and Bye, W.A., 1984, The Membranous epithelial (M) cells and the mucosal immune system. Ann.Rev.Med., 35:95

PARTICULATE SYSTEMS FOR SITE SPECIFIC DRUG DELIVERY

S.S. Davis and L. Illum

Dept of Pharmaceutical Sciences, University of Nottingham
University Park, Nottingham, NG7 2RD, UK

INTRODUCTION

The Drug Delivery Research Group at the University of Nottingham has for a number of years been interested in the development of a wide variety of systems for site specific drug delivery to include micro and nanospheres, polymers and prodrugs. This paper will focus especially on the use of colloidal carriers such as nanospheres, emulsions and liposomes to achieve site specific delivery of drugs to sites in the body such as the lungs, liver, spleen, the general circulation, the lymphatics, the bone marrow and the joint cavities after parenteral administration. Microspheres administered by non-parenteral routes of administration will further have the ability of targeting to the lungs, the nasal cavity and the gastrointestinal tract. The use of the gastrointestinal tract is of special importance for vaccine delivery and for the exploitation of bioadhesive particles to modify gastrointestinal transit.

The terms "drug targeting" and "site specific delivery" are used indiscriminately in the literature for any carrier system that obtains some degree of accumulation in specific sites in the body. It would be useful to define precisely what is meant by these terms. For example, how much of the administered dose should reach the target site in order that delivery systems may be given the accolade of a drug targeting system? Should it be that greater than 90% or 50% of the dose reaches the designated site, or would 10% at the target site be acceptable? Some published papers have claimed drug targeting even though calculations show that on close examination less that 1% of the dose reached the designated site (Mizushima et al, 1982). This may well be "site specific delivery" but is not, in our opinion, "drug targeting". We consider that the term "drug targeting" should be reserved for those situations where more than 50% of the administered dose reaches the target site. An alternative is to consider the ratio of target drug concentrations to those at non-target sites. Such a drug targeting index has been proposed by Boddy and Aarons, (1991). From a clinical standpoint, the important factor is the improvement in the therapeutic index of the drug; such that a higher dose of the drug can be administered before the manifestation of adverse side effects and reactions or that a smaller dose of drug will result in the same therapeutic effect.

Site specific delivery or targeting can be achieved at different levels of sophistication. A primary level of targeting can involve a specific organ, for example, the heart, liver, lung etc. A second order of targeting can be to a specific cell type within that organ, for example, for an organ such as the liver, it is possible to identify a whole range of cell types that may constitute appropriate target sites. These include Kupffer cells, hepatocytes and endothelial cells. It is possible to define a further tertiary level of targeting

Targeting of Drugs 4, Edited by G. Gregoriadis *et al.*
Plenum Press, New York, 1994

where intracellular structures such as the lysosomal compartment, cytoplasm, golgi apparatus and the nucleus are the sites of interest. The latter site has particular relevance in new developments such as gene therapy (Vanchieri, 1993).

Drug targeting can be characterised as being either passive or active (Tomlinson and Davis, 1986). In passive targeting, the delivery system reaches its target site by the operation of natural mechanisms. These mechanisms include filtration of particles by narrow capillary structures, or capture of particles by tissue macrophages. Passive targeting can be controlled or modified by changing particle size or alteration of surface structures. In the latter case, the surface of a particle can have an important role in determining the uptake of critical plasma components in processes known as "opsonization" and "dysopsonization" (Moghimi and Patel, 1992). These will be discussed further below. Active targeting can be achieved by the attachment of groups that render the particle recognised by body tissues or cell structures. Common examples include the use of sugar groups (lectin mediated uptake) as well as mononclonal antibodies and their fragments. More recently, our group has identified the possibility of using polymeric coatings such as poloxamers, in order to achieve targeting to endothelial cells in the bone marrow by exploiting recognition processes (Illum and Davis, 1987; Porter et al, 1992). It is believed that the polymer predisposes the particle to the adsorption of a specific blood component which then provides the necessary structure for a recognition event at the endothelium in the bone marrow.

COLLOIDAL CARRIERS

When considering aspects of drug delivery including drug targeting, it is often instructive to consider how the human body itself achieves similar tasks. An excellent example is low density lipoprotein (LDL). These small particulates are involved in cholesterol metabolism. The core of the particle consists of cholesterol ester. The surface of the particle comprises a layer of phospholipid and interestingly, a recognition moiety in the form of apoprotein (B) which provides the necessary element of targeting. Not surprisingly, attempts have been made to produce such particles semi-synthetically with a certain degree of success (Vitols et al, 1985; Van Berkel et al, 1993).

This paper will concentrate on the general principles that influence the targeting of colloidal particles in the body. Both parenteral administration (vascular and lymphatic targeting) as well as targeting within the gastrointestinal tract will be considered. Most of the data presented are from studies on model particles in the form of non-degradable polystyrene latices. However, the results presented are equally applicable to other particle systems such as degradable microspheres, emulsions and liposomes. Examples have been chosen to demonstrate how a proper understanding of the physicochemical properties of small particles, and the relevant aspects of cell biology and molecular biology can be used to produce targeting systems for different body sites. They also serve to provide insight into how particle engineering can be used to dictate performance events in biology. The critical properties of the particle are not, only particle size and surface charge, but other less well definable characteristics such as surface hydrophobicity.

RELEVANT ASPECTS OF BIOLOGY

Before embarking on the development of a novel targeting system, it is always instructive to have a clear understanding of the relevant aspects of the biological environment (Posnansky and Juliano, 1984). Not surprisingly, there are many barriers that can prevent the successful targeting of a particulate system to its intended site of action. It is also critical to realise that considerable differences can occur between the normal and pathological state. For example the reticuloendothelial system, that has an important role in body defence can demonstrate very different characteristics under certain pathological conditions as compared to the case for the normal or healthy individual.

When attempting the targeting of particles to the systemic circulation, it is important to have an appreciation of the endothelial barrier that lines the blood vessels. In most tissues the endothelial barrier is intact and therefore there are no opportunities for extravasation of particles, however small they may be. Even if a particle was able to pass through endothelial structures, the basement membrane would constitute another significant barrier. In certain tissues, such as endocrine glands, etc. gaps exist within the endothelial wall. However, these gaps are covered by a membrane and the basement membrane is intact. It is unlikely therefore that particulates could find their way out of the circulation from such tissues to extravascular sites. In the liver, the gaps in the endothelial lining are of quite a significant size (eg 100-150 nm). Here, particles may be able to pass through to reach the underlining parenchyma, since the basement membrane is partial or absent (Wisse and Leeuw, 1984). Certainly, so far as the liver is concerned, it is now well established that small particles can escape from the circulation in this way. However, it needs to be remembered that the blood sinusoids of the liver also contain phagocytic Kupffer cells. These resident macrophages have an impressive capacity to recognise and capture rapidly those particulate systems that are recognised as being foreign. As a consequence, the Kupffer cells in the liver represents a major site for the uptake of injected particles that have unnatural characteristics. This includes not only injected delivery systems but also invasive microorganisms. Only when the particle is disguised and recognised as self does it have the ability to avoid macrophage capture and achieve extravasation.

Thus, the opportunities for extravasation of particles from the circulation under normal conditions is very limited. However, in pathological states, such as at sites of inflammation and in certain tumours, not only is the endothelial layer disrupted but so may be the basement membrane. These pathologies can provide an opportunity for particles to escape to such target sites, either by simple processes of diffusion or more likely by being carried by cells (diapedesis) (Illum et al, 1989).

INTRAVENOUS INJECTION OF COLLOIDAL CARRIERS

The processes that occur upon administration of a particulate system into the blood circulation are critical in determining subsequent events that can include particle recognition and targeting. Immediately upon injection, all particles will be conditioned in some way through the rapid adsorption of blood components (Moghimi and Patel, 1992; Leonard and Vroman, 1991). This conditioning process will be highly dependent upon the surface nature of the particle. In general terms, hydrophobic particles will adsorb considerable amounts of protein, whereas hydrophilic particles have a reduced capacity for protein uptake. This process of adsorption is known as "opsonization" and it is a critical stage in particle recognition. The adsorption of certain blood components (including complement, gamma globulin, fibronectin) can render a particle easily recognised by macrophages. The Kupffer cells of the liver have a battery of receptors that can recognise a wide variety of opsonic factors (Toth and Thomas, 1992).

It is believed until recently that one strategy in reducing macrophage uptake was to minimise this opsonisation process. Reduced uptake of plasma components onto particles is an important factor in controlling macrophage uptake, but it is now appreciated that certain hydrophilic particle surfaces can selectively encourage the adsorption of plasma components that can act as so-called dysopsonins that prevent recognition of particles by macrophages (Moghimi and Patel, 1992; Moghimi et al, 1993). Thus, as with many other biological processes, the mechanisms affecting particle uptake by macrophages are controlled by a balance of competing (opsonins and dysopsonins) factors and processes.

A wide variety of blood components may be adsorbed to the surface of particles injected into the circulation. This uptake is largetly dictated by the nature of the particle surface (Figure 1) (Norman et al, 1992; Norman et al, 1993). Such particle conditioning is a competitive and dynamic process. The competitive uptake of plasma components to foreign surfaces has been investigated in detail in the field of biomaterials. Vroman and colleagues have suggested that the component to be adsorbed initially to a foreign surface will be albumin since this is present at a high concentration (Leonard and Vroman, 1991). However, with time, other components, that have a higher affinity for the surface but are present at lower concentrations will gradually displace the albumin. Consequently, Vroman has proposed that a whole series of adsorption, displacement reactions can occur when a

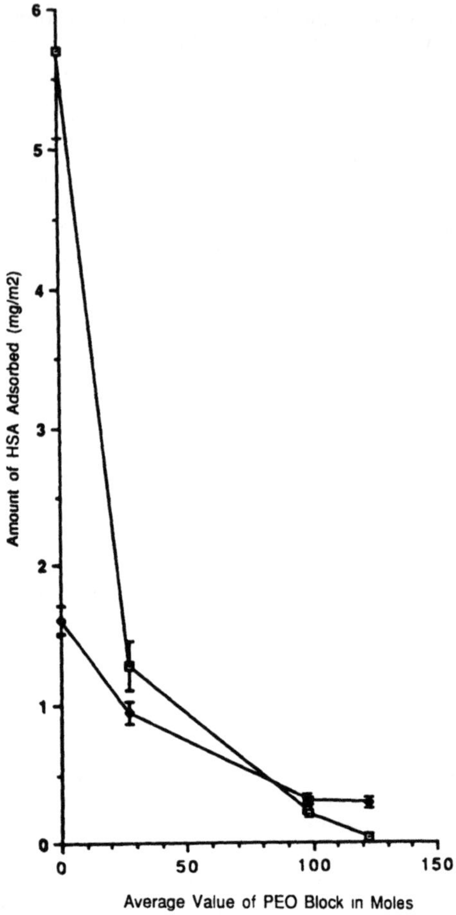

Fig.1 The effect of PEO chain length on the adsorption of human serum albumin to polystyrene particles coated with poloxamers and poloxamines.
▣ : measured by desorption/SDS-PAGE
♦ : measured by radiolabelling of HSA

biomaterial (implant) is exposed to plasma (Leonard and Vroman, 1991). Similar events no doubt occur with particulates injected into the bloodstream but the time scales can be short before a particle reaches a target site whereas an injected circulating particle will be presenting its changing surface to many sites, possibly on frequent occasions.

THE LIVER AS A TARGET

From the foregoing description, it will be appreciated that a simple way of targeting to the liver can be achieved by the injection of small particles that are rapidly opsonized and as a result are recognised as foreign by Kupffer cells. This approach has indeed been shown to work in various experimental studies (Davis and Illum, 1986). Our own work shows that labelled particles injected into an animal model such as the rabbit, are captured both rapidly and efficiently (Davis et al, 1993). In pharmacokinetic terms, this is a process of "total extraction". While such rapid capture represents a barrier to targeting particulates to other sites, it does represent a useful strategy for the treatment of diseases that involve the liver and in particular the Kupffer cells thereof. For example, in certain parasitic diseases, such as leishmaniasis, the causative organism is resident in the macrophages (Alving, 1983). Consequently, great benefit can be gained by delivering antiparasitic drugs within particles and for these to be targeted passively to the liver. The study of Alving and others, who used antimony-containing drugs entrapped in liposomes, is classical in this regard (Alving, 1983). The liposomal delivery system was able to provide total clearance of the disease in a hamster model at a fraction of the dose that was needed for the free drug.

PARTICLES WITH EXTENDED CIRCULATION TIMES

The potential barrier function of reticuloendothelial capture in particle targeting has been discussed by many authors in the past. For example, in 1984, Juliano and Posnanski (Posnanski and Juliano, 1984) commented that the reticuloendothelial barrier cannot be entirely avoided. "However, it was conceivable that a detailed understanding of the subtle recognition properties of macrophages will eventually allow one to construct particles or drug delivery systems that can evade uptake by reticuloendothelial cells". In a series of studies we have shown that it is indeed possible to overcome the reticuloendothelial barrier and effectively "fool" the body into accepting foreign particles as natural. This has been achieved by providing particles with a surface coating that not only presents a hydrophilic aspect to the biological environment but also provides what is known as a steric barrier in the form of extended polymer chains (Fig. 2) (Illum et al, 1987; Illum and Davis, 1987). It is known from theories of colloid science that such sterically stabilized particles demonstrate good stability toward particle-particle interaction and such concepts have their biological equivalent. Our work and that of Andrade and de Gennes (Porter et al, 1992; Moghimi et al, 1993; Jeon et al, 1991) has provided a more detailed mechanistic understanding of the processes involved. A hydrophilic stabilised surface not only greatly reduces the uptake of host proteins but apparently encourages the uptake of other key factors that can act as dysopsonins, ie. blood components that discourage particle capture.

Our early studies were conducted with various block copolymers of polyoxyethylene (PEO), polyoxypropylene (POP), adsorbed to the surface of labelled polystyrene particles, by using block copolymers, that were well adsorbed to the surface of the particle through their hydrophobic (POP) block. By providing the surface with an appropriate thickness and

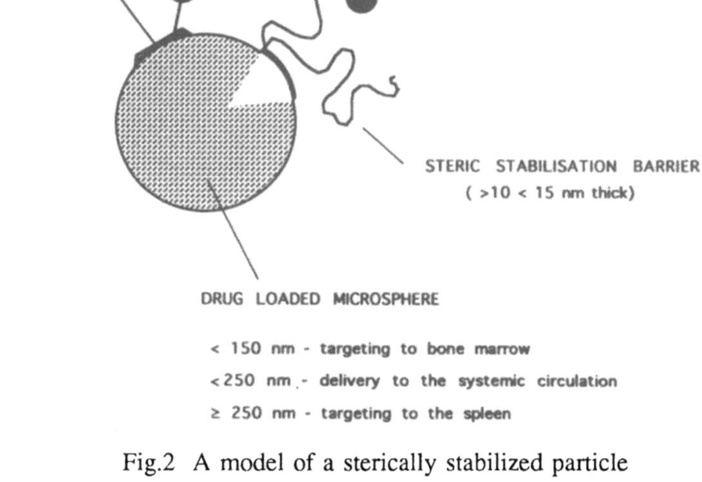

Fig.2 A model of a sterically stabilized particle

density of polyoxyethylene groups, we discovered that the resultant particles demonstrated long circulation times and were poorly taken up by macrophages in vitro or as in vivo (Davis et al, 1993). The same concept has now been used to provide liposomes with long circulating times by the incorporation of polyethylene glycol groups that render the surface of the particle hydrophilic and sterically stabilised (Klibanov et al, 1990).

Particles with long circulation times can be used to deliver drugs to targets within the vascular compartment or to allow particles the opportunity to reach extravascular pathologies in the form of tumours and inflammation (Illum et al, 1989; Illum et al, 1987). Particle constructs can also be prepared that have both a sterically stabilizing layer and attached homing ligands in the form of sugar or monoclonal antibodies. Our work with adsorbed block copolymers has now been extended to include studies with particles that have surface modifying moieties grafted to their surface thereby eliminating the possibility of desorption of the stabilizing polymer layer in vivo (Brindley et al, 1992; Brindley et al, in press; Dunn et al, in press; Harper et al, 1991). These systems have also allowed us to investigate the importance of the structure of the surface layer; for example, chain length and surface density of attached polyoxyethylene groups. Coated biodegradable particles produced from materials such as polylactide-coglycolide and polymalic acid have also been investigated (Stolnik et al, 1993).

We have been intrigued as to the nature of the various opsonic and particularly dysopsonic factors that can control the recognition and non-recognition processes for injected particles. So far we have established that the dysopsonic factors are multicomponent in nature, and involve both high and low molecular weight materials (Muir et al, 1991). The design of a particle surface that encourage the selective uptake of specific factors that control particle fate is a current research objective.

TARGETING TO THE BONE MARROW

In screening a wide variety of block copolymers for their ability to provide non-recognition of particles by macrophages, we were surprised to find that certain block copolymer coatings all containing PEO not only prevented liver uptake but directed particles to the bone marrow (Illum and Davis, 1987; Porter et al, 1992). Particle size was a key factor, in that particles had to have a mean size equal to or below about 150 nm in order to display this bone marrow targeting phenomenon (Fig. 3). In our rabbit model, more than 50% of the dose was found to be contained within the active bone marrow. At first, this uptake was thought to be mediated by macrophages, notwithstanding the fact that the particles were able to avoid recognition by the macrophages in the liver. Subsequent detailed histological examination has revealed that the bone macrophages are not involved

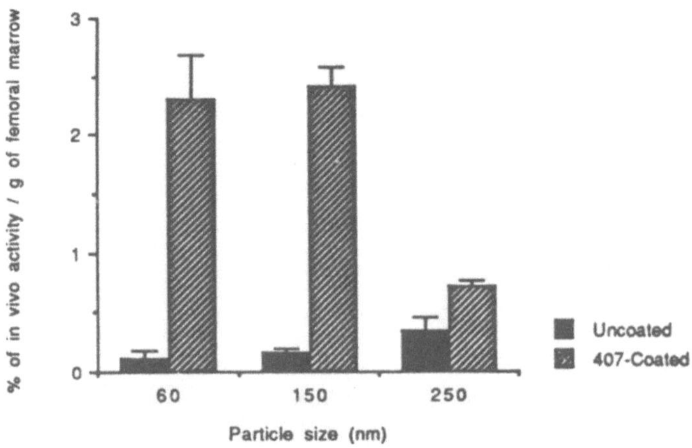

Fig. 3 The effect of particle size on the targeting of poloxamer 407 coated particles

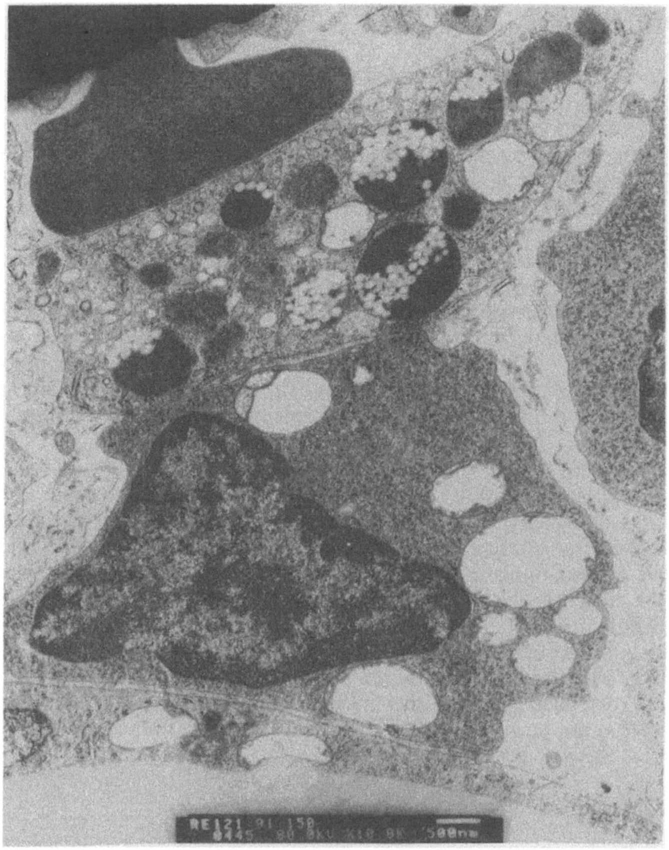

Fig. 4 Polystyrene particles coated with poloxamer 407 captured in the bone marrow endothelial cells in rabbits

and the particles are captured exclusively by endothelial cells lining the sinusoids of the bone marrow (Fig. 4). The particles are apparently recognised and taken up by a receptor-mediated process and are located within structures known as "dense bodies" that are lysosomal in nature. We have yet to determine whether the dense body is the final site for uptake of the particles or whether they are able to translocate to underlying tissues. Such additional processing stages have obvious implications for the use of the particulate system in various targeting opportunities.

Interestingly, the ability of particles coated with certain block copolymers to be recognised by the bone marrow endothelia is highly specific and is demontrated by only a few polymeric materials and is critically dependent upon molecular weight as well as particle size (Porter et al, 1992). A present hypothesis is that the particle surface encourages the uptake of a critical plasma component and it is this plasma component that is subsequently recognised by the bone marrow endothelial cells. We believe that this is the first demonstration of active targeting through the use of a purely physicochemical mechanism (Porter et al, 1992).

A more detailed understanding of the underlying processes would clearly be of benefit. It is now known that the endothelia lining the various tissues and organs of the body can demonstrate considerable heterogeneity which could provide opportunities for targeting similar particle constructs to other defined sites in the vascular compartment.

SPLEEN TARGETING

As described above, small particles with appropriate hydrophilic and sterically stabilised surfaces can demonstrate long circulation times. However, we find that sterically stabilized particles of an intermediate size, for example 250 nm and larger, while demonstrating little uptake by the liver, are eventually captured in the rat model by the spleen (Fig. 5) (Moghimi et al, 1991). We have shown that this uptake by the spleen occurs through a process of initial filtration rather than macrophage capture. Indeed, as discussed for the bone marrow, it would be surprising if there was a macrophage involvement in uptake since the particle surface has been modified to prevent such an occurrence. However, while filtration is the responsible mechanism for removing hydrophilic particles of intermediate size from the circulation, the particles eventually find their way into macrophages by 24 hours. We are yet to elucidate the mechanisms for this process of translocation. It could be that the adsorbed layer of polymer is gradually removed in the spleen, thereby allowing subsequent macrophage presentation. Enhanced delivery of particles to the macrophages of the spleen could have a number of therapeutic applications; in particular the design of novel vaccine systems.

On repeated administration of particles, we find that it is possible to exhaust the responsible processes whatever they might be. After administration of particles coated with the block copolymer poloxamine 908 to rats over a period of 4 days, we find that the particles, instead of remaining in the circulation, or being captured by the spleen, are taken up instead by the liver. If the treated animals are allowed a period of recovery, and once again given the polymer coated particles, these are once more to be found in the circulation, rather than in being directed to the liver. These data strongly suggest that the lack of recognition of the polymer coated particles by the liver macrophages is mediated by the uptake of critical dysopsonic factors that may be depleted upon repeated administration of such particles (Moghimi and Patel, 1989).

LYMPHATIC TARGETING

A survey of available literature indicates that while considerable attention has been given to the intravenous administration of colloidal particles, much less attention has been given to lymphatic delivery. Indeed some years ago, Weinstein and Lesserman (1984) commented that too much attention was then being paid to the intravenous route and too little attention focused on delivery via regional lymph nodes and to injection into body compartments such as the peritoneal cavity. We have recently set up a programme to investigate lymphatic delivery where the same strategy of applying concepts of particle

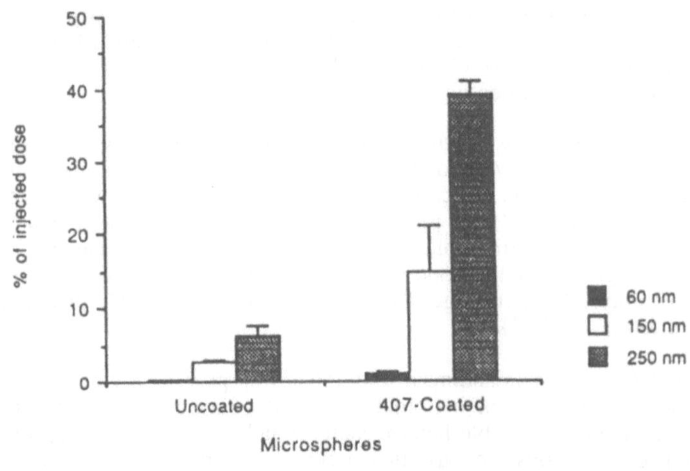

Fig. 5 Effect of particle size on spleen uptake of polystyrene particles coated with poloxamer 407, 24 hours after intravenous administration in rats

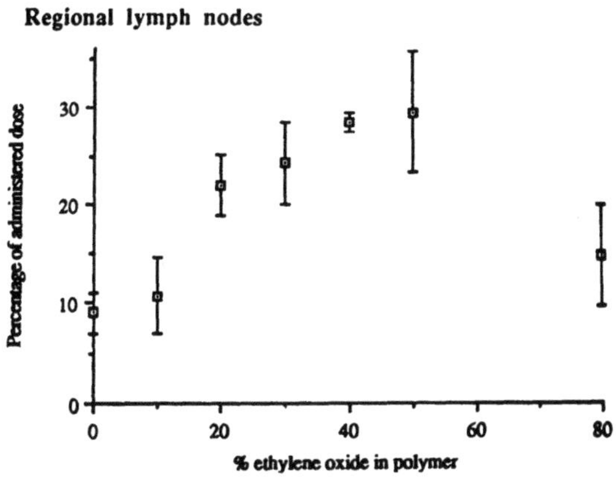

Fig. 6 Relationship between ethylene oxide content of poloxamers and their effect on the uptake by the regional lymph nodes of 60 nm polystyrene microspheres 24 hours after subcutaneous administration into the foot pad of the rat

engineering to dictate performance events are being exploited. The target has been the regional lymph nodes following subcutaneous administration of various coated particles. The work has again demonstrated that particle size and surface nature are both critical in determining enhanced lymphatic uptake. Particles less than 100 nm in size are required for the effective targeting to the regional lymph nodes of the rat after subcutaneous injection. Hydrophobic particles are largely retained at the injection site and are moved only slowly and to a low extent to the regional lymph nodes. In contrast, particles with very hydrophilic surfaces (ie. those that provide long circulation times for intravenously administered particles) leave the injection site rapidly but pass through the regional lymph nodes and find their way into the systemic circulation (Fig. 6). Particles with intermediate hydrophilic/hydrophobic characteristics are moved from the site of injection and are sequestered in the lymph nodes, by the lymph node macrophages. By studying a wide range of block copolymers of the poloxamine and poloxamer series, we have been able to discover novel relationships between polymer characteristics and lymph node uptake, as well as to define more general properties such as critical particle surface hydrophobicity. Some of the systems currently under development can deliver more than 30% of the dose to the primary, secondary and tertiary nodes following subcutaneous administration (Moghimi et al, 1994). This figure greatly exceeds the quantity that has been achieved previously. Targeting to the lymphatics has a number of therapeutic applications to include imaging and diagnostic procedures, radiation therapy as well as drug delivery.

THE GASTROINTESTINAL TRACT

Particle delivery within the gastrointestinal tract can be conducted using the same general principles that have been developed for the parenteral administration of particles. Once again, particle engineering can be used to achieve a better uptake at designated sites. Unfortunately, so far as the gastrointestinal tract is concerned, there is currently confusion as to the mechanisms responsible for particle processing and above all, a lack of realism as what may be achieved in terms of transmucosal transport in the human gastrointestinal tract. While it is now well established that small particles are well taken up by specific cells in the gastrointestinal tract, the extent of uptake in relation to administered dose is extremely small. As a result such uptake has little or no relevance to the delivery of drugs (for example therapeutic peptides and proteins) but has an important role in the development of vaccines and immunotherapies (Davis, 1993).

Within the gastrointestinal tract, there are specialised regions known as "Peyer's patches". These patches contain specific cells known as M-cells which have a role of

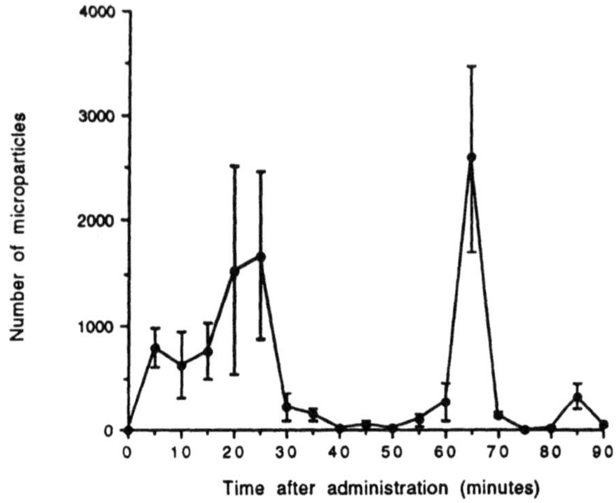

Fig. 7 The detection of microparticles in mesenteric lymph of rats, particle size 1 micron

sampling the gastrointestinal contents. M-cells are different from the normal absorptive cells, not only in terms of their surface (for instance, an absence of lysosomal material that could lead to the degradation of sensitive materials) but also the lack of an overlying layer of mucus. M-cells can take up both particulate and dissolved materials and present these to underlying lymphoid tissue (O'Hagan, 1990). This normal physiological process has importance in the development of tolerance to ingested food materials. For the case of particulate carriers it represents an opportunity for the design of mucosal vaccines. When an antigen is entrapped into a particle and taken up by the M-cells, it can be presented to the underlying lymphoid tissue. Subsequent stages of particle processing and presentation can lead to the production of significant quantities of secretory IgA at mucosal sites as well as the appearance of IgA and IgG within the serum.

The extent of particle uptake that occurs under normal conditions has been a matter of much discussion in the literature and still remains a contentious issue. Some, including ourselves, believe that such uptake is very small, whereas other groups have reported astonishing figures of more than 20% of the dose of administered particles being taken up from the gastrointestinal tract. Almost invariably such studies have involved the use of indirect methods of assessing particle uptake. We have recently studied this phenomenon by the direct counting of particles appearing in the mesenteric lymph after oral dosing (Jenkins et al, 1994). Fluorescent polystyrene particles were administered to the rat intestine and particles within mesenteric lymph were measured by fluorescence activated cell sorting. Particles of different sizes were studied . Particles were detected in the lymph but quantities found were extremely small (Fig. 7). Interestingly the particles seem to be transported in a phased manner which may indicate the presence of some form of specific mechanism; eg. a receptor that is recycled. Our calculations show that the maximum uptake of particles represents much less than 0.1% of the administered material and is size dependant. Such low uptakes render particulate systems of little importance for drug delivery to the systemic circulation but are more than sufficient for mucosal vaccination. Concepts of particle engineering can be used to produce systems that are optimal in their performance. Particle size, surface, nature and surface charge, together with protective coatings are being examined. The incorporation of the antigen is another important factor. Whether the antigen should be contained in the bulk of the particle or should be external (adsorbed or grafted) has yet to be ascertained. A combined effect may be beneficial. Antigen release characteristics could also be critical. In some instances a sustained release would be beneficial (eg. to achieve tolerance) whereas for vaccine development, pulsatile release may well be more appropriate.

CONCLUSION

Particles in the colloidal size range can have a variety of uses in drug targeting. Site specific delivery can be achieved through the application of concepts of particle engineering, where particle size, surface charge and surgace properties are optimized to achieve defined biological responses.

REFERENCES

Alving, C.R., 1983, Delivery of liposome-encapsulated drugs to macrophages, Pharm. Ther.,22:407

van Berkel, T.J.C., Kruijt, J.K., Harkes, L., Nagelkerke, J.F., Spanjer, H. and Kempen, H.-J. M., 1993, Receptor-dependent targeting

Boddy, A.V. and Aarons, L.J., 1991, Application of a linear recirculation model to drug targeting, J.Pharmacokinet.Biopharm., 19:355

Brindley, A., Davies, M.C., Lynn, R.A.P, Davis, S.S., Hearn, J. and Watts, J.F., 1992, The surface characterization of model charged and sterically stabilized polymer colloids by SSIMS and X.P.S., Polymer Reports, 33:1112

Brindley, A., Davis, S.S., Davies, M.C. and Watts, J.F., Polystyrene colloids with surface grafted poly(ethylene oxide) as model systems for site specific drug delivery. Part 1: Preparation and surface chemical characterisation using SIMS and XPS, J.Colloid Interf.Sci., in press

Davis, S.S. and Illum, L., 1986, Colloidal delivery systems: opportunities and challenges, in: "Site-specific drug delivery", E. Tomlinson and S.S. Davis, eds., John Wiley & Sons, Chichester

Davis, S.S., Illum, L., Moghimi, S.M., Davies, M.C., Porter, C.J.H., Muir, I.S., Brindley, A., Christy, N.M., Norman, M.E., Williams, P. and Dunn, S., 1993, Microspheres for targeting drugs to specific body sites, J.Control.Rel., 24:157

Davis, S.S., 1993, Oral delivery of peptides and proteins, in: "Perspectives in Medicinal Chemistry", B. Testa, E. Kyburz, W. Fuher and R. Giger, eds., VCH, Weinheim

Dunn, S.E., Brindley, A., Davis, S.S., Davies, M.C. and Illum, L., Polystyrene -poly(ethylene glycol) (PS-PEG 2000) particles as model systems for site specific drug delivery. Part 2. The effect of PEG surface density on the in vivo cell interaction and in vivo biodistribution, Pharm.Res., in press

Harper, G.R., Davis, M.C., Davis, S.S., Tadros, Th.F., Taylor, D.C., Irving, M.P. and Waters, J.A., 1991, Steric stabilization of microspheres with grafted polyethylene oxide reduces phagocytosis by rat Kupffer cells in vitro, Biomaterials, 12:695

Illum, L. and Davis, S.S., 1987, Targeting of colloidal particles to the bone marrow, Life Sci., 40:1553

Illum, L., Wright, J. and Davis, S.S., 1989, Targeting of microspheres to sites of inflammation, Int.J.Pharm., 52:221

Illum, L., Davis, S.S., Muller, R.H., Mak, E. and West, P., 1987, The organ distribution and circulation time of intravenously injected colloidal carriers sterically stabilized with a block co-polymer-poloxamine 908, Life Sci., 40:367

Illum, L., Davis, S.S., 1984, The organ uptake of intravenously administered colloidal particles can be altered using a non-ionic surfactant (Poloxamer 338), FEBS Lett., 167:79

Jeon, S.I., Lee, J.H., Andrade, J.D. and de Gennes, P.G., 1991, Protein-surface interactions in the presence of polyenthylene oxide, J.Colloid Interf.Sci., 142:149

Jenkins, P.G., Howard, K.A., Blackhall, N.W., Thomas, N.W., Davis, S.S. and O'Hagan, D.T., 1994, Microparticulate absorption from the rat intestine, J.Control.Rel., 29:339

Klibanov, A.L., Maruyama, K., Torchillin, V.P. and Huang, L., 1990, Amphipathic polyenthyleneglycols effectively prolong the circulation time of liposomes, FEBS Lett., 268:235

Leonard, E.F. and Vroman, L., 1991, Is the Vroman effect of importance in the interaction of blood with artificial materials?, J.Biomater.Sci.Polym.Ed., 3:95

Mizushima, Y., Hamano, T. and Yokoyama, K., 1982, Use of a lipid emulsion as a novel carrier for corticosteroids, J.Pharm.Pharmacol., 34:49

Moghimi, S.M. and Patel, H.M., 1992, Opsonophagocytosis of liposomes by peritoneal macrophages and bone marrow reticuloendothelial cells, Biochim.Biophys.Acta, 1135:269

Moghimi, S.M., Hawley, A.E., Christy, N.M., Gray, T., Illum, L. and Davis, S.S., 1994, Surface engineered nanospheres with enhanced drainage into lymphatics and uptake by macrophages of the regional lymph nodes, FEBS Lett., 344:25

Moghimi, S.M., Muir, I.S., Illum, L., Davis, S.S. and Kolb-Bachofen, V., 1993, Coating particles with a block co-polymer (poloxamer 908) suppresses opsonisation but permits the activity of dysopsonins in the serum, Biochim.Biophys.Acta, 1179:157

Moghimi, S.M., Porter, C.J.H., Muir, I.S., Illum, L. and Davis, S.S., 1991, Non -phagocytic uptake of intravenously injected microspheres in rat spleen: Influence of particle size and hydrophilic coating, Biochem.Biophys.Res.Commun., 177:861

Moghimi, S.M. and Patel, H.M., 1989, Differential properties of organ-specific serum opsonins for liver and spleen macrophages, Biochim.Biophys.Acta, 984:379

Muir, I.S., Moghimi, S.M., Illum, L., Davis, S.S. and Davies, M.C., 1991, The effect of block copolymer on the uptake of model polystyrene microspheres by Kupffer cells - in vitro and in vivo studies, Biochem.Soc.Trans., 19:3295

Norman, M.E., Williams, P. and Illum, L., 1992, Human serum albumin as a probe for surface conditioning (opsonization) of block copolymer-coated microspheres, Biomaterials, 13:841

Norman, M.E., Williams, P. and Illum, L., 1993, The adsorption of plasma and serum proteins to block copolymer coated microspheres, Biomaterials, 14:193

O'Hagan, D.T., 1990, Intestinal translocation of particulates - implications for drug and antigen delivery, Adv.Drug Deliv., 5:265

Porter, C.J.H., Moghimi, S.M., Illum, L. and Davis, S.S., 1992, The polyoxyethylene/ polyoxypropylene block co-polymer poloxamer-407 selectively redirects intravenously injected microspheres to sinusoidal endothelial cells of rabbit bone marrow, FEBS Lett., 305:62

Posnansky, M.J. and Juliano, R.L., 1984, Biological approaches to the controlled delivery of drugs; a critical review, Pharmacol.Rev., 36:277

Stolnik, S., Davies, M.C., Illum, L., Davis, S.S., Boustra, M. and Vert, M., 1993, The preparation of sub 200 nm biodegradable colloidal carriers from poly(β-malic acid-co-benzyl malate) copolymers and their surface modification with Poloxamer and Poloxamine surfactants, J.Control.Rel., in press

Tomlinson, E. and Davis, S.S., 1986, "Site Specific Drug Delivery", eds., Wiley, Chichester

Toth, C.A. and Thomas, P., 1992, Liver endocytosis and Kupffer cells, Hepatology, 16:255

Vanchieri, C., 1993, Opportunities opening up for Gene Therapy, J.Nat.Cancer Inst., 85:90

Vitols, S.G., Masquelier, M. and Peterson, C.O., 1985, Selective uptake of a toxic lipophilic anthracycline derivative by the low-density lipoprotein receptor pathway in cultured fibroblasts, J.Med.Chem., 28:451

Weinstein, J.N. and Leserman, L.D., 1984, Liposomes as drug carriers in cancer chemotherapy, Pharmacol.Ther., 24:207

Wisse, E. and De Leeuw, A.M., 1984, Structural elements determining transport and exchange processes in the liver, in: "Microspheres and Drug Therapy, Pharmaceutical, Immunological and Medical Aspects", S.S. Davis, L. Illum, J.G. McVie and E. Tomlinson, eds., Elsevier, Amsterdam

Participants of the NATO Advanced Studies Institute "Targeting of Drugs: Advances in System Constructs" held in Cape Sounion Beach, Greece during 24 June - 5 July 1993. The Organizing Committee included Gregory Gregoriadis (ASI Director and Chairman), G. Poste (ASI Co-Director), P. Couvreur, G. Deliconstantinos (local Committee) and Brenda McCormack (ASI Co-ordinator)

CONTRIBUTORS

T.M. Allen, Department of Pharmacology, University of Alberta, Edmonton, AB T6G 2H7, Canada

A.C. Allison, Dawa Corporation, Belmont, CA, USA

I.A.J.M. Bakker-Woudenberg, Dept of Clinical Microbiology, Erasmus University, Rotterdam, The Netherlands

D.R. Bard, Strangeways Research Laboratory, Worts' Causeway, Cambridge, CB1 4RN, UK

J. van Beeumen, Vakgroep Biochemie, Fysiologie en Microbiologie, Ledeganksstraat, 35, 9000 Ghent, Belgium

G. Blume, Medizinische Biophysik, Urologische Klinik und Poliklinik der Technischen Universität, München, Germany

P. Breton, Laboratoires UPSA, Laboratoire de Galénique, 128 rue Danton, 92506 Rueil-Malmaison Cedex, France

A. Casabianca, Institute of Biological Chemistry, "G. Fornaini", University of Urbino 61029, Urbino, Italy

L. Chiarantini, Institute of Biological Chemistry, "G. Fornaini", University of Urbino 61029, Urbino, Italy

J.S. Cohen, Department of Pharmacology, Georgetown University, Washington DC 20007, USA

D.J.A. Crommelin, Department of Pharmaceutics, Utrecht University, PO Box 80 082, 3508 TB Utrecht, The Netherlands

P. Couvreur, Laboratoire de Pharmacie Galénique et de Biopharmacie, URA CNRS 1218, Université Paris XI, 92296 Chatenay-Malabry, France

S.S. Davis, Department of Pharmaceutical Sciences, University of Nottingham, University Park, Nottingham, NG7 2RD, UK

B. Deveese, Vakgroep Biochemie, Fysiologie en Microbiologie, Ledeganksstraat, 35, 9000 Ghent, Belgium

A. Epenetos, Tumour Targeting Laboratory, Imperial Cancer Research Fund, Oncology Unit, Department of Clinical Oncology, Royal Postgraduate Medical School, Hammersmith Hospital, DuCane Rd, London, W12 0HS, UK

E.M. Eugui, Syntex Research, Palo Alto, CA, USA

A.T. Florence, Centre for Drug Delivery Research, The School of Pharmacy, University of London, Brunswick Square, London WC1N 1AX, UK

A. Fraternale, Institute of Biological Chemistry, "G. Fornaini", University of Urbino 61029, Urbino, Italy

G. Gregoriadis, Centre for Drug Delivery Research, The School of Pharmacy, University of London, Brunswick Square, London WC1N 1AX, UK

H. Haisma, Department of Medical Oncology, Free University Hospital, Amsterdam, The Netherlands

A.M. Hillery, Centre for Drug Delivery Research, The School of Pharmacy, University of London, Brunswick Square, London WC1N 1AX, UK

N. Hussain, Centre for Drug Delivery Research, The School of Pharmacy, University of London, Brunswick Square, London WC1N 1AX, UK

L. Illum, Department of Pharmaceutical Sciences, University of Nottingham, University Park, Nottingham, NG7 2RD, UK

F. Jacquemotte, Institut Meurice, Avenue Emile Gryzon, 1, B-1070, Brussels, Belgium

P.U. Jani, Centre for Drug Delivery Research, The School of Pharmacy, University of London, Brunswick Square, London WC1N 1AX, UK

J. Jankun, Deparment of Chemistry, The University of Toledo, Toledo, Ohio 43606, USA

C.G. Knight, Strangeways Research Laboratory, Worts' Causeway, Cambridge, CB1 4RN, UK

F. Lescure, Laboratoires UPSA, Laboratoire de Galénique, 128 rue Danton, 92506 Rueil-Malmaison Cedex, France

B. McCormack, Centre for Drug Delivery Research, The School of Pharmacy, University of London, Brunswick Square, London WC1N 1AX, UK

M. Magnani, Institute of Biological Chemistry, "G. Fornaini", University of Urbino 61029, Urbino, Italy

A. Maraveyas, Tumour Targeting Laboratory, Imperial Cancer Research Fund, Oncology Unit, Department of Clinical Oncology, Royal Postgraduate Medical School, Hammersmith Hospital, DuCane Rd, London, W12 0HS, UK

L. Marchal-Heussier, Laboratoires UPSA, Laboratoire de Galénique, 128 rue Danton, 92506 Rueil-Malmaison Cedex, France

R. Mayer, Laboratoire de Biochimie des Glycoconjugués et Lectines Endogènes, Université d'Orléans et Centre de Biophysique Moléculaire, CNRS, 1, Rue Haute, 45071 Orléans, France

D.K.F. Meijer, Department of Pharmaceutical Pharmacology and Clinical Pharmacy, Section of Pharmacokinetics and Drug Delivery, Groningen Inst for Drug Studies, University Centre for Pharmacy, Ant. Deusinglaan 2, 9713 AW Groningen, The Netherlands

P. Midoux, Laboratoire de Biochimie des Glycoconjugués et Lectines Endogènes, Université d'Orléans et Centre de Biophysique Moléculaire, CNRS, 1, Rue Haute, 45071 Orléans, France

M. Monsigny, Laboratoire de Biochimie des Glycoconjugués et Lectines Endogènes, Université d'Orléans et Centre de Biophysique Moléculaire, CNRS, 1, Rue Haute, 45071 Orléans, France

U.K. Nassander, Department of Pharmaceutics, Utrecht University, PO Box 80 082, 3508 TB Utrecht, The Netherlands

F. Puisieux, Laboratoire de Pharmacie Galénique et de Biopharmacie, URA CNRS 1218, Université Paris XI, 92296 Chatenay-Malabry, France

A.-C. Roche, Laboratoire de Biochimie des Glycoconjugués et Lectines Endogènes, Université d'Orléans et Centre de Biophysique Moléculaire, CNRS, 1, Rue Haute, 45071 Orléans, France

L. Rossi, Institute of Biological Chemistry, "G. Fornaini", University of Urbino 61029, Urbino, Italy

D. Roy, Laboratoires UPSA, Laboratoire de Galénique, 128 rue Danton, 92506 Rueil-Malmaison Cedex, France

C. Seguin, Laboratoires UPSA, Laboratoire de Galénique, 128 rue Danton, 92506 Rueil-Malmaison Cedex, France

A.H. Sehon, Department of Immunology, The University of Manitoba, Winnipeg, Canada R3E 0W3

G. Storm, Department of Pharmaceutics, Utrecht University, PO Box 80 082, 3508 TB Utrecht, The Netherlands

M.H. Vingerhoeds, Department of Pharmaceutics, Utrecht University, PO Box 80 082, 3508 TB Utrecht, The Netherlands

M.C. Woodle, Liposome Technology Inc., Menlo Park, CA, USA

E.P. Wraight, Department of Nuclear Medicine, Addenbrooke's Hospital, Hills Road, Cambridge, CB2 2QQ, UK

M. Monsigny, Laboratoire de Biochimie des Glycoconjugués et Lectines Endogènes, Université d'Orléans et Centre de Biophysique Moléculaire, CNRS, ... Orléans, France

INDEX

MIX
Papier aus verantwortungsvollen Quellen
Paper from responsible sources
FSC® C105338

If you have any concerns about our products,
you can contact us on
ProductSafety@springernature.com

In case Publisher is established outside the EU,
the EU authorized representative is:
Springer Nature Customer Service Center GmbH
Europaplatz 3, 69115 Heidelberg, Germany

Printed by Libri Plureos GmbH
in Hamburg, Germany